Audio, Video and Data Telecommunications

Audio, Video and Data Telecommunications

DAVID PETERSEN

The McGraw-Hill Companies

London · New York · St Louis · San Francisco · Auckland
Bogotá · Caracas · Lisbon · Madrid · Mexico
Milan · Montreal · New Delhi · Panama · Paris · San Juan
São Paulo · Singapore · Sydney · Tokyo · Toronto

Published by
McGraw-Hill Publishing Company
Shoppenhangers Road, Maidenhead, Berkshire, SL6 2QL, England
Telephone 01628 23432
Fax 01628 770224

British Library Cataloguing in Publication Data

Petersen, David
Audio, video and data telecommunications
I. Title
621.382

ISBN 0-07-707427-0

Library of Congress Cataloging-in-Publication Data

Petersen, David
Audio, video and data telecommunications/David Petersen.
p. cm.
Includes bibliographical references and index.
ISBN 0-07-707427-0
1. Telecommunication. I. Title.
TK5101.P42 1992 92-9766
621.382—dc20 CIP

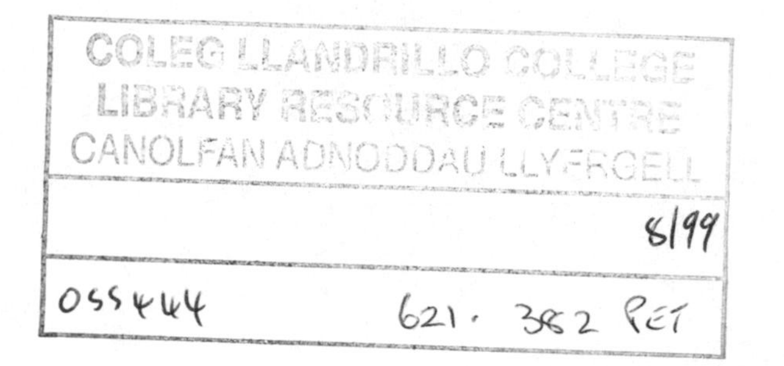

McGraw-Hill 
A Division of The McGraw-Hill Companies

Reprinted 1997

Typeset by Interprint Ltd, Malta
Printed and bound in Great Britain by The Basingstoke Press (75) Limited
Printed on permanent paper in compliance with ISO Standard 9706

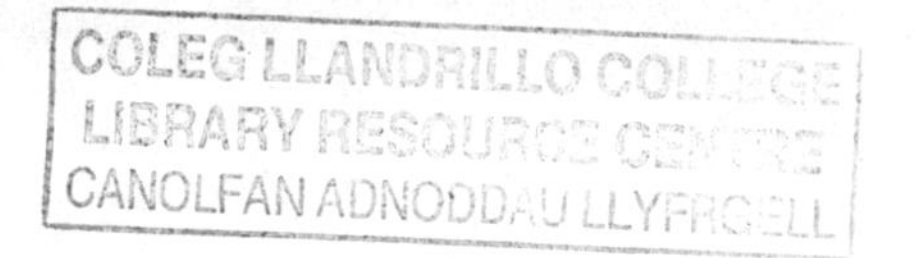

... of making many books there is no end;
and much study is a weariness of the flesh.
Ecclesiastes 12, 12

Contents

Preface

The final decade of this millennium should see many further advances in information technology (IT), a subject area that brings together the techniques of computing and telecommunications. Computing involves information processing. Telecommunications is the near-instantaneous exchange of information of all kinds such as speech, pictures and computer data, between separate locations. Future generations may well look back on the late 20th century as the 'information age'.

The principles of computing and telecommunications have been known for many years, but their latest forms of exploitation have had to await advances in very-large-scale integrated (VLSI) circuits and other electronic devices. The increasing use of telecommunications, made economically possible by this hardware, continues to stimulate great interest and justifies a new book on the subject, which takes account of new applications.

Telecommunications theory may be expressed with great precision and elegance using fairly sophisticated mathematics, so, for the expert author, the temptation to rely on this approach is almost irresistible. Such treatment, however, may obscure the principles for the beginner and discourage further study. The many books that have been written on the subject are usually aimed at either degree or diploma students. In the former, theory dominates and engineering techniques tend to be played down. In the latter, the more practical approach tends to produce just a collection of facts. This book tries to combine the best of these two approaches for the benefit of the student with no previous knowledge of the subject. Such an approach should also appeal to practising engineers who wish to revise their knowledge.

This book covers briefly the main principles of telecommunications theory and relates them to modern practice. It does not attempt to be comprehensive but aims to be user-friendly. It tries to unravel the conceptual threads of the subject in a logical and narrative sequence, using illustrative examples and exercises along the way. The first two chapters set the scene. Chapter 3 deals with the appropriate background mathematical theory. Chapters 4 to 7 describe, with the aid of practical examples, the engineering principles and techniques used in telecommunications. Chapter 8 describes noise and signal errors, and considers their effect on the recovery of information from the signal. Chapter 9 introduces information theory. Chapters 10 and 11 show how the principles and techniques, outlined in the previous chapters, are applied in modern telephone networks and in data communications. These two chapters are, by far, the largest in the book, showing the emphasis placed on the application of long-standing principles. A final chapter summarizes the book and speculates about the future. With the exception of the first two chapters and the summary, the chapter order is somewhat arbitrary and, since a book provides random access to its contents, students often skip about at will, referring backwards and forwards. There are many equally logical orders in which the subject may be tackled.

The scope and level of treatment should satisfy the requirements of the early years of undergraduate courses in British universities and polytechnics, where telecommunications is taught as a main subject in electrical and electronic engineering degree courses. It will also serve as a useful introduction to specialist options normally offered in the final year of such courses. Much of the material will be found suitable for those following higher diploma courses. The overall level of treatment assumes that mathematics and physics have been studied to A-level and a little knowledge of electronics is also assumed.

Each chapter begins with a set of objectives and an introduction. It is hoped by this means to give the reader some preliminary explanation, and an idea of where he or she is going. After all, if you do not know where you are going, you may end up in the wrong place! Each chapter closes with a summary, and some self-assessment questions (SAQs), which provide for immediate revision and highlight those areas which require more study before proceeding further. Throughout the text, use is made of exercises, which are designed to encourage the reader to think through the significance of the points being made and develop some of these a little further. Responses are given to all the exercises at the end of each chapter, and the answers to the SAQs may be checked from the text. This encouragement of student activity and the feedback provided by the responses and answers should make this book especially suitable for self-study and distance learning. When used in this way, the reader is advised to follow the material in the order in which it is presented.

Extended numerical questions, much beloved by college examiners, have not been included. Such questions, in the author's view, tend to reflect the interests and idiosyncrasies of the examiner, so the student is advised to use past question papers for the course being pursued. However, many examination-type questions can be found in the textbooks listed after the final chapter.

In preparing this book, I am greatly indebted to John Reeves, a former colleague at Portsmouth Polytechnic, for his advice and many useful suggestions. I would also like to acknowledge the help of Graham Freebody of BT in the preparation of Chapter 10. Finally, I am grateful to my wife Lilian for her understanding and encouragement.

David Petersen

1

Telecommunication signals

Getting information across

Objectives

When you have completed this chapter you will be able to:

- Define telecommunications
- Outline the principles of line and radio techniques
- Define what is meant by a signal
- Categorize signals into audio, video and data types
- Define the main features of signal waveforms
- Distinguish between analogue and digital signals
- Describe various forms of analogue and digital signal
- Describe complex waveforms in terms of their sinusoidal components

'Telecommunication' is literally 'distance communication'. In practice, the term is restricted to mean communication using electrical energy to carry information. For example, over a distance, we may see with a television, hear with a telephone and write with a telegraph. All these 'teledevices' enable information to be exchanged over any distance, great or small.

What is meant by 'information' is probably well understood intuitively by most people. Information is what makes us aware of something we did not already know—it is the difference between our knowledge at one instant and the next. This definition will have to be refined later, but it will do initially.

Telecommunication is a special form of communication. Conversation and writing are regarded as examples of communication, but not telecommunication, although significant distance is involved. The signals used in communication consist of information mainly in the form of speech, script or drawings, although body language and smell, for example, could also be included. In telecommunication the signals are energy, in the form of an electric current or an electromagnetic wave, encoded so that information is carried by the energy flow.

The aim of this chapter is to describe the main types of signal used in telecommunications. This provides a basis for the remaining chapters, which are concerned with the means of carrying these signals in order to convey information with maximum efficiency and minimum cost, and processing the signals so that they can be used in a variety of applications.

1.1 Forms of communication

Primitive attempts at communication were probably crude grunts (speech) and signs (hand signals). These early attempts gradually evolved into our sophisticated modern languages, the letters and words of which are codes that can convey large amounts of information far more effectively than grunts and signs. The use of diagrams and pictures (graphics) has also developed, and plays an important role in the communication of ideas. Mathematical notation is another extremely powerful and precise means of communicating certain concepts.

Initially, communication occurred over very short range—shouting distance at most. Communicating farther was difficult and various methods were adopted: the North American Indians used smoke signals, while, in more-developed Europe, flags, beacons and semaphores were used. All these methods relied on signals being carried by light waves and were therefore still, like the sound waves of speech, restricted in range. Today, much communication takes place over greater distances than the range of sight because telephone wires, optical fibres and radio waves are used to carry information.

Wire techniques have their origins in the early railway signalling systems, developed in the 19th century, which used electrical signals, routed along wires between the signal boxes, where they rang bells and moved pointers. They are also a development of the telegraph service in which messages and data were sent world-wide over wire links between punched-paper tape readers and teleprinters. Today, information, usually in the form of speech, pictures or computer data, may be disseminated or exchanged over any distance from a few metres (as between computers in the same room) to many millions of metres (as in space probes). The systems that perform these tasks use *electric currents* or *electromagnetic waves* to carry energy (information) between a transmitter and a receiver. The energy travels along a *wire* or *radio* (wireless) link or along a *glass fibre*. Figure 1.1 shows a telecommunications system using a line (wire or fibre) link.

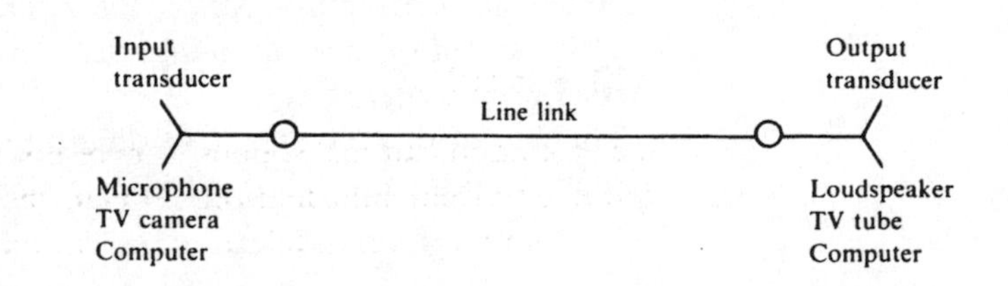

Figure 1.1. A line-linked telecommunications system.

The first task in any telecommunication system is to convert the information to be transmitted, whether it is in the form of speech, pictures or computer data, into electrical energy. This is done by a suitable *transducer*, such as a microphone, TV camera or computer. In a wire system, this energy, known as the *signal*, takes the form of electric current, which travels along a conductor. In an optical system, light is guided through a glass fibre. In radio, the signal is sent through space as an electromagnetic wave launched by a transmitting aerial. The wave is collected by a receiving aerial as in Fig. 1.2.

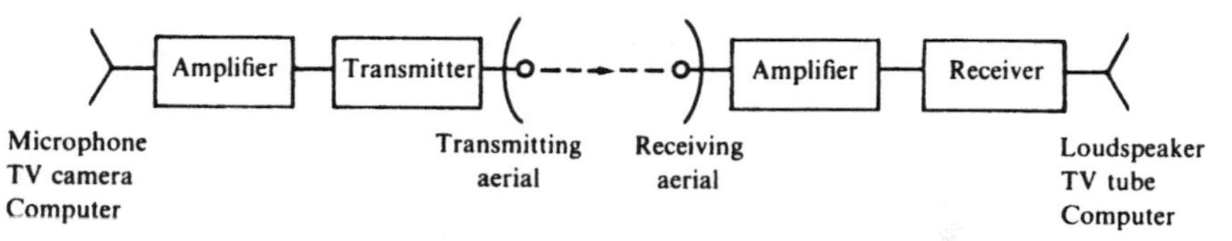

Figure 1.2. A radio-linked system.

In all cases, the received signal must be converted by a transducer such as a loudspeaker, TV tube or computer interface circuit so that the transmitted information is recoverable. The recovered information must be a reasonably close approximation to the original. In this respect, the fidelity criteria become more stringent as we go from low- to high-quality reproduction.

The transmitting aerial (alias antenna or dish) used in line-of-sight links needs to be very directional, so that a large proportion of the transmitted energy is beamed to reach the receiver and is not wasted by dispersion. A two-way radio system would have a transmitter and receiver at each end of the link and the same aerial would then serve for both transmission and reception. Little dispersion occurs if the radio wave is enclosed in a copper tube or *waveguide*, but the cost of waveguides restricts their use to short links. Broadcast systems, unlike line-of-sight systems, transmit power over a wide area and are often only one-way or unidirectional. The antennae used to receive broadcasts are usually selective and directional, so that they tend to pick up the wanted signal rather than the many other electrical signals that abound in space.

Much signal power is lost in a long radio link, so the source power must be increased, by an amplifier, before transmission. Signal power outputs of the order of a few hundred watts for radiophones and up to 1 MW for ultra-high-frequency TV broadcasting are typical. The available power at the receiving aerial is usually so small that much amplification is required before the signal information can be recovered. An extreme example of a long link occurs in satellite TV, where the signal path is about 36 000 km.

1.2 Signals

Telecommunications signals for reception by humans represent either aural or visual information. Touch, taste and smell have not yet been telecommunicated. In telecommunication between computers, on the other hand, digital signals (in the form of binary pulses of current) are used. These three forms of signal, audio, video and data, will now be considered in turn.

Audio signals

When we speak, our vocal cords vibrate, causing the surrounding air to vibrate, which, in turn, makes the eardrum of our listener vibrate. The link in a speech communications system is the air between the speaker and the listener. If a vacuum existed between the two, there would be no link. There must be something which can vibrate to form the link.

Exercise 1.1
Check on page 19

The North American Indians sometimes tapped railway lines in order to send information rather than use smoke signals. What scientific principles were involved in this technique and what was the advantage of communicating this way?

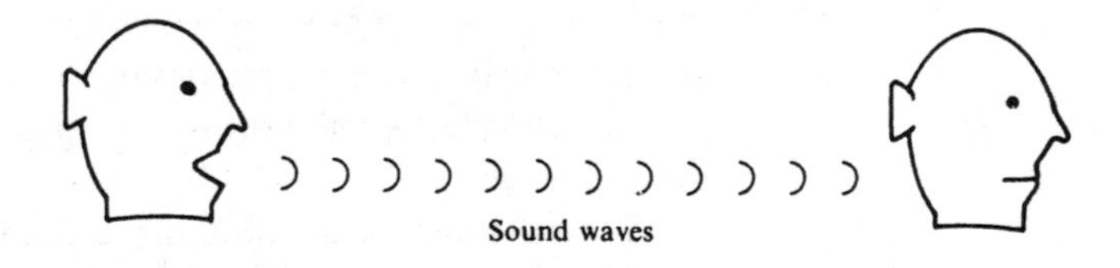

Figure 1.3. A speech communications system.

As the sound link, shown in Fig. 1.3 lengthens, the speaker has to shout in order to make the listener understand what's being said against a background of *noise.* Noise is any unwanted signal, and one person's noise may be another's signal. Furthermore, if the distance between speaker and listener becomes too great, the listener would not even hear the speaker because the sound energy reaching the ear would be too small. The sound level has gone below the *threshold of hearing*—a condition in which the listener's eardrum is not vibrating sufficiently to produce a sensation. When this condition is reached, there will be no exchange of information, even at negligible noise level. Figure 1.4(a) shows the waveforms of two tones of differing volume or loudness.

In telecommunications, the term 'tone' usually refers to a sound made up of a single frequency, and this is the meaning assumed in this book. In music, a tone is an 'interval' between two notes whose frequency ratio is about 9/8.

In Fig. 1.4(a), A_1 and A_2 are the *amplitudes* of the tones. The greater the amplitude, the louder is the sound. An amplifier, for example, increases the amplitude of a signal and so increases its loudness. The sensation of loudness is subjective and depends on air pressure changes on the eardrum. These pressure changes are related to the power carried by the sound waves. In an average conversation the average power is about $1\ \mu\text{W/m}^2$, and the threshold of hearing occurs at about

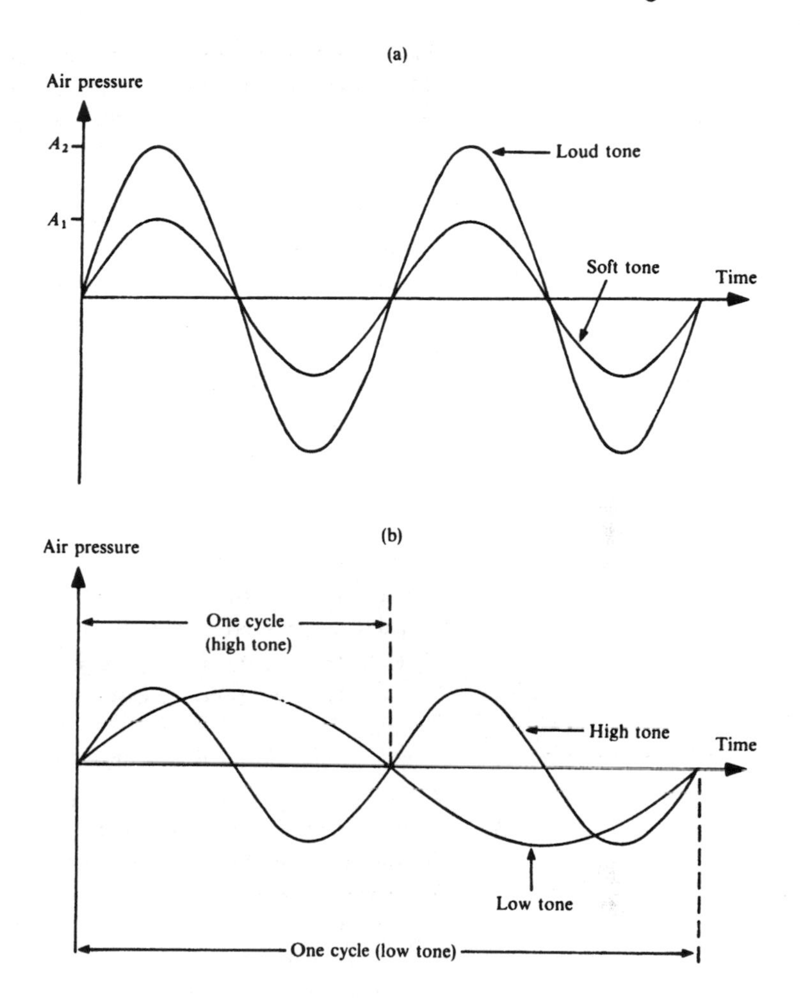

Figure 1.4. Waveforms for two different tones: (a) loud and soft tones; (b) high and low tones.

1 pW/m^2. The way in which loudness is quantitatively expressed is described briefly in Appendix A.

Exercise 1.2
Check on page 19

1. Estimate the sound power incident on a telephone handset diaphragm of 20 mm diameter during a normal conversation.
2. Assuming that the diaphragm is 10% efficient at converting sound into electrical energy, estimate the current amplitude of a pure tone signal of 1 V amplitude at the speaker's end of the line.

Another subjective feature of sound that must be considered is *pitch*. This is the audio sensation of 'height' or 'depth'. For example, a woman's voice generally sounds 'higher' than a man's voice. The characteristic of sound that determines pitch is *frequency*, which is the number of complete vibrations per second or hertz (Hz). The greater the frequency, the higher the pitch. In the example of Fig. 1.4(b), the frequency of the higher tone is double that of the lower. The upper pitched sound would be perceived as being eight notes (one *octave*) higher in a major or minor musical scale. Small freqency ranges are often measured in such multiples of 2 or octaves. The human voice, for instance, covers a range of about 64 to 1024 Hz—a *bandwidth* of approximately four octaves—from bass to treble. The range of frequency that the human ear can detect is about 20 Hz to 20 kHz. This much larger bandwidth is more conveniently measured in multiples of 10 or *decades*. Further details on frequency and pitch appear in Appendix A.

The waveforms in Fig. 1.4 may be expressed as a sine function thus:

$$y = A\sin(2\pi ft)$$

where y is the displacement at time t, A is the amplitude of the vibration and f is the frequency. An analogous electric signal current may be expressed in a similar way:

$$i = I_m \sin(2\pi ft) \tag{1.1}$$

where I_m is the maximum or peak value of the signal current i. In telecommunications this signal would carry the sound information, which, in this case, is that of a tone of frequency f.

The time interval for one cycle is the *period* (T). In one complete vibration, $2\pi ft$ would go from 0 to 2π radians. So $2\pi fT$ must equal 2π and hence, $fT = 1$.

A *periodic* signal is one that repeats itself from one period to the next, one in which i at time t is equal to i at time $t + T$, that is $i(t) = i(t + T)$.

If, in Fig. 1.4, the graph had used distance instead of time along the horizontal axis, then the distance along this axis before the wave repeated itself would be the *wavelength* (λ). This is illustrated in Fig. 1.5.

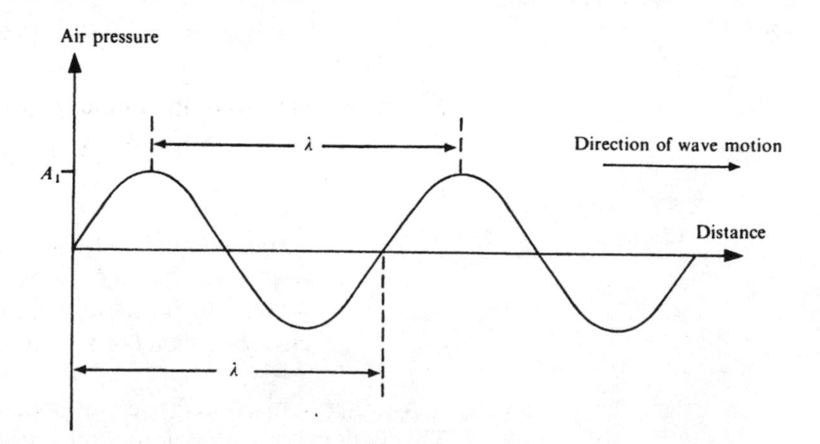

Figure 1.5. Graph showing wavelength.

Waves may be described by their wavelength instead of their frequency. For example, the higher-frequency sound wave in Fig. 1.4(b) has a frequency (f) twice that of the lower-frequency wave and, therefore, has half the wavelength (λ). A frequency of f would mean that the wave would travel a distance of f wavelengths in one second. The distance travelled in one second is the *velocity* (v), so

$$v=\lambda \times f$$

Alternatively, it could be said that a wave travels one wavelength in one period, so

$$v=\lambda/T$$

For sound the velocity is about 330 m/s in air, 1460 m/s in water and 5000 m/s in steel. The velocity of all electromagnetic waves in free space is the same—about 3×10^8 m/s, a value usually denoted by c because it is thought to be constant everywhere in the universe.

Exercise 1.3
Check on page 19

Calculate the wavelengths, corresponding to the following frequencies, when the various waves or signals stated are propagated through air:

1. A sound wave of 660 Hz
2. A radio signal of 100 MHz
3. A radar signal of 10 GHz
4. An infrared signal at 20 THz
5. A visible light signal at 600 THz

(Here is a reminder of thc dccimal multipliers used: M (mega) $=10^6$, G (giga) $=10^9$, T (tera) $=10^{12}$.)

Exercise 1.4
Check on page 19

Deduce the signal time delay that occurs in a 5000 km communications link for:

1. An electrical signal
2. A sound signal in steel
3. A sound signal in air

With these delays in mind, comment on the suitability of acoustic transmission for two-way communication.

When sounds of different amplitude and frequency are combined, we perceive a difference in what we hear according to the mix of frequencies, that is, the frequency and relative amplitudes of the sinusoidal components. This is what enables a distinction to be made between different musical instruments or voices. Since the variations in the mix are almost infinite, each voice is very characteristic, having its own distinctive quality or *timbre*.

Speech is made up of vowels and consonants. The vowels are the longer sounds that give carrying power to the voice and the consonants are shorter sounds that break up the vowel sounds to give more meaning to speech. Typical waveforms for these two types of sound are shown in Fig. 1.6.

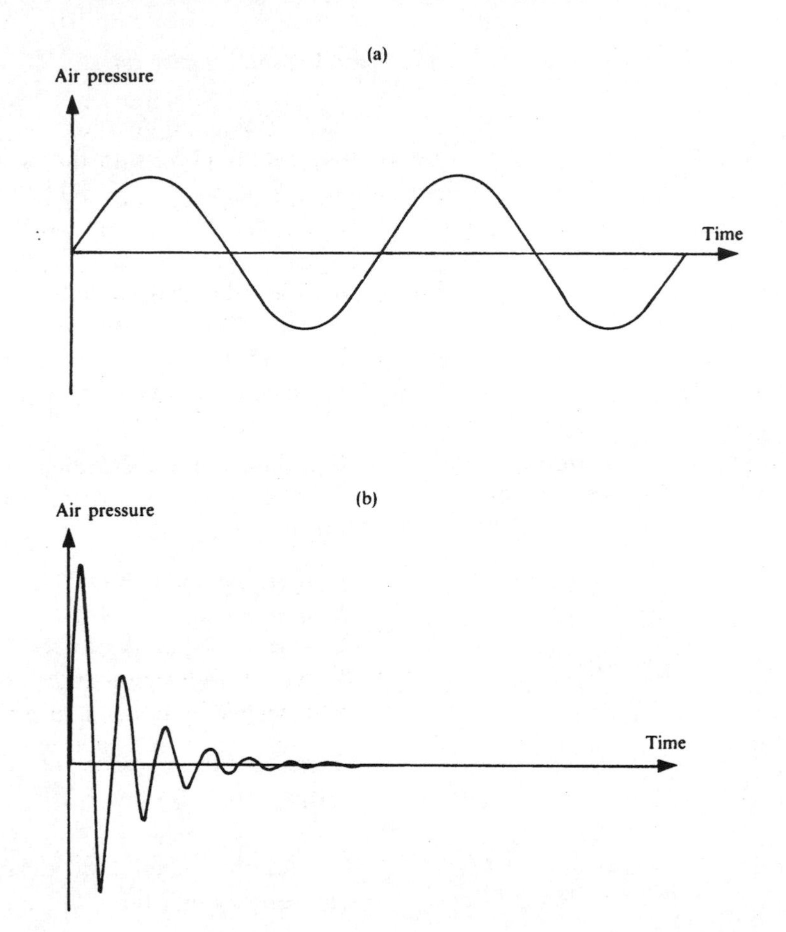

Figure 1.6. Waveforms for vowels and consonants: (a) vowel waveform; (b) consonant waveform.

The point made by Fig. 1.6 is that consonants have higher-frequency sound components than vowels. If the frequency range of a telecommunications link is curtailed, the effect is to lose the consonants and hence the intelligibility of the speech. In practice, it has been found that a range of frequencies of just over 3 kHz gives acceptable commercial speech quality, so the public telephone network has adopted a range of 300 to 3400 Hz—a bandwidth of 3.1 kHz or about one decade. As this bandwidth is reduced, intelligibility is lost—which is another way of saying that signal information is reduced when the bandwidth is reduced. It follows from this that transmission bandwidth savings can be made by sacrificing fidelity in the received signal, while maintaining its intelligibility. This principle is used in voice coders or *vocoders*, which are used in digitized speech transmission (see Sec. 1.3).

Now that the main characteristics of audio signals have been described, we go on to consider video signals.

Video signals

'One picture is worth a thousand words' is an adage which suggests that video signals will contain much more information and so have much greater bandwidth than their audio counterparts. For the purposes of telecommunications, a picture is made up of a matrix of picture elements called *pixels* or *pels*. These elements vary in light intensity according to the picture. Figure 1.7 shows how alphanumeric characters might be made up with an 8×9 matrix of pixels.

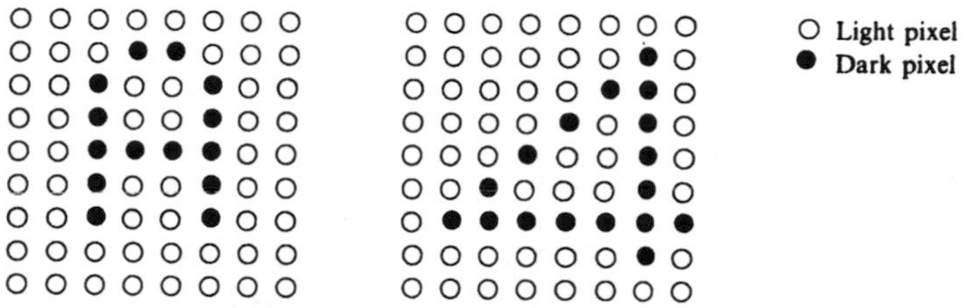

○ Light pixel
● Dark pixel

Figure 1.7. Matrix representation of alphanumeric characters.

Documents may be telecommunicated via the public telephone network using facsimile (FAX). FAX uses an electronic scanner to generate a black-on-white image of the characters to be transmitted, as exemplified in Fig. 1.8. A binary unit of information is the *bit*, eight bits making up

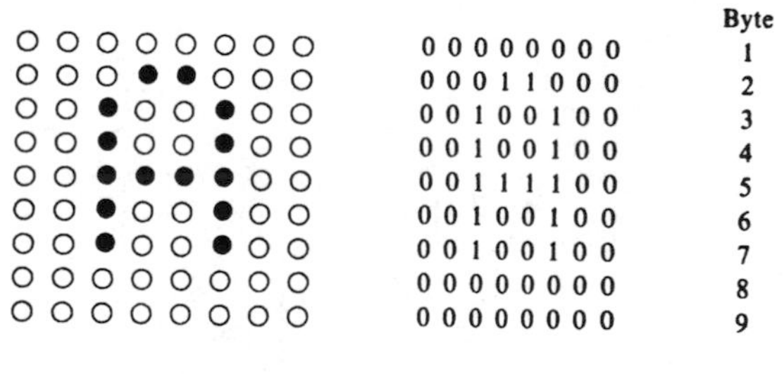

	Byte
0 0 0 0 0 0 0 0	1
0 0 0 1 1 0 0 0	2
0 0 1 0 0 1 0 0	3
0 0 1 0 0 1 0 0	4
0 0 1 1 1 1 0 0	5
0 0 1 0 0 1 0 0	6
0 0 1 0 0 1 0 0	7
0 0 0 0 0 0 0 0	8
0 0 0 0 0 0 0 0	9

Figure 1.8. Binary representation of an alphanumeric character.

one *byte*. As each line of a character is scanned, a series of bytes is produced. In this example, nine bytes might be transmitted sequentially along the telephone link to represent one character, and each byte would be used to build up a line of the matrix at the receiver. The FAX system may be represented as in Fig. 1.9. The signal waveform for byte 5, for example, of the image shown in Fig. 1.8 would take the form shown in Fig. 1.10.

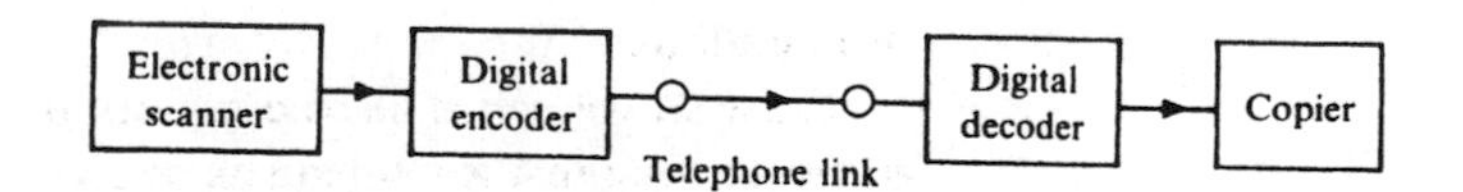

Figure 1.9. The FAX system.

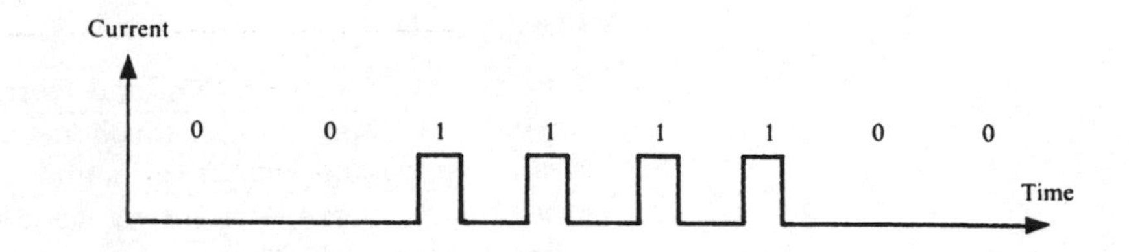

Figure 1.10. A binary signal waveform for one line of a single character.

Normally a document would consist of many lines of characters and whole lines at a time would be scanned rather than individual characters. Reproduction is effected by the copier scanning the paper and placing a dot on the paper for each '1' and leaving a space for each '0'. In the copier, thermal printing is used in which the dot matrix is printed on the surface of heat-sensitive paper by heating elements. A complete line of text takes about one second to scan, so a typical A4 page of 60 lines is reproduced in about one minute.

Pictures, unlike documents, have a light intensity or *luminance* that varies continuously from white through to black, so a continuously variable, rather than binary, signal would be needed to control the intensity of a TV image. The luminance signal has to produce a voltage for the grid of a *cathode ray tube* (CRT) so that the energy of the electrons impinging on the screen is varied. The beginning of each line is marked by a synchronizing (sync) pulse. A typical TV video signal for four lines of scan would take the form shown in Fig. 1.11.

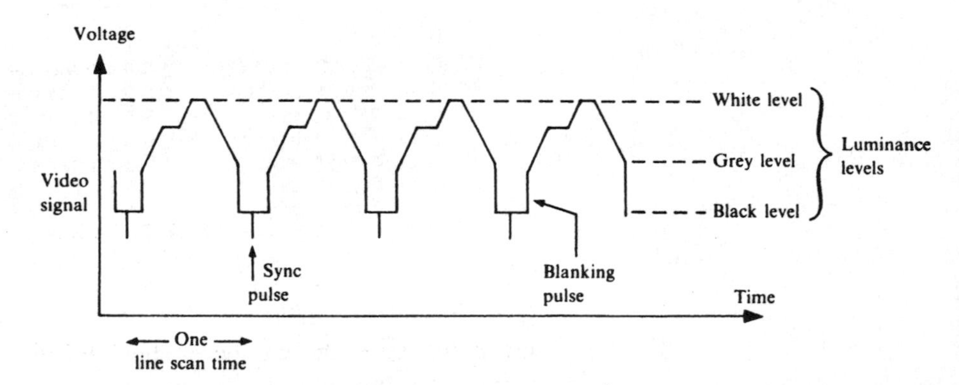

Figure 1.11. A typical video signal waveform.

Exercise 1.5
Check on page 20

Draw the waveform obtained during a line scan of the image shown in Fig. 1.12.

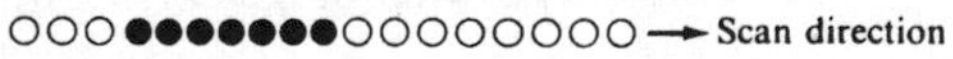

○○○●●●●●●●○○○○○○○○ → Scan direction

Figure 1.12. A simple image.

The TV receiver separates the sync pulse from the luminance information and uses the pulse to trigger the receiver timebase generator. A timebase waveform is needed to produce the electron beam scan pattern or *raster* across the CRT screen as shown in Fig. 1.13. At the end of each scan

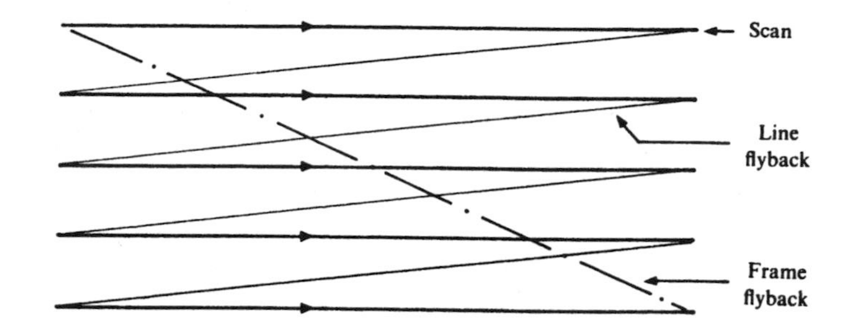

Figure 1.13. A simple raster.

the electron beam must fly back to begin the next scan, and during this very short period the energy of the electrons must be reduced so the screen appears blank.

In a colour TV system, *chrominance* information would also be included in the video waveform. The moving images of television are produced by usng repeated rapid scanning to encode a succession of image signals and persistence of vision does the rest. TV systems will be explained more fully in Chapter 6.

Data signals

Data signals consist of electrical pulses of the kind shown in Fig. 1.10. Binary-coded pulses are the most common, but other codes are also used in telecommunications for reasons that will emerge in later chapters. These forms of signal are mainly associated with computer communications, hence the term 'data signals', but they are also used for voice communications. We consider these signals in more detail in the next section.

1.3 Analogue and digital signals

The terms 'analogue' and 'digital' have been mentioned several times and we have now reached the stage where we can try to define what is meant by these terms and show how the same information may be carried in either form.

If an electrical signal has the same waveform as, for instance, a speech signal, the electrical signal is then *analogous* to the speech signal. The electrical current is proportional to the air pressure of the speech signal. An electrical analogue of a tone of constant amplitude is shown in Fig. 1.14.

Certain engineering problems were first solved with the aid of electrical circuit models or simulations in which voltage was analogous to the problem variables. This was the principle of the, now almost obsolete, analogue computer. Analogue signals are continuous in time and can assume any magnitude within their amplitude. It is this *continuity* in time

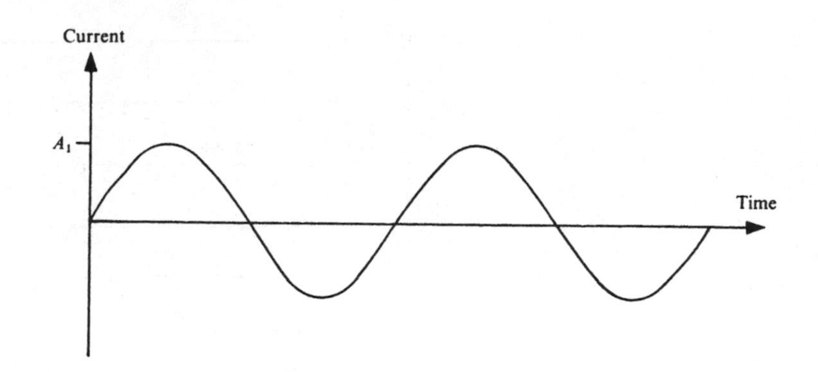

Figure 1.14. The electrical analogue signal of a tone.

and magnitude that is implied by the term 'analogue' as it is now used. Digital signals, on the other hand, are discontinuous in time and have only *discrete* values of magnitude—very often only two, as is the case with digital computers. These two-valued digital signals are known, more precisely, as *binary* signals, but the general term 'digital' tends to be used. It should be appreciated that digital signals are very easy to process and their information is easily recovered in the presence of noise—think of Morse code. This is not the case with analogue signals, as will be explained in Chapters 7 and 8. Consequently many systems convert analogue signals into digital form prior to transmission. Let us see how an analogue signal in the form of a single tone might be digitized by considering Fig. 1.15.

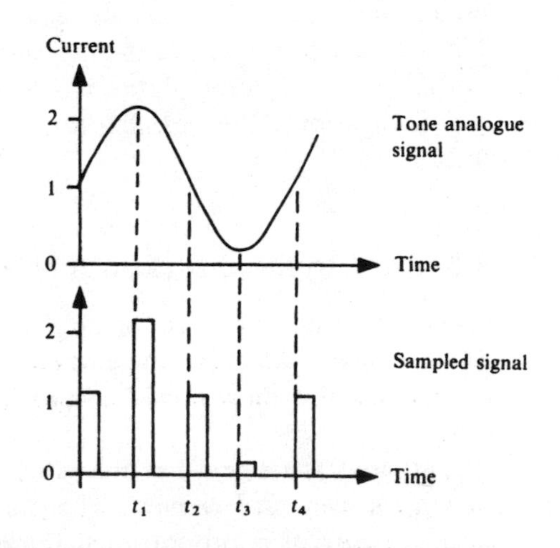

Figure 1.15. An analogue signal and its sampled version.

The amplitude of the analogue signal in Fig. 1.15 is *sampled* at regular intervals t_1, t_2, etc., giving samples with amplitudes of about 1.2, 2.2 and 0.2. The sample values depend on when the samples are taken and can have any of the instantaneous values assumed by the analogue signal. The sampled waveform is digital in time but analogue in amplitude.

In order to obtain a waveform that is digital in magnitude, sample values are encoded, usually into binary-coded pulses, as exemplified in Fig.

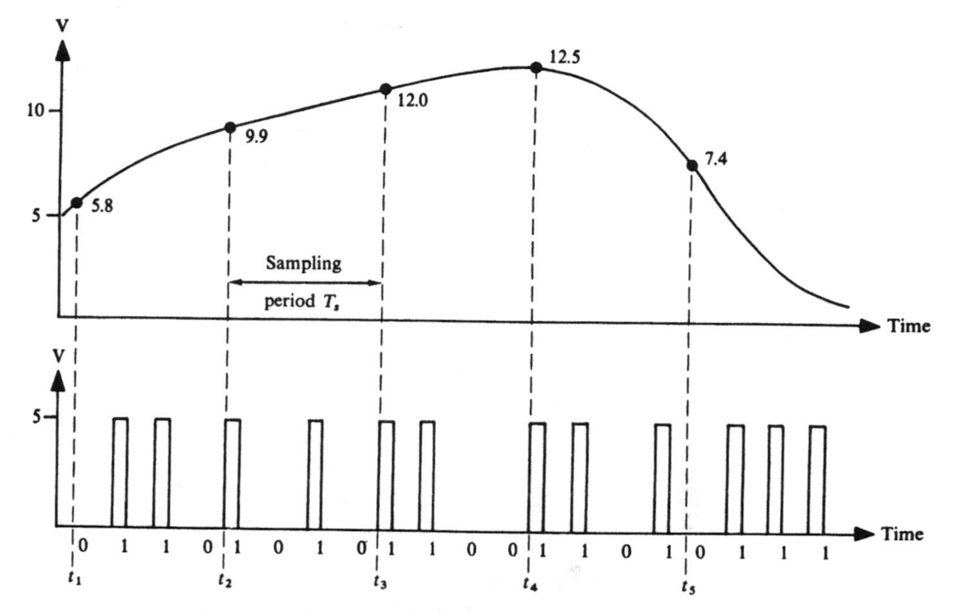

Figure 1.16. An analogue signal and its pulse-coded version.

1.16. In the example of Fig. 1.16, the first sampled amplitude value of the analogue signal is, say, 5.8 V, which encodes into more bits than the four allowed in this example. The nearest we can get to 5.8 with four bits is 0110, giving an error of 0.2 V. The next sample is about 9.9 V, which encodes to 1010 with an error of 0.1 V. With four-bit coding, the signal samples over the range 0 to 15 V can only be resolved to the nearest whole number. The signal values have been *quantized* into integer values by the four-bit conversion. Each four-bit word is transmitted as a current pulse pattern in a *time slot* T_s. Note that the fourth sample in Fig. 1.16 is 12.5 V, which encodes as either 1100 or 1101 with the maximum *quantization error* of 0.5 V in either case. In this example the quantum happens to be 1. Words of more bits would reduce the quantum and the quantization error.

Quantization error has to be accepted as the price paid for the advantages obtained when digital transmission is used (these are described in Chapter 7). Standard integrated circuits (ICs) known as *analogue-to-digital converters* (ADCs) are available for signal processing. This important method of transmitting analogue information, known as *pulse code modulation* (PCM), is covered in more detail in Chapter 7. The original analogue signal can be reconstituted, albeit with some error, from the binary-coded signal by using *digital-to-analogue converter* (DAC) ICs. The closeness of the recovered analogue signal to the original improves as the number of bits increases.

It is shown in Chapter 7 that the sampling rate must be greater than twice the frequency of the analogue signal being sampled. In the compact disc (CD) system, which also uses PCM, the sampling rate is 44.1 kHz, giving a frequency response in excess of the audio range of 20 kHz.

Sixteen-bit coding is used, giving $2^{16} = 65\,536$ discrete signal levels. If these levels were not binary encoded, we would have a '65 536-ary' signal and not a 'binary' signal! In general, a digital signal consisting of M levels is known as an *M-ary signal.*

Bandwidth savings can be made by reducing the number of quantized levels (or bits), but fidelity suffers. Vocoders, which are used in some digitized speech systems, use bit rates that are reduced to a minimum level consistent with intelligibility in order to save transmission bandwidth.

Exercise 1.6
Check on page 20

Distinguish between:

1. Analogue and digital signals
2. Binary and digital signals
3. FAX and TV signals

1.4 Some characteristics of digital signals

The pulses shown in Fig. 1.16 are known as binary units or bits. They will have to be sufficiently brief so that they can fit into the sampling period. The number of bits per second (b/s) is usually referred to as the *bit rate* rather than the frequency. A standardized bit rate of 64 kb/s, for example, is used in many digital telephone networks. Computer terminals linked together on a local area network (LAN) might use 10 Mb/s. The pulse period (T) is the time shown in Fig. 1.17.

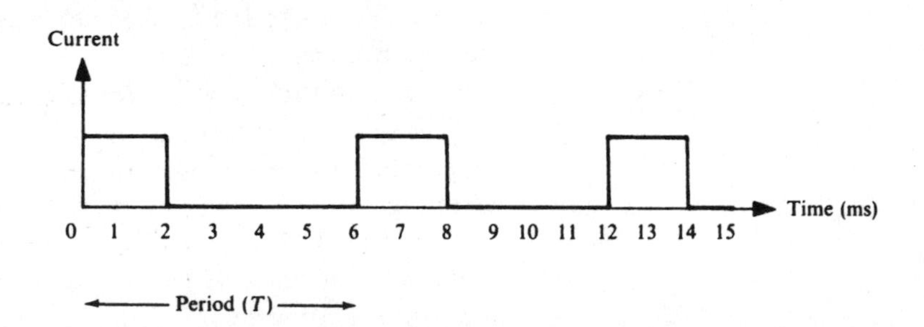

Figure 1.17. The pulse period.

The bit rate (pulse frequency) is the reciprocal of the period, as it would be for any periodic signal. The duration of a pulse is known as the *pulse width* (W). It is the time interval during which the pulse level is high or low. Figure 1.18 shows pulses of various rates and widths. In Fig. 1.18(a), the pulse width is half the period, so that the space between pulses is equal to the pulse width. That is, there is a one-to-one *mark/space ratio.* In Fig. 1.18(b), the pulse width is a third of the period, so the mark/space ratio is 1:2.

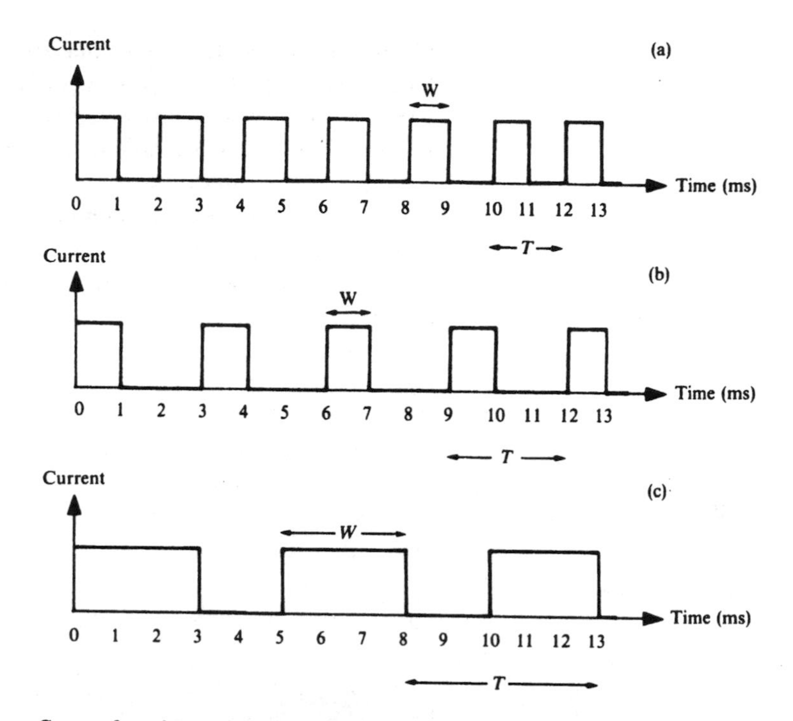

Figure 1.18. The pulse width.

Exercise 1.7
Check on page 20

State, for the pulse signal in Fig. 1.18(c), the values of the period, the pulse width, the mark/space ratio and the bit rate.

Types of binary waveform

Three important forms of binary signal are shown in Fig. 1.19. In this example, we have a clock signal that establishes the timing and the bit rate

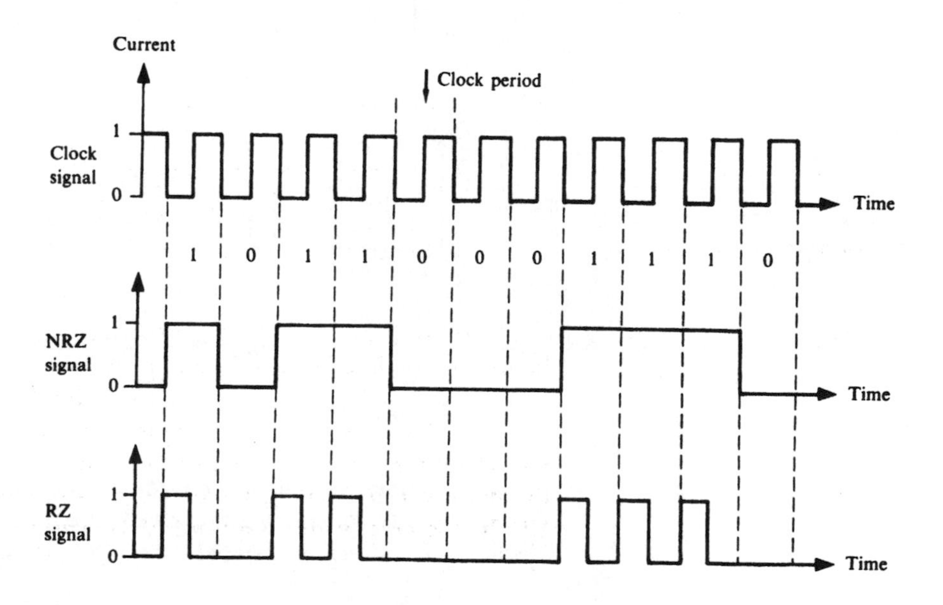

Figure 1.19. Clock, NRZ and RZ binary pulse waveforms.

of the system. A binary sequence 10110001110 is represented by a current pulse every time the digit 1 occurs. The resulting current waveform may have the shape of the *non-return-to-zero* (NRZ) signal in which the current stays high during the whole of a clock period. Alternatively, it may have the shape of the *return-to-zero* (RZ) signal in which the current returns from high to zero during each clock period containing 1. Both the NRZ and RZ signals are synchronized to the clock signal, so they are described as *synchronous.* Normally, their transitions occur when the clock pulse goes from high to low (that is, on the trailing edge of the clock pulse). The NRZ signal requires less bandwidth than the RZ signal because it has less transitions, but it loses its timing information more readily than the RZ signal as the number of consecutive 0s or 1s increases. The RZ signal has this drawback for consecutive 0s only.

A form of coding that overcomes the loss of timing due to a long sequence of 0s or 1s uses transitions instead of levels to represent the binary digits. For example, a transition from high to low could represent 1 and a transition from low to high could represent 0, giving the *biphase* RZ signal waveform shown in Fig. 1.20.

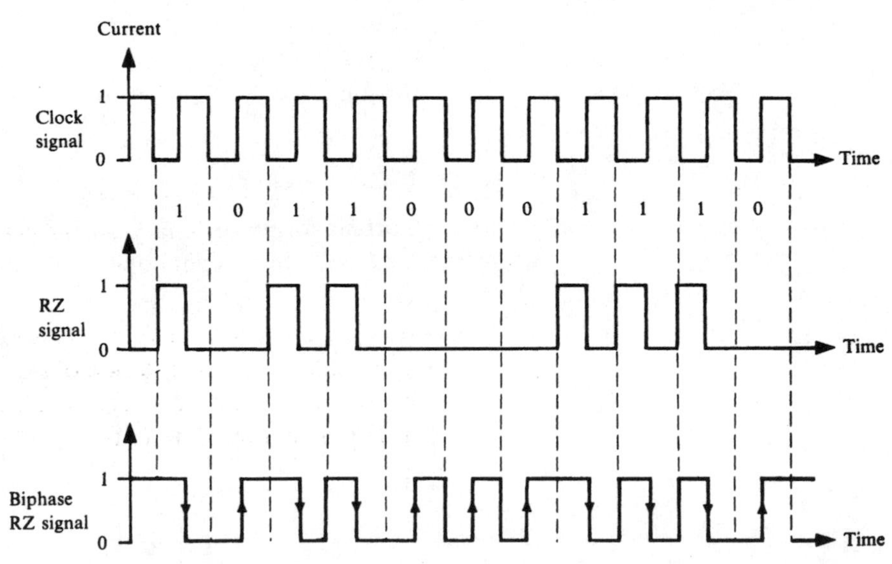

Figure 1.20. Two forms of RZ binary signals.

Exercise 1.8
Check on page 20

Draw pulse waveforms of the kind shown in Figs 1.19 and 1.20 to show NRZ, RZ and biphase RZ signals for the binary sequence 01010111000.

The examples of pulses seen so far are *unipolar* or *single-current.* If binary 0 is represented by a negative current instead of zero, the result is a *bipolar* or *double-current* signal. Both types of signal are used.

1.5 Signals and their frequency content

All periodic signals, that is, those described by $y(t)=y(t+T)$, are made up of a number of sinusoidal components or *harmonics*, which go to make up a *Fourier series* of the following general form:

$$y(t)=A_1\sin(2\pi f_1 t+\phi_1)+\cdots A_n\sin(2\pi f_n t+\phi_n)+\cdots \qquad (1.2)$$

where A_n, f_n and ϕ_n are the amplitude, frequency and phase, respectively, of the nth harmonic. The first term of the series is called the *fundamental* rather than first harmonic. The cosine function provides a similar alternative description of a complex waveform as a summation or *superposition* of cosinusoidal components of various amplitude, frequency and phase. For periodic waveforms, the frequencies are integer-related and would be represented as spectral lines in the frequency spectrum of the signal. Figure 1.21 shows the spectrum of Eq. (1.2).

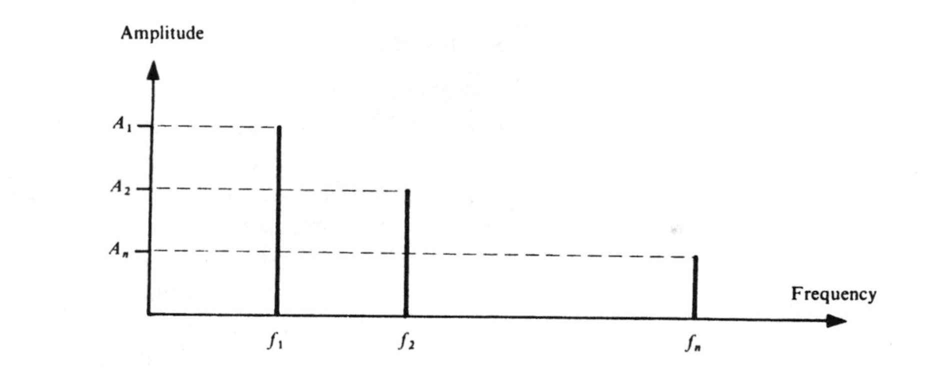

Figure 1.21. A frequency line spectrum.

A single pulse of width W has a continuous spectrum whose envelope is shown in Fig. 1.22. The representation of signals in terms of their frequency components plays an important part in telecommunications theory and is covered in more detail in Chapter 3.

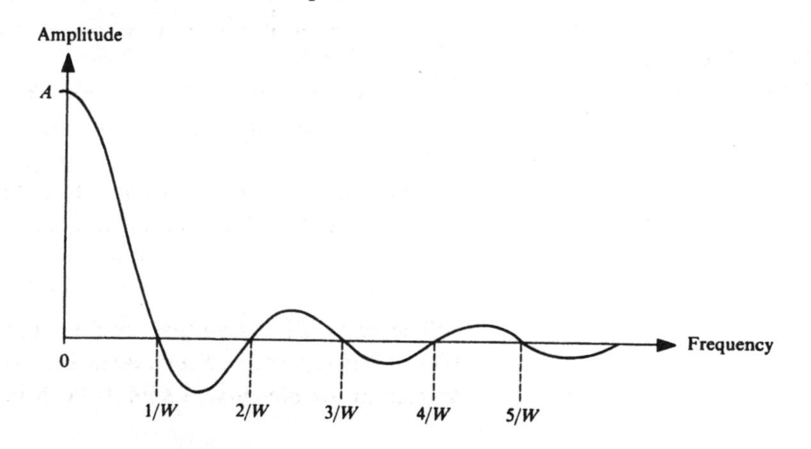

Figure 1.22. The continuous spectrum envelope of a single pulse.

Transmitting information by means of digital signals has many advantages. The availability of cheap ADCs, DACs and microprocessors has enabled digital techniques to play an ever-increasing role in telecommunication engineering.

Digital signals are made up of binary digits, which make up a word. The rate of transmission of these words may be up to many millions per second, giving bit rates of hundreds of Mb/s. The shape of the pulses making up the bit stream does not need to be preserved in order to preserve the information content of a signal. It is only necessary to discern the presence or absence of current in order for a receiver to recover information from the signal. Thus the transmission bandwidth required for a digital signal is of the order of the bit rate of the data stream. This will be higher than that required to transmit the original analogue signal. For example, an analogue telephone speech signal requires a bandwidth of 3.1 kHz, but when it is transmitted by PCM at 64 kb/s a bandwidth of about 64 kHz is required.

Summary

This chapter has provided a description of the main categories of signal used in telecommunications: audio, video and data. A signal consists of energy, which is used to carry information in the form of speech, pictures or computer information. If this energy is electromagnetic, the signal can be sent great distances over line or radio links, the information being carried by encoding the signal in an analogue and/or digital form.

The history of telecommunications began with simple signals such as were used in the railway telegraph. Morse code was used in wire and radio links for long-distance telecommunication and in flash-light links over short distances. These rather crude digital signals carried very little information but were effective in conveying this information against a noisy background. As technology developed, analogue electrical signals were used to carry more information, first along wires and then over the air as radio waves. Much of this information was often obscured by electrical noise from electrical storms, electrical gadgets and other people's signals.

We have now come full circle in that many modern signals have reverted to digital form in order to overcome the noise problem. Improved technology has, at the same time, enabled these signals to carry much more information, at much lower cost, than their digital ancestors, and the use of Morse code has recently been discontinued. The latest technique for sending messages has actually gone back to using light as the carrier. However, instead of using Morse-coded light-beam flashes through the atmosphere, the light signal is guided through cables consisting of hundreds of glass fibres each carrying many pulse-coded messages. Large networks of these cables now exist and provide the means of conveying huge amounts of information at great speed over enormous distances, with almost complete freedom from noise. The same digital techniques have enabled radio to be used in space probe and satellite communications.

A signal is a stimulus, which produces an effect or sensation at a receiver. The signal may be a pressure variation, which will produce a sensation of sound or touch. In telecommunications, the signal in the transmission medium (the channel) is electromagnetic, and it may be detected when it is transduced into a sensible form. A signal that contains information varies with time in an unpredictable manner. When we sense how it is varying, we have received information. Information is encoded in the signal in a manner that suits the communications medium. In telecommunications the medium will be a cable or radio link, probably carrying many communication channels, in which the signal will become distorted by channel time delay and non-linearity, and obscured by noise. These effects are covered in subsequent chapters.

► Checks on exercises

1.1 Two principles were involved. First, when a railway line was tapped, the steel was set into vibration and a sound wave travelled along the line. By using a railway line to guide the sound, most of the sound energy was kept in the line and did not disperse. This meant that the range of the signals might, according to the length of the line and the weather conditions, be greater than the smoke signals (radio waves are sent along waveguides for similar reasons). Secondly, the messages conveyed by the signal were encoded as a rhythmic pattern of sounds—an early form of digital audio signal!

1.2

1. A 20 mm diaphragm would have an area of $\pi \times (10^{-2})^2\ \text{m}^2$, so the incident power would be $10^{-6} \times \pi \times 10^{-4}$, which is about 300 pW.
2. If the diaphragm is 10% efficient then the electrical power in the line would be about 30 pW. Thus $\frac{1}{2} v_m i_m = 30$ pW, so $i_m = 60$ pA.

1.3

1. The relationship that must be used is $v = \lambda \times f = 330$ m/s. So $\lambda = 330/660 = 0.5$ m.

The relationship that must be used in the remaining cases where electromagnetic waves are involved is $c = \lambda \times f = 3 \times 10^8$ m/s. So $\lambda = (3 \times 10^8)/f$. This gives wavelengths of the following values:

2. 3 m for 100 MHz.
3. 30 mm for 10 GHz.
4. 15 μm for 20 THz.
5. 500 nm (nanometres) for 600 THz ($n = 10^{-9}$).

1.4

1. An electrical signal would be delayed by 5000 km/(3×10^8 m/s) = 17 ms.

2. A sound signal in steel would be delayed by 5000 km/(5000 m/s) = 1000 s.
3. A sound signal in air would be delayed by 5000 km/(330 m/s) = 15 000 s.

Acoustic transmission for two-way communication over such long links is impracticable because of the long time gap between transmission and reception.

1.5 The waveform obtained during a line scan of the image is shown in Fig. 1.23.

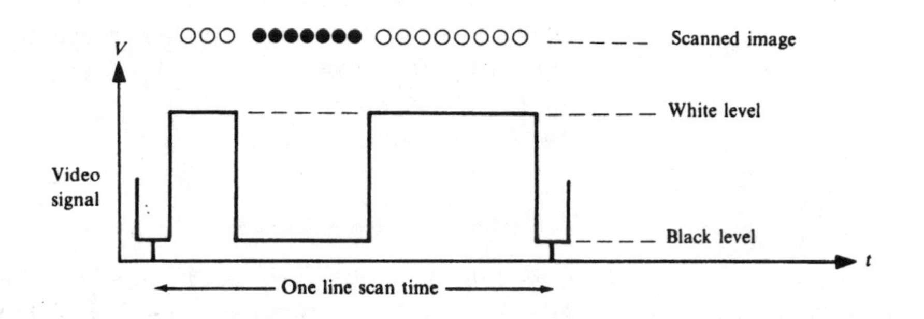

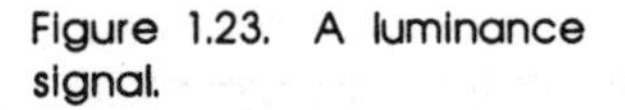
Figure 1.23. A luminance signal.

1.6
1. An analogue signal can have any value within its amplitude. A digital signal has a limited number of discrete values only.
2. A binary signal has only two values, so it is one type of digital signal.
3. A FAX signal is a binary signal because it carries only black (1) or white (0) luminance information. A TV signal must carry all shades from black, through grey, to white. This continuous variation in luminance classifies the TV signal as analogue.

1.7 For the pulse signal in Fig. 1.18(c), the period is 5 ms, the pulse width is 3 ms, the mark/space ratio is 3/2 and the bit rate is 1/(5 ms) = 200 b/s.

1.8 The NRZ, RZ and biphase RZ signal waveforms for the sequence 01010111000 are shown in Fig. 1.24. The biphase code, alias the *Manchester code*, has double-width pulses whenever 1s and 0s are adjacent and single-width pulses at repetitions, so timing is preserved over long sequences of 0s or 1s. Other forms of code exist that maintain timing information in spite of long sequences of 0s. The *high-density bipolar 3* (HDB3) code, described in Chapter 7, is an important example.

Self-assessment questions

1.1 What term would be used to describe measuring at a distance?

1.2 What are the main features of electrical energy that make it preferable to other forms of energy for telecommunications?

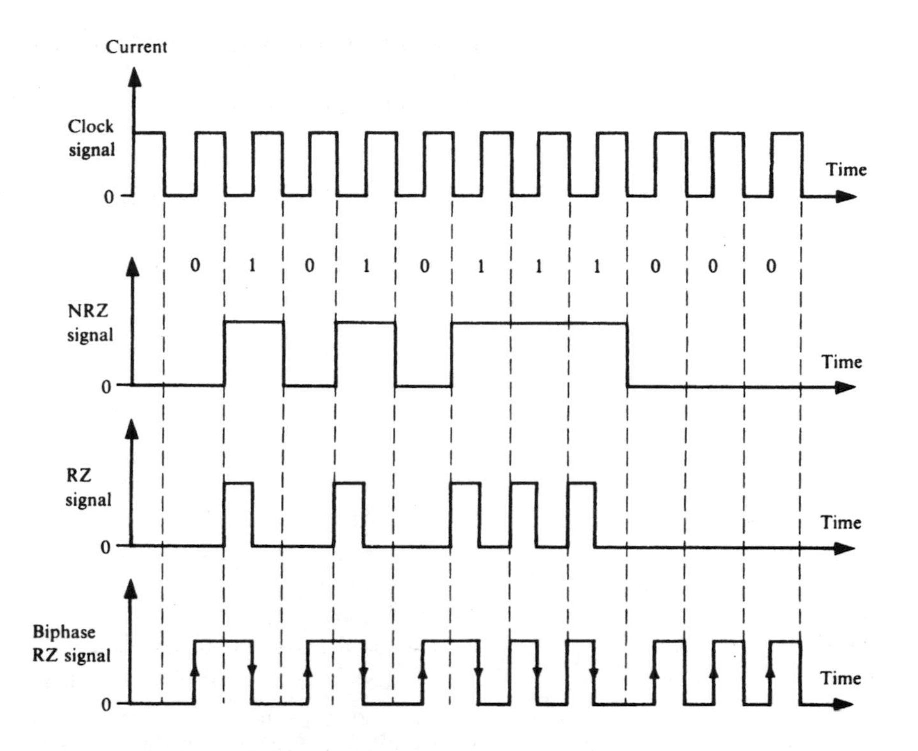

Figure 1.24. NRZ, RZ and biphase RZ binary pulses.

1.3 State which factors determine the minimum signal energy required for intelligible reception of an audio signal.
1.4 What properties of a sound signal determine pitch and timbre?
1.5 Why has the telephone network adopted the frequency range 300 to 3400 Hz?
1.6 Why do video signals have a greater bandwidth than audio signals?
1.7 Show, with the aid of a diagram, how the character α would be transmitted along a telephone line.
1.8 What are the functions of the sync pulse and luminance signal in TV?
1.9 What advantage does a digital signal like Morse code have over a speech signal in long-distance telecommunication?
1.10 State which of the following are analogue and which are digital: temperature, heat flow, atomic spectra, the number of pages in a book.
1.11 Write down the definition of a periodic waveform.
1.12 Write out the Fourier series for the first four components of a signal that consists of odd-order harmonics only, all in the same phase.
1.13 State the difference between the spectra of periodic and aperiodic signals.

2

Channels and networks

Carrying and routing signals

Objectives

When you have completed this chapter you will be able to:

- Describe various forms of line and radio link
- Define what is meant by a telecommunications channel
- Describe how more information may be carried along a link by multiplexing
- Define duplex, half-duplex and simplex transmission
- Define a network
- State why switching is used in telephone networks
- Describe the principle of cross-point switching
- Describe line concentration and expansion
- Describe the facilities offered by a private branch network
- State the function of local area networks
- Distinguish between local, wide and metropolitan area networks

In the last chapter it was established that telecommunication required a line or radio link to carry information, in the form of electrical energy, between a transmitter and a receiver.

The aim of this chapter is to describe various kinds of link and how electrical signals are carried, how these links may be organized into networks and how efficient use may be made of links and networks.

2.1 Communication by line

Telecommunication by line involves the use of a length of some material such as metal or glass to *guide* the signal between a sender and a receiver. The material takes the form of long, continuous strands called *wires* or *fibres*.

Wire links

In the simplest form of speech communication, the voice signal is converted, by a microphone (MIC), into an analogous (voice-shaped) electrical signal so that it can be sent along wires to a loudspeaker (LS), which converts the signal current back into the original voice signal. This process is outlined in Fig. 2.1. Notice that two wires are required in the link between the MIC and LS so that an electrical circuit or *loop* is formed to provide for current flow. If a tone was applied to a MIC, the signal current would have the waveform shown in Fig. 1.14.

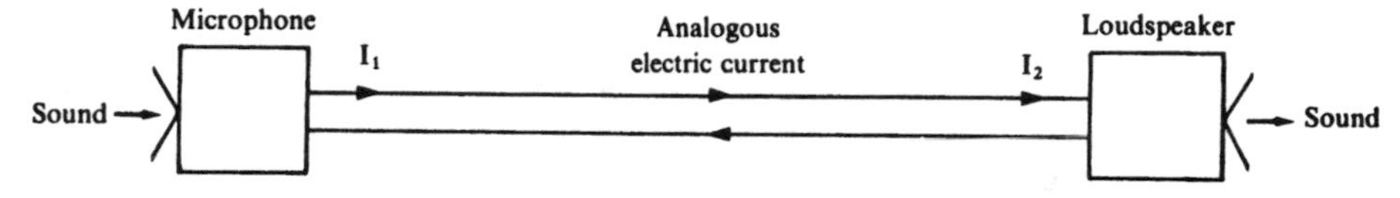

Figure 2.1. The microphone loudspeaker circuit.

The LS consists of a diaphragm, which is vibrated by an electromagnet whose coils carry the voice signal current. The MIC is normally a parallel-plate capacitor type with a metal diaphragm forming one of the plates. The small change in voltage across the plates, which occurs when the capacitance is varied by the sound pressure on the diaphragm, requires amplification if sound of acceptable quality is to be received over links of more than a few metres. Before the advent of cheap electronic amplifiers, microphones used carbon granules, which formed part of the loop. These were pressed down under a diaphragm, which caused their area of contact and hence their resistance to vary as the pressure on the granules varied. A battery, situated in the local telephone exchange, was included in the loop to provide a current whose value varied with the sound-induced resistance variation. This arrangement gave an electrical signal that did not require amplification, but the sound quality was poor due to the 'crackling' granules. Carbon granule MICs became obsolete in the late 1970s. The modern handset contains an electromagnetic LS, an electrostatic MIC and an amplifier powered by batteries situated in the local exchange.

Signal currents are conveyed by copper or steel conductors (wires). When wires are grouped together they form *cables*. The simplest form of cable is a pair of insulated wires, such as those used to connect a subscriber's premises to the public telephone network. A local telephone link is made up of cables consisting of many hundreds of these insulated 'pairs' all within the same protective cover. Each wire pair provides an information path or *channel*. We shall see later in the chapter how one pair may accommodate many channels.

All wires absorb energy from the signal as it progresses down the line, so that the signal power at the receiving end of the line is less than that at the sending end. The ratio of the transmitted signal power to the received signal power is the *loss* or *attenuation* (A):

$$A = I_1^2 R_1 / I_2^2 R_2$$

It is normally assumed that $R_1 = R_2$, although this is not always the case in practice, so that:

$$A = (I_1/I_2)^2 = 20 \log(I_1/I_2)\ \mathrm{dB}$$

(see Appendix A). A similar expression for attenuation in terms of voltage is also valid, namely:

$$A = 20 \log(V_1/V_2)\ \mathrm{dB}$$

A cable may be represented by a cascade of T sections as shown in Fig. 2.2. The cable parameters are: the wire resistance R (Ω/m), the self-

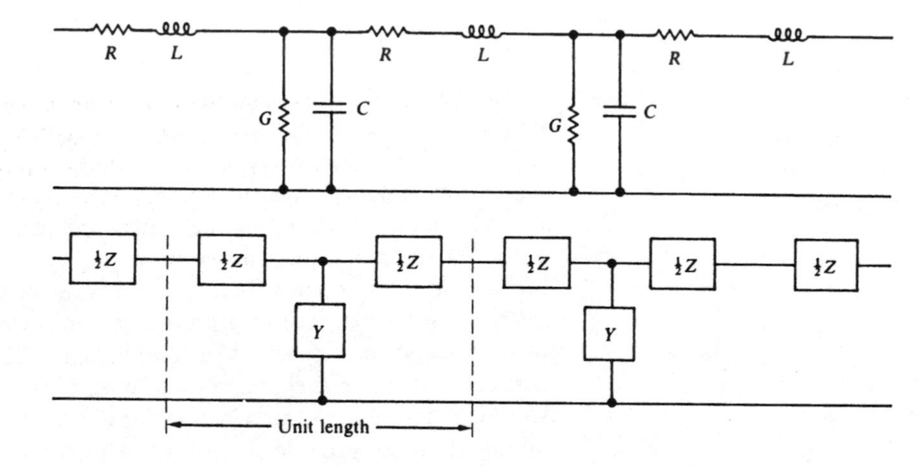

Figure 2.2. Representations of a cable pair.

inductance L (H/m), the leakage conductance G (S/m) and the capacitance C (F/m); for a sinusoidal signal of angular frequency $\omega = 2\pi f$

$$Z = R + \mathrm{j}\omega L \quad \text{and} \quad Y = G + \mathrm{j}\omega C$$

When a sinusoidal signal is input to a cable, the current and voltage waves travel along the cable until they reach the receiving end. The signal power that is not absorbed by the end loading is then reflected back along the cable. The only way to prevent reflection is to have either an infinitely long cable or a cable loaded by an impedance equal to the *characteristic impedance* (Z_0). Z_0 is defined as shown in Fig. 2.3. The characteristic impedance of a cable is the impedance that would be measured at the input of an infinitely long cable. Such a cable can be simulated by terminating a finite cable with a load Z_0. The characteristic impedance is given, from Fig. 2.3, by the expression:

$$Z_0 = Z\delta x + Z_0/[Y\delta x(Z_0 + 1/Y\delta x)]$$

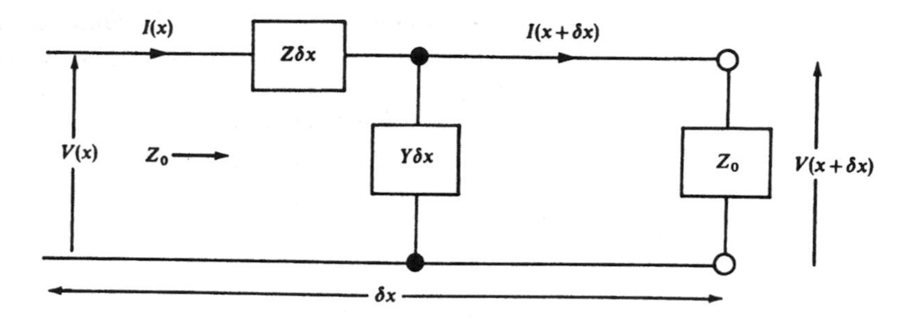

Figure 2.3. Definition of characteristic impedance.

so

$$Z_0^2 Y = Z + ZZ_0 Y\delta x$$

As $\delta x \to 0$,

$$Z_0^2 \to Z/Y = (R + j\omega L)/(G + j\omega C)$$

for a sinusoidal input signal.

Exercise 2.1
Check on page 48

1. The inductance and capacitance per unit length of an isolated copper pair is given respectively by $L = \mu \ln(d/r)/\pi$ and $C = \pi\varepsilon/\ln(d/r)$, where μ and ε are the permeability and permittivity, respectively, of the insulating material, r is the radius of the wire and d is the distance between the wires. If $r = 0.5$ mm, $d = 3$ mm, $\varepsilon = 10^{-9}/36\pi$ F/m and $\mu = 4\pi \times 10^{-7}$ H/m, calculate C and L.
2. Calculate the characteristic impedance of the twin cable if R and G are negligible.

Distortion produced by lines

The characteristic impedance of a line varies between $\sqrt{(R/G)}$ at $\omega = 0$ and $\sqrt{(L/C)}$ at $\omega = \infty$ as shown in Fig. 2.4(a). The output signal will thus vary in amplitude and phase with frequency. To keep the characteristic impedance of the line constant throughout the working frequency range, R/G

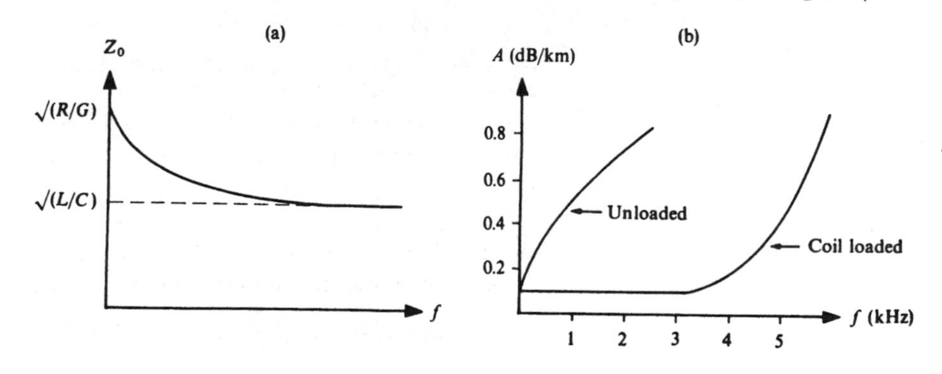

Figure 2.4. Variation of line Z_0 and attenuation with frequency.

must be made equal to L/C. That is, $LG=CR$. This means that, since L is low and C is high, L or G must be increased. It would be foolish to increase G because this would increase the line power loss, so L must be increased. This is done by connecting *loading coils* at intervals along the line to give the effect shown in Fig. 2.4(b).

Looking back to Fig. 2.3 we see that

$$V(x)-V(x+\delta x)=\delta V=I(x)Z\delta x$$

so, as $\delta x\to 0$,

$$\delta I/\delta x\to \mathrm{d}I/\mathrm{d}x=YV$$

and

$$I(x)-I(x+\delta x)=\delta I=V(x)Y\delta x$$

so, as $\delta x\to 0$,

$$\delta I/\delta x\to \mathrm{d}I/\mathrm{d}x=YV$$

Hence

$$\mathrm{d}^2I/\mathrm{d}x^2=Z\,\mathrm{d}I/\mathrm{d}x=ZYV=\gamma^2V \tag{2.1}$$

where the *propagation constant* is given by

$$\gamma=\surd(ZY)=\surd[(R+\mathrm{j}\omega L)(G+\mathrm{j}\omega C)]=\alpha+\mathrm{j}\beta$$

When R and G are negligible, $\beta=\omega\surd(LC)=$ phase delay/unit length of line.

A solution to Eq. (2.1) is

$$V(x)=A\,\mathrm{e}^{-\gamma x}+B\,\mathrm{e}^{\gamma x} \tag{2.2}$$

as also are $V(x)=A\,\mathrm{e}^{-\gamma x}$ and $V(x)=B\,\mathrm{e}^{\gamma x}$. Now $V(0)=A\,\mathrm{e}^{-0}$ so $A=V(0)$. Hence

$$V(x)=V(0)\,\mathrm{e}^{-(\alpha+\mathrm{j}\beta)x}=V(0)\,\mathrm{e}^{-\alpha x}\,\mathrm{e}^{-\mathrm{j}\beta x}$$
$$=V(0)\,\mathrm{e}^{-\alpha x}(\cos\beta-\mathrm{j}\sin\beta)$$

If the input voltage varies sinusoidally with time then $V(0)$ is written as $v(0)$ and is given by

$$v(0)=V_p\sin(\omega t)=\mathrm{Im}\ V_p\,\mathrm{e}^{\mathrm{j}\omega t}$$

where Im ≡ imaginary part of $V_p\,\mathrm{e}^{\mathrm{j}\omega t}$. Hence

$$\begin{aligned}v(x)&=\mathrm{Im}\ V(0)\,\mathrm{e}^{-\alpha x}\,\mathrm{e}^{-\mathrm{j}\beta x}\\&=\mathrm{Im}\ V_p\,\mathrm{e}^{\mathrm{j}\omega t}\,\mathrm{e}^{-\alpha x}\,\mathrm{e}^{-\mathrm{j}\beta x}\\&=\mathrm{Im}\ V_p\,\mathrm{e}^{-\alpha x}\,\mathrm{e}^{\mathrm{j}(\omega t-\beta x)}\\&=V_p\,\mathrm{e}^{-\alpha x}\sin(\omega t-\beta x)\\&=V_p\,\mathrm{e}^{-\alpha x}\sin[\omega(t-\beta x/\omega)]\\&=V_p\,\mathrm{e}^{-\alpha x}\sin[\omega(t-t')]\end{aligned}$$

When $t'=T$, $x=\lambda$ so $T=\beta\lambda/\omega$. Hence wave velocity $(\lambda/T)=1/\sqrt{(LC)}$. As the sine wave travels along the line there is an exponential reduction in its amplitude (see Fig. 2.5)

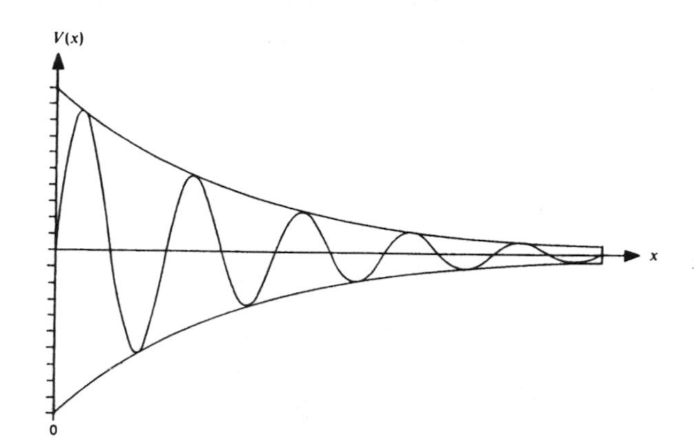

Figure 2.5 Sine wave attenuation along cable.

$$\text{attenuation } A=V(x)/V(0)=e^{-\alpha x}$$

so $e^{-\alpha}$=attenuation per unit length, where the *attenuation constant* $\alpha=-\ln[V(1)/V(0)]$ nepers. Also

$$\begin{aligned}\text{attenuation/unit length in dB} &= 20\log[V(1)/V(0)]\\ &= 20\log(e)^{-\alpha}=-8.7\alpha\end{aligned}$$

Obviously, the attenuation of a cable increases with length. What is not so obvious is that attenuation also increases as the frequency goes up. This is partly because a cable radiates *electromagnetic waves* when current passes, and the higher the current frequency, the more power is radiated. This accounts for some of the cable loss. The radiation also gives rise to unwanted signals or *noise* in adjacent signal paths—an effect known as *cross-talk*.

A more important cable loss is due to the *skin effect* in which the current flows mainly on the outer 'skin' of a wire instead of being uniformly distributed over the whole cross-sectional area. This is because a greater number of flux linkages exist in the centre of a conductor than at the surface. As the current frequency increases, the effective area (a_{eff}) reduces. This increases the wire resistance (R) because $R=\rho l/a_{eff}$, where ρ is the resistivity of the material and l the wire length. The conducting skin depth (δ) is given by $\delta=\sqrt{(\rho/\pi f\mu)}$, where f is the frequency. For copper, $\delta=6.6$ mm at 100 Hz, 66 μm at 10 kHz and only 0.66 μm at 1 MHz, so the effect is appreciable in the upper audio range.

Coaxial cables

Not much can be done about skin effect, but radiation loss and cross-talk may be reduced by using *coaxial* cable, of the kind shown in Fig. 2.6. The electromagnetic fields are almost entirely contained within the outer conductor of the cable, which is normally earthed. This considerably reduces radiated power loss from the cable compared with that which would occur, at the same frequency, in a twin. Coaxial cables are used when the signal frequency exceeds about 30 MHz and are very familiar through their use in interconnecting TV sets and aerials. They are also used in networks linking data-handling equipment. They are subject to the same kinds of losses and phase shifts as any line, and these have to be made good by *repeaters* when, for example, they are used in under-sea telecommunication links. One of the longest such links, which runs between Vancouver in Canada and Sydney, Australia—a distance of almost 8000 miles—has about 1000 repeaters, which compensate the loss and correct the phase distortion in each eight-mile section of cable between repeaters.

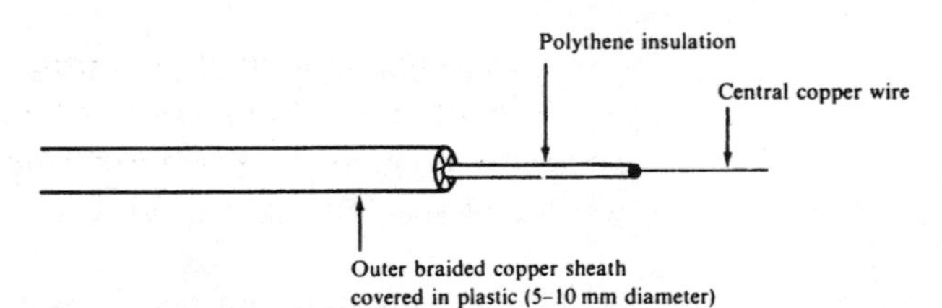

Figure 2.6. The construction of a coaxial cable.

Exercise 2.2
Check on page 48

1. The capacitance per unit length of a coaxial cable is given by $C=2\pi\varepsilon/\ln(r_o/r_i)$ where ε is the permittivity of the insulating material, r_i is the radius of the inner wire and r_o is the radius of the outer conductor. The approximate inductance per unit length of such a cable is given by $L=\mu\ln(r_o/r_i)/2\pi$ where μ is the permeability of the insulating material. If $r_i=0.3$ mm, $r_o=3$ mm, $\varepsilon=10^{-9}/36\pi$ F/m, $\mu=4\pi\times10^{-7}$ H/m and R and G are negligible at the operating frequency, calculate the characteristic impedance and the velocity of the signal along the line.
2. If the wire loop resistance is 40 Ω/km and the leakage conductance is 2 μS/km, calculate Z_0 and the attenuation per kilometre at zero frequency.

Waveguides

The attenuation of a coaxial cable increases with frequency rise, as does that of any cable. The signal energy is lost mainly in the insulation between the inner and outer conductors—a form of *dielectric loss* incurred as the line capacitance is repeatedly charged and discharged at the signal frequency. In practice, the loss becomes significant above about 3 GHz. For this reason, another form of transmission medium is used consisting of a hollow tube down which the signal energy, in the form of an electromagnetic wave, is sent. The signal conductor is now thought of as a *wave* guide rather than a *current* guide, although there is signal current in the wall of the waveguide. The guide may be thought of as a coaxial cable with the inner conductor dispensed with. The frequency above which it is more appropriate to think of waves rather than currents is somewhat vague, but 3 GHz may be regarded as the break frequency.

The electromagnetic wave is launched into a guide by a loop of wire or *probe* and the wave progresses by being reflected from side to side between the walls of the guide. A waveguide is a metal tube, of either rectangular or circular cross-section, with inner surfaces that are polished to facilitate internal reflection. In effect, a waveguide provides a confined space in which a super-high-frequency radio wave is propagated. Energy losses occur mainly due to the non-perfectly reflecting walls. In order to minimize losses, the guide must be made so that it is an accurately machined, rigid structure, and high-quality joints are needed to link sections. If the guide is to go round corners, precision bends must be used. These factors make a waveguide very expensive and means that they are normally used only for short links. For example, they may be used to connect transmitters and receivers to their antennae but they would not be used on telecommunications trunk routes. A much cheaper form of 'waveguide' in the form of a *glass fibre* is available for long-distance links.

Glass-fibre cables

A glass or optical fibre is a fine strand (about 100 μm diameter) of very high-purity glass made up of a glass core of refractive index μ_1 surrounded by a glass cladding of lower refractive index μ_2 as shown in Fig. 2.7. The fibre propagates light waves by internal reflections in a similar manner to that of the waveguide—the light beams swinging from side to side as they are internally reflected at the core/cladding interface. The purpose of the cladding is to provide a change of refractive index and thus a reflective surface within the fibre. The glass used for both the core and the cladding

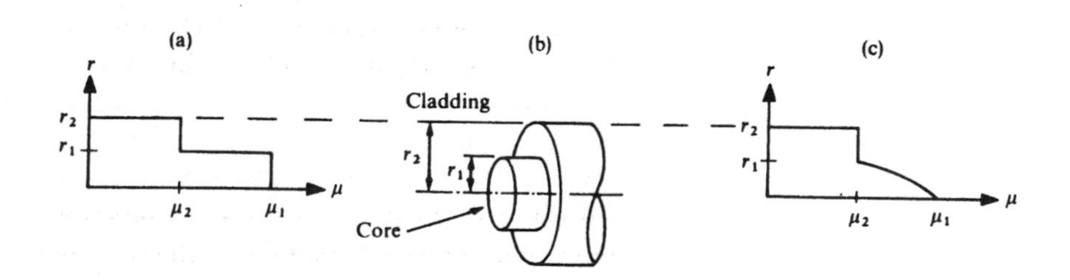

Figure 2.7. Glass or optical fibre construction: (a) stepped index; (b) fibre construction; (c) graded index.

must be of very high quality because any defects or impurities will cause scattering. Fibres are now being manufactured of such high quality that very little energy is lost during propagation. Many fibres are enclosed in protective sheaths to produce optical fibre cables capable of providing an enormous number of channels.

There are two basic types of optical fibre; the *stepped-index* and the *graded-index* types. In the former, there is a step change in the refractive index of the fibre at the core/cladding interface as shown in Fig. 2.7(a). The graded-index fibre, shown in Fig. 2.7(c), is similar to the stepped-index fibre but the refractive index profile changes gradually. The effect of this is to cause refraction rather than reflection so the light is bent away from the fibre surface. These fibres tend to operate with one beam only, unlike the stepped-index type, so giving less distortion.

Light signals are generated and launched into a fibre by means of either a *laser* or a *light-emitting diode* (LED). These devices emit light as a result of applied signal voltages and electron movement. Their action depends on the principle that, when energy is applied to certain materials, some of the atoms absorb energy and enter an unstable excited state. The atom spontaneously reverts back to its stable state, emitting its recently acquired energy in the form of light. The frequency of the light depends on the material used. The laser gives more powerful radiation than a LED. Its name is an acronym derived from a description of its action: Light Amplification by Stimulated Emission of Radiation.

Although lasers and LEDs can produce visible light, most present-day fibre systems use standardized signals of wavelengths 0.8, 1.3 and 1.55 μm (all of which are in the near-infrared band) because dispersion in glass is minimal over this band. Note that a wavelength of 1 μm is equivalent to a frequency of $3 \times 10^8/10^{-6} = 300$ THz. At the receiving end of the fibre a light-sensitive detector is required to convert the light back into the electrical signal. A hypothetical fibre-optic link for speech telecommunication is shown in Fig. 2.8.

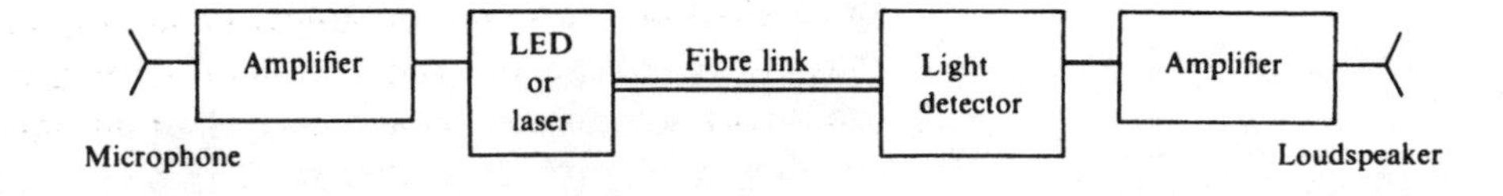

Figure 2.8. A hypothetical fibre-optic speech telecommunications system.

Note that, with an optic link, there is only a single fibre connection. There is no need to 'complete the circuit' by using twin wires as is done when electric current is used for telecommunication.

In practice, speech is not telecommunicated exactly in the way suggested by Fig. 2.8 because the frequency of light is so much greater than that of speech. In order to telecommunicate along a fibre, the light signal must be *encoded* or *modulated* by the speech signal before transmission, in order to give a suitable signal, containing the required information, which the fibre can carry. The information is recovered at the receiver by the opposite process of *decoding* or *demodulation.* In effect, the voice baseband frequency is translated or shifted to a suitable optical frequency. This technique gives a bonus of being able to send many voice signals along the fibre simultaneously. The same idea applies to any link that uses high-frequency transmission, such as a coaxial cable or radio wave. Instead of having one baseband *channel* of communication to each wire, we are now able to *multiplex* many channels on one link. The concept of a physical information link (wire or radio) is now being extended to the concept of an information channel, many of which may be accommodated on one link.

2.2 Multiplexing

Commercial-standard speech requires a bandwidth of only 3.1 kHz, so many messages could be sent at once on a link if separate frequency channels were used. This is an example of multiplexing.

In order to send a large number of independent telephone messages from one point to another, a multi-wire link with a pair of wires for each channel (what might be called *space multiplexing*) is usually used if the distance is short. As the distance increases, it becomes more economical to use one wire to carry many signals, each contained within an allotted frequency band or channel. When wire links are used with one wire for each communication channel, speech is sent in the form of electric signals covering the same frequency band as the speech: this is called *baseband* transmission. When light or radio waves are used the speech band must be shifted to the higher frequencies of these waves. This allows many speech signals to be sent simultaneously along the same link by *frequency-division multiplexing* (FDM) as shown in Fig. 2.9. In FDM each channel is allocated a portion of the available frequency band (a frequency slot) and has exclusive use of this band all the time. Each channel carries a baseband signal, which has been modulated as described in Chapter 4.

An alternative method of multiplexing, using time rather than frequency sharing, is used for the digital signals of PCM speech and computer data transmission. This is *time-division multiplexing* (TDM), in which each

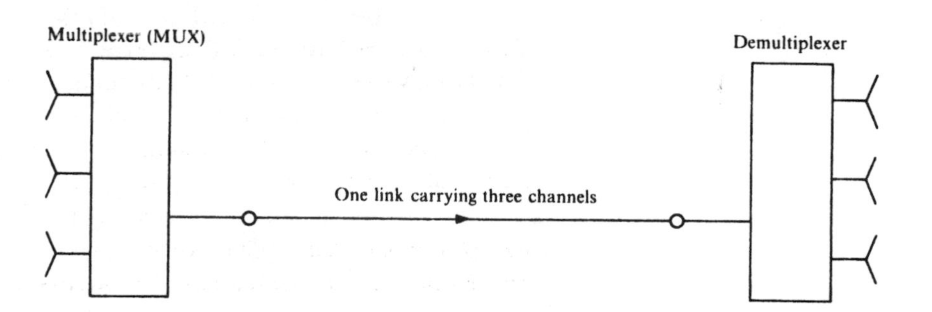

Figure 2.9. A basic multiplexing system.

channel is allocated the whole of the link bandwidth for specific periods known as *time slots.* In 32-channel PCM speech, for example, there are 32 time slots each containing an 8-bit sample. These 32 slots constitute a *frame* and the frames must be repeated at the sampling rate. The two multiplexing methods are compared in Fig. 2.10.

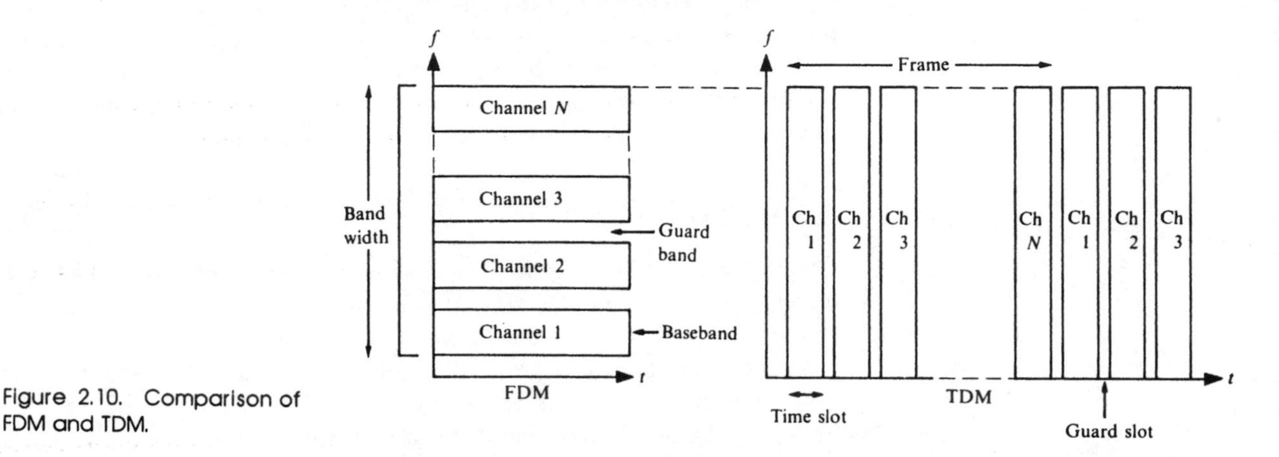

Figure 2.10. Comparison of FDM and TDM.

In each form of multiplexing, 'guard bands' or 'guard slots' separate adjacent channels. Analogue signals are continuous in time so it is not possible to use TDM with such signals. TDM can only be used with pulse signals, the pulses occupying specific time slots corresponding to the time between samples of the same signal. FDM and TDM enable many independent signals to be sent with great economy, despite the high cost of special cables, fibres and radio equipment compared with copper wires. Modulation and multiplexing are fundamental to telecommunications and will be considered in greater detail in later chapters.

The number of channels available using FDM is determined by the bandwidth. This becomes less of a constraint in the higher-frequency bands where the baseband is much smaller, relatively, than the available bandwidth. However, the number of channels available using TDM is limited by the minimum time slot that can be allowed to give the equipment time to operate. To increase the channel capacity of PCM telephone signals, several levels of TDM are used. For example, the CCITT scheme time-division multiplexes four 32-channel TDM signals to give a first level of 128 channels. Batches of four of these 128-channel groups can be multiplexed to give a second level of 512 channels, and so on, as in Fig. 2.11. The

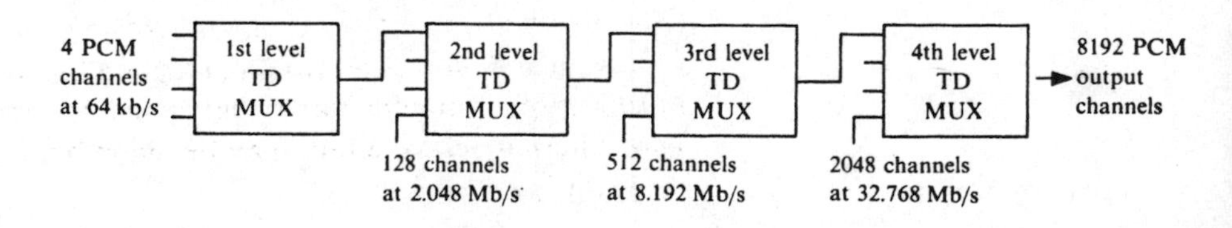

Figure 2.11. CCITT scheme for multi-level PCM TDM.

output, after four levels of TDM, contains 8192 channels at 131·072 Mb/s. In practice, the bit rate at each level has to be increased by slightly more than a factor of 4 because extra bits have to be added for frame alignment, which is necessary for demultiplexing. However, the bit rates are usually expressed in the rounded values 2, 8 and 32 Mb/s for convenience.

2.3 Radio systems

Having looked at the principles of signal transmission through wires, guides and fibres, we now consider 'wireless' transmission.

Signal energy can be carried by radio waves without there being a physical link or 'guide' between the transmitter and receiver. Radio waves are electromagnetic waves covering the spectrum from below 300 Hz to 3 THz. The radio-frequency spectrum has been conveniently divided into bands as shown in Table 2.1.

Table 2.1. Division of the radio-frequency spectrum and some applications

Frequency band	Frequency band name	Abbreviation	Main application
3 Hz – 30 Hz	Extremely low	ELF	Submarine communication
30 Hz – 300 Hz	Ultra-low	ULF	Submarine communication
300 Hz – 3 kHz	Infra-low	ILF	Baseband telephony
3 Hz – 30 kHz	Very low	VLF	Telegraphy
30 kHz – 300 kHz	Low	LF	Navigation and radio
300 kHz – 3 MHz	Medium	MF	AM broadcasting
3 MHz – 30 MHz	High	HF	AM world-wide radio
30 MHz – 300 MHz	Very high	VHF	FM broadcasting
300 MHz – 3 GHz	Ultra-high	UHF	TV broadcasting
3 GHz – 30 GHz	Super-high	SHF	Radar and satellite communications
30 GHz – 300 GHz	Extremely high	EHF	Radar and radio astronomy
300 GHz – 3 THz	Tremendously high	THF	Research
3 THz – 30 THz	Infrared	IR	Optical communication

The IR band has been included because, although not a radio band, it is used for optical telecommunications in fibres so it is 'wirelike' telecommunication. The VLF and ILF bands also use 'wire' links. ULF waves penetrate water to a depth of about 10 m and are therefore used for radio communication with submarines. Research is being done on the use of powerful transmitters at ELF for submarine communications at greater depth.

Radio waves are produced by *aerials* or *antennae*, which can be of many different kinds: from *omnidirectional* or *isotropic* types, radiating equally in all directions, to highly directional ones. A radio wave can travel between antennae as either a *space wave*, a *surface wave* or a *sky wave*. Alternatively, the link may be made by *wave scatter* or via a *satellite* as shown in Fig. 2.12.

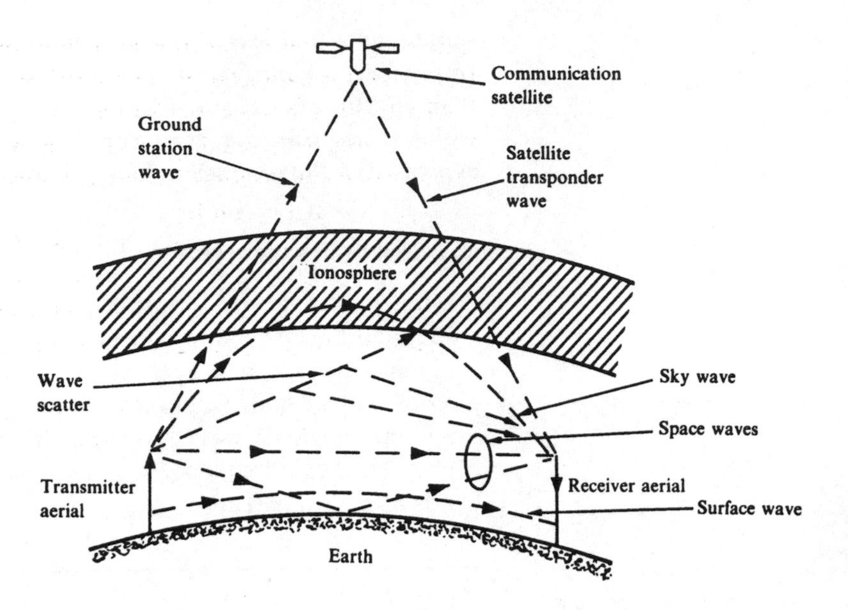

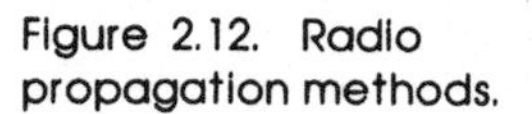
Figure 2.12. Radio propagation methods.

The *space wave* travels in the space between transmitter and receiver and has two components: one travelling almost directly, the other reflecting off the Earth's surface. Since the lower part of the Earth's atmosphere causes some refraction of the radio waves, this form of transmission can extend beyond the horizon.

The *surface wave* is guided by the surface of the Earth—the ground/air interface acting as an enormous waveguide in the VLF and LF bands.

The *sky wave* is directed towards the *ionosphere*. This is a region in the upper atmosphere, 100 km or more above the Earth's surface, where ionized gases exist. This ionization is caused by light from the Sun, so the height of the ionosphere and its intensity of ionization have diurnal and seasonal variations. Radio waves are *refracted* by this layer, and at frequencies in the MF and HF ranges the refraction is sufficient to return the waves to Earth.

The *scatter* mode of propagation is similar to that of the sky wave but happens at VHF and UHF. The waves are directed towards the *troposphere*, the region of the Earth's atmosphere immediately adjacent to the surface of the Earth and extending upwards for some tens of kilometres. Changes in the properties of the various tropospheric layers, such as density and moisture, give rise to scattering of radio signals. Some of the scattered waves return to Earth, giving an effect similar to ionospheric reflection but with very much greater energy loss. By using directional aerials and large transmitted power, reliable communication links well beyond the radio horizon can be established.

The use of communications satellites orbiting the Earth is a fairly recent technique for radio telecommunications. A satellite receives signals, amplifies them and then transmits, on a lower frequency, back to Earth. Satellite systems work in the UHF and SHF bands and provide a large number of channels for TV, long-distance telephony and data communications, by FDM or TDM.

Space waves find use in sound and TV broadcasting, multi-channel telephony systems and mobile radio systems operating in the VHF and UHF bands. Surface waves are used for world-wide telecommunications in the LF band and sound broadcasting in the MF band. Sky waves are used for long-distance and radio-telephony and sound broadcasting in the HF band. Scatter occurs in the UHF and SHF bands and provides multi-channel radio-telephone links.

As the frequency increases, the radio signals can be made extremely directional. Aerial systems using large parabolic 'dishes' or reflectors produce very narrow beams comparable with powerful searchlights.

2.4 Aerial systems

The function of an aerial or antenna is to convert electric current into electromagnetic (EM) waves and vice versa—it is a current/EM wave transducer. The wavelengths used in radio communication vary widely from about 30 000 km (10 Hz) down to 0.3 mm (1 THz). Aerials vary in size from a small fraction to many wavelengths and can take the form of slung wires, wire loops, poles, rods, wire grids, slot arrays and dishes according to the operating frequency band and the application. The basic element of many aerials is the *half-wave dipole* shown in Fig. 2.13.

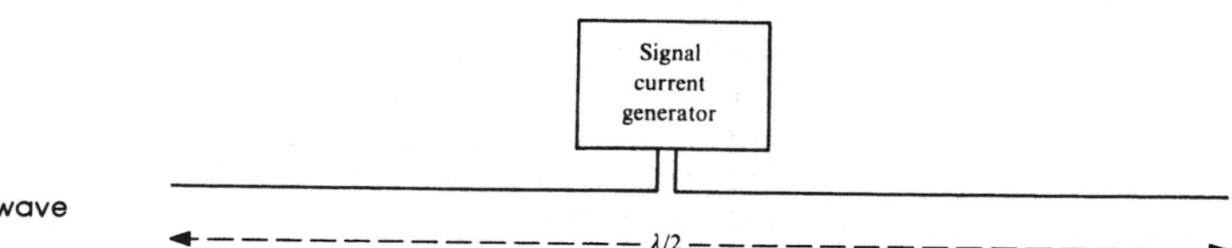

Figure 2.13. A half-wave dipole.

In a transmitting system, a signal current generator is connected to two rigid conductors or poles. The overall length of the resulting dipole is equal to half the wavelength of the radiated wave. A simple idea of how the dipole works may be obtained by imagining the signal current flowing alternately from end to end through the generator. This current produces an electric field (E) parallel to the dipole and a magnetic field (H) concentric with the dipole.

The current flowing towards the open end of a pole must equal the returning current. Hence the electromagnetic field at the ends of the

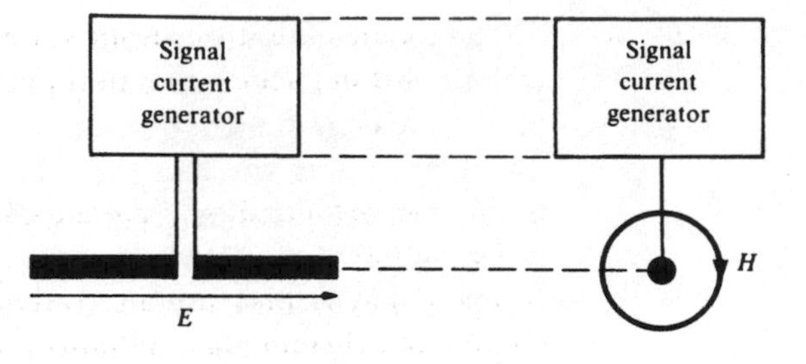

Figure 2.14. The electromagnetic field around a dipole.

dipole shown in Fig. 2.14 is zero. The current at the generator end of the poles changes value as the generated current varies. The current distribution along the dipole at maximum generator current is shown in Fig. 2.15.

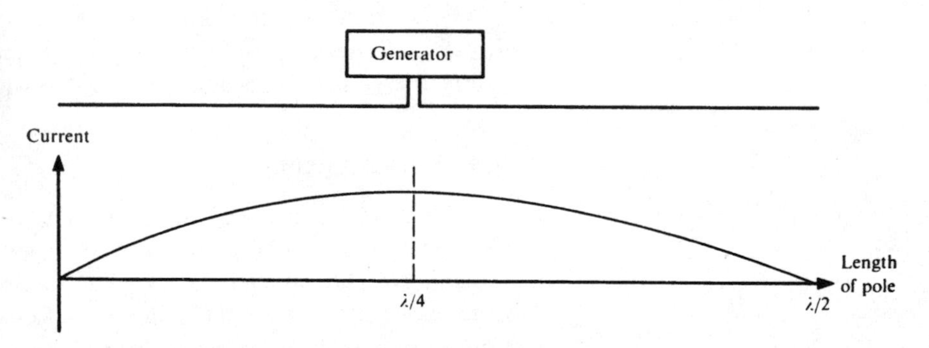

Figure 2.15. Current distribution in a dipole at maximum generator current.

If the generator current falls instantaneously to zero, there will be zero current throughout the length of the dipole. If the generator current reverses, the instantaneous current distribution will be as shown in Fig. 2.16.

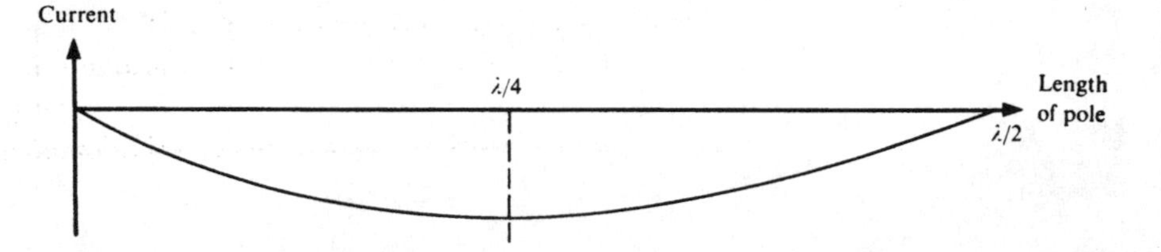

Figure 2.16. Dipole current distribution when generator current reverses.

The frequency at which the current ebbs and flows depends on how long it takes it to travel to the end of a pole and back again—the shorter the pole, the higher the frequency. Changes in polarity of the generator must roughly coincide with the changes in direction of the current in order to sustain the flow, so the length of a pole must be related to the generator frequency. This suggests that the current amplitude will be maximized when the generator frequency exactly matches the natural current flow in the dipole—a condition of *resonance*, which occurs if the dipole length is $\lambda/2$ or any multiple of this.

Exercise 2.3
Check on page 49

Calculate the various dipole aerial lengths required for electromagnetic wave propagation at 10 kHz, 300 MHz, 10 GHz and 3 THz.

A *monopole* aerial is often used, which is, in effect, a dipole formed by a $\lambda/4$ pole and its reflection from the 'ground' plane on which it stands. An electromagnetic field radiates outwards from a dipole at the frequency of the applied current. The radiation pattern may be displayed in the form of a *polar diagram*, which is a polar plot of the signal strength for all directions in a particular plane using the aerial position as the origin. The horizontal and vertical radiation patterns for a vertical mono/dipole are given in Fig. 2.17.

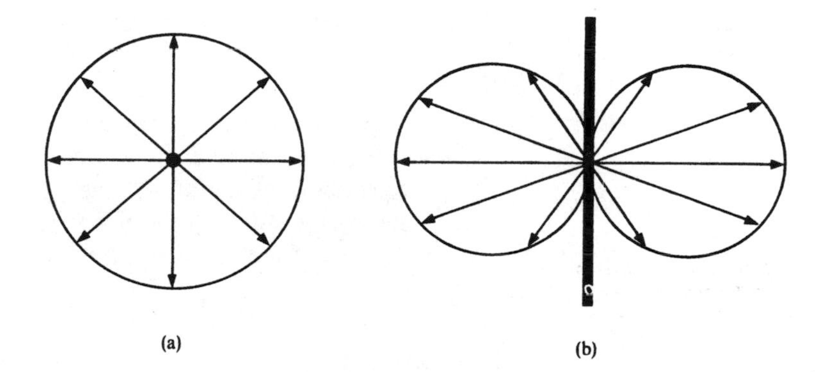

Figure 2.17. Polar diagrams for a half-wave dipole: (a) mono/dipole horizontal plane; (b) dipole vertical plane.

The horizontal polar diagram shows the dipole aerial to be isotropic in this plane, whereas the vertical polar diagram shows directionality. In order to increase directionality in the horizontal plane, extra conducting elements are combined with the dipole to produce an aerial system known as a *Yagi array* illustrated in Fig. 2.18.

By using an array of extra conducting elements (a reflector and directors), the radiation is concentrated in one particular direction along the line of the array. A measure of this *directivity* is the *beamwidth*, which is

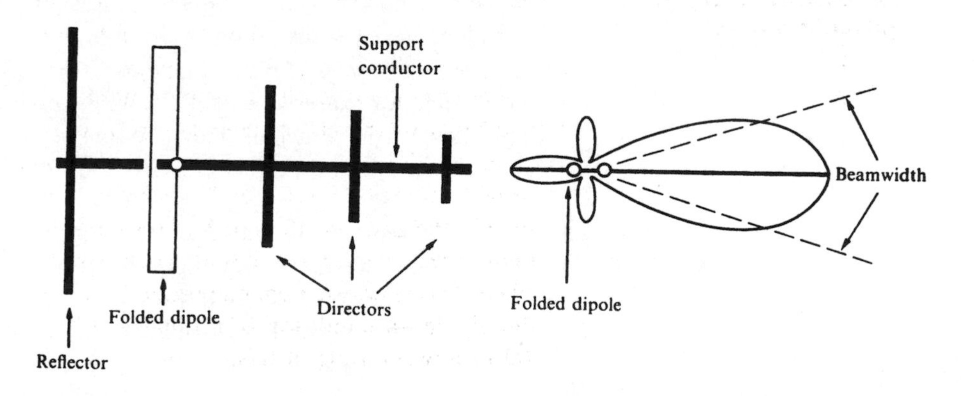

Figure 2.18. A vertical Yagi array and its horizontal polar diagram.

defined as the angle between the directions in which half the maximum power is transmitted or received. The *gain* of an aerial is the ratio of the maximum power it transmits or receives divided by that of an isotropic aerial. The smaller the beamwidth, the greater the gain. The gain is increased by adding more elements. It should be appreciated that gain in this context does not mean signal amplification in the electronic sense, where 'gain' is a measure of the increase in the signal power achieved by the equipment as it converts input supply power into output signal power.

Exercise 2.4
Check on page 49

A broadcast transmitter is sited at point A and uses a vertical half-wave dipole. A similar aerial, situated at point B, 8 km from A, receives a signal in the aerial that is just sufficient to give acceptable reception. If a similar receiver is sited at point C, 16 km from A, and a Yagi array is pointed at A, what is the minimum gain required by the Yagi aerial?

The input impedance of an aerial is given by $Z_i = R_i + jX_i$ where the reactance X_i represents energy stored in the electric field immediately surrounding the aerial—the *near field*. The radiated power of an aerial is 'seen' by the signal generator as radiation resistance R_{rad}. The total input resistance of the aerial is given by

$$R_i = R_{rad} + R_{ohmic}$$

where R_{ohmic} is the conduction resistance. The more input power that an aerial can convert into radiated power the better, so aerial efficiency is given by

$$\eta = R_{rad}/(R_{rad} + R_{ohmic})$$

It can be shown that, for a wire aerial of length L, R_{rad} is proportional to L^2 whereas R_{ohmic} is proportional to L, so efficiency reduces considerably as L decreases.

Radio transmission at low frequency is inefficient, because it is not possible to match the generator and aerial impedances, and thus effect maximum power transfer between the signal source and the aerial. Even with long aerial wires, slung between high towers in a large green-field site, transmitter powers of many kilowatts are required. The LW transmitter at Droitwich, for example, uses 400 kW. Receiving aerials at these frequencies are often made of a coil (many loops) wrapped around a ferrite core in order to improve their power conversion efficiency.

Exercise 2.5
Check on page 49

Compare the ratio of the highest to the lowest frequency of commercial speech in baseband (300 Hz to 3.4 kHz) to that when it is translated (shifted) through a range of 100 MHz. What improvement in aerial performance due to this translation from base to VHF band is suggested by this comparison?

As the wavelength decreases it becomes easier to construct more efficient and directional aerials. We reach the ultimate in this respect in the SHF band, where the microwave aerials often consist of flared waveguides that launch the signal via a reflecting parabolic dish, which projects a very narrow radio beam into space rather like a car headlight. These dishes are a common sight in the countryside, where they can be seen mounted on towers, often on hill tops, to give line-of-sight communication with another dish using space wave propagation. The Post Office Tower, in London, bristles with such dishes as well as Yagi arrays. Figure 2.19 shows a typical terrestrial microwave link.

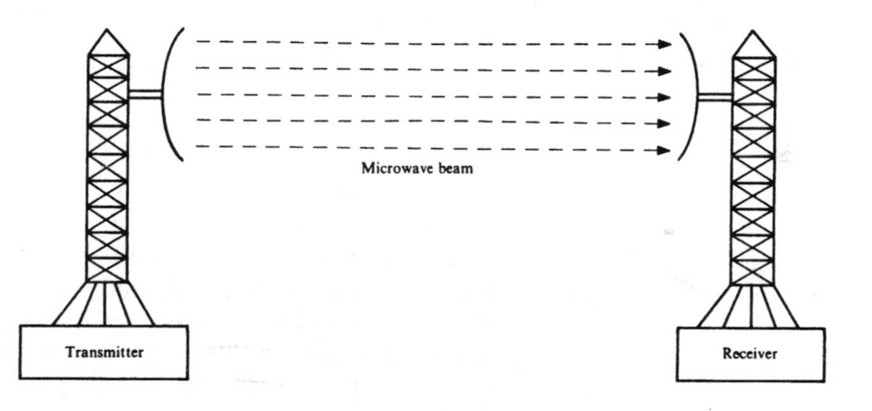

Figure 2.19. A microwave link.

Although microwaves cannot go through hills and buildings, there is little atmospheric absorption in certain SHF bands, so these dishes can be as far apart as the geography of the land will allow. The high frequency of microwave links enable many low-frequency communication channels to be provided by using FDM.

The natural extension of the terrestrial link for international communications is the satellite link. A satellite is placed, usually in a *geostationary* orbit (in which it moves round at the same speed as the Earth rotates), above the equator and a transmitting dish is permanently focused on it. The satellite carries a similar dish and the signals received from Earth are amplified, lowered in frequency by the equipment on board and transmitted back to a receiver on Earth. Direct broadcasting by satellite (DBS) also uses this principle, although in this case the returned signal is dispersed over a wide area.

In terrestrial broadcasting a single aerial system is designed to transmit TV and sound programmes over a large geographical area. In this case a single tower or mast will carry several aerials corresponding to the broadcast frequency bands as illustrated in Fig. 2.20.

The dipole aerial for TV broadcasting is shorter than that used for radio because the TV channels are likely to be in the UHF band while the radio channels, if they use frequency modulation (FM), will be in the VHF band.

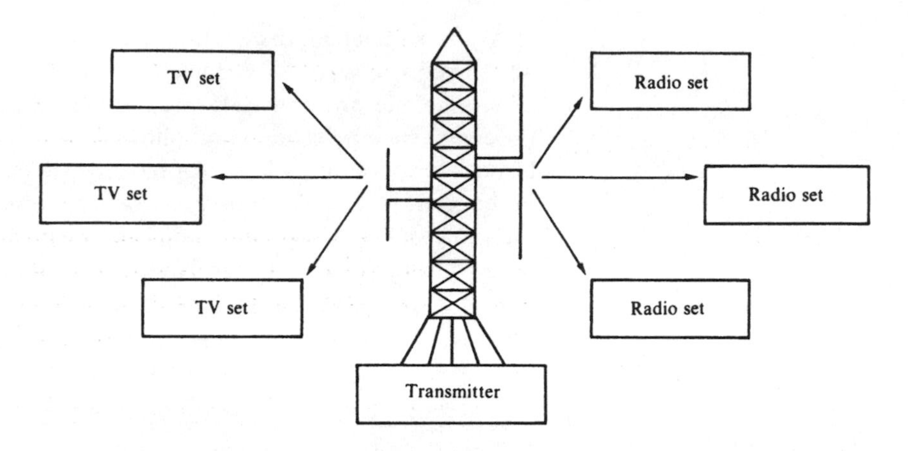

Figure 2.20. A TV and radio broadcasting system.

A TV channel has to accommodate both vision and sound within the same frequency channel, so a TV set consists of two receivers, one for vision and the other for sound. Each receiver uses a different carrier frequency in the same channel. In the UK, the video signal is sent using amplitude modulation (AM) while the monophonic sound signal is transmitted using FM because this gives better sound quality than AM—an FM receiver is fairly immune to AM signal interference. For example, sound interference on the TV picture is quite noticeable but not the converse. This can be observed by slightly mistuning a TV set. High-fidelity stereo sound, using digital encoding, is becoming available throughout the UK using a system, developed by the BBC, called NICAM (an acronym for Near Instantaneously Companded Audio Multiplexing).

2.5 One-way and two-way telecommunication

Radio and TV services are one-way: you simply listen or view without replying. Others, such as the telephone service, are bidirectional because a conversation is required. One way of achieving two-way transmission would be to duplicate the basic system of Fig. 2.1 as shown in Fig. 2.21. A system that carries information in both directions simultaneously is called a *full-duplex* system.

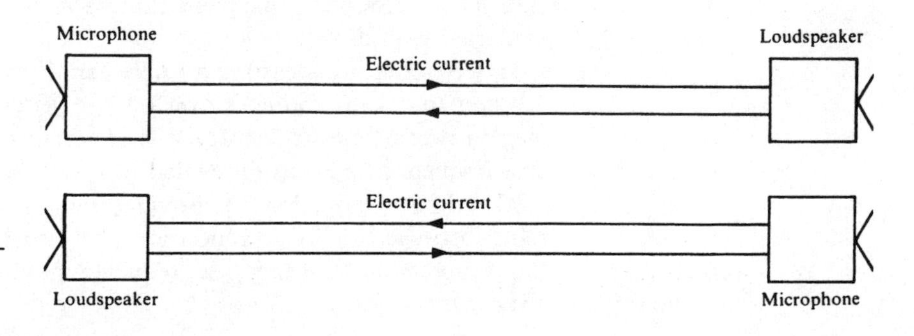

Figure 2.21. A two-way four-wire telecommunication system.

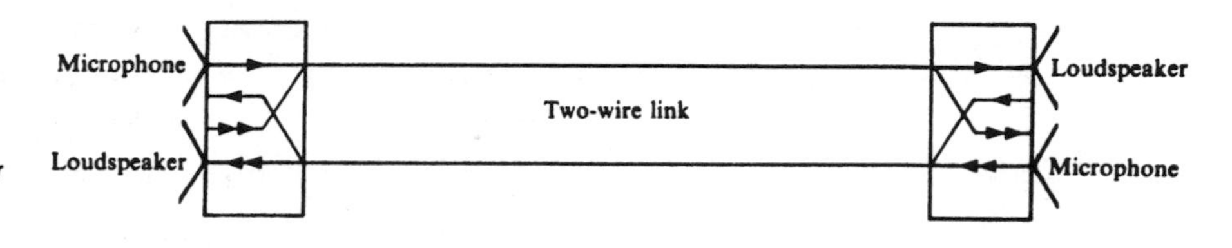

Figure 2.22. A two-wire direct-linked arrangement for two telephones.

A telephone is normally connected to the public network via two wires, so that we have effectively the arrangement shown in Fig. 2.22. The same two wires carry signals in both directions, at the same time if necessary, so duplex operation is achieved. Over long-distance telephone links the presence of repeaters, which are unidirectional, means that a four-wire link must be used to achieve duplex operation. In polite conversation one person listens while the other talks—this is called *half-duplex* transmission. It is a method favoured in some car radio systems such as taxis and takes the form shown in Fig. 2.23. A half-duplex system carries information in both directions, but not simultaneously. A one-way system is termed *simplex*.

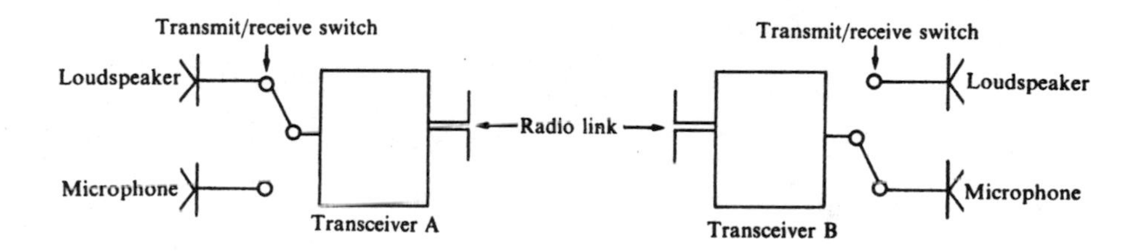

Figure 2.23. A half-duplex radio-telephone link.

2.6 Networks

Radio and TV broadcasting services cover the whole country by means of a *network* of transmitters, which are linked to studios where the programmes are produced. Two telephones can carry signals in either direction using a two-wire link. In order to enable many telephones to intercommunicate, it is necessary to set up a network of two-wire links as exemplified in Fig. 2.24. Direct linking, in which each telephone is permanently connected with every other, is impracticable for a large number of telephones.

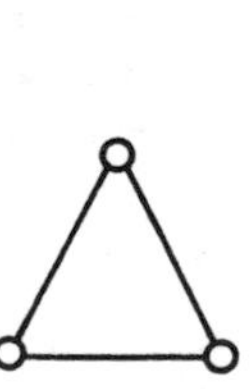

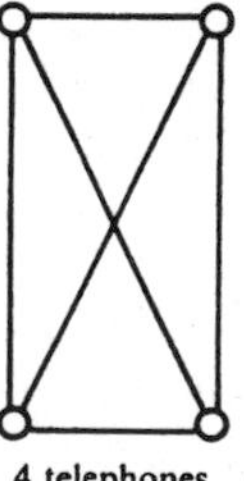

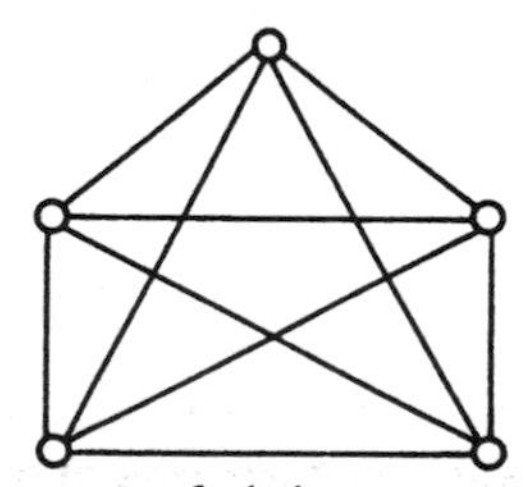

Figure 2.24. Networks of directly linked telephones.

Exercise 2.6
Check on page 49

Calculate the number of links needed to interconnect every telephone in a network consisting of n telephones.

2.7 Public telephone network switching

In order to overcome the problem of having to provide a huge number of cable links to connect the network fully, each telephone in a local area is connected to an *exchange* where any two telephones can be linked together on demand. What this entails, in simple terms, is illustrated in Fig. 2.25. This shows the links that must be made in the

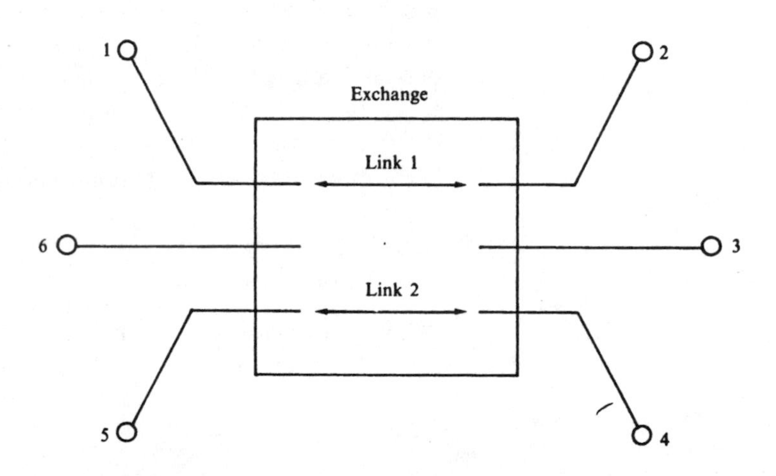

Figure 2.25. A star-connected telephone network.

exchange to interconnect telephones 1 and 2 and telephones 4 and 5 in a star-configured network. In the original early exchanges these links were made manually by a human operator sitting at a switchboard consisting of row upon row of sockets each connected to a telephone in the manner shown in Fig. 2.26.

In Fig. 2.26, telephones 1 and 5 are shown linked by a cable terminated at each end by a jack plug. Making these connections is a

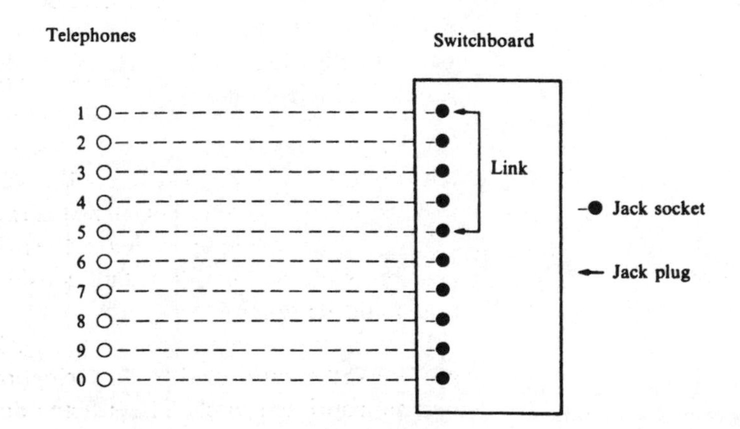

Figure 2.26. A switchboard serving 10 telephones.

labour-intensive task and would be impracticable in the large exchanges that are now commonplace in the public telephone network because of the enormous present-day telephone usage. A modernized form of manual connection is still used in many private exchanges.

The first non-manual method of making connections was invented in 1897 by Almon B. Strowger. This uses electromechanical switches or *selectors* to find a connecting path through the exchange to interlink local subscribers or connect callers to the trunk routes that link the main switching centres (see Chapter 10). Some local exchanges using Strowger switching still exist but they are being replaced by digital exchanges. All the main switching centres are digital.

Cross-point switching

Modern exchanges make use of *cross-point* or *space switching* in which a *matrix* of orthogonal wires are linked by closing a switch (×) at their cross-points as shown in Fig. 2.27. The switching is effected by AND logic gates as shown in the inset of Fig. 2.27. So, for example, telephones 1 and 5 would be connected by closing the switch (enabling the AND gate) at the cross-point of row 5, column 1.

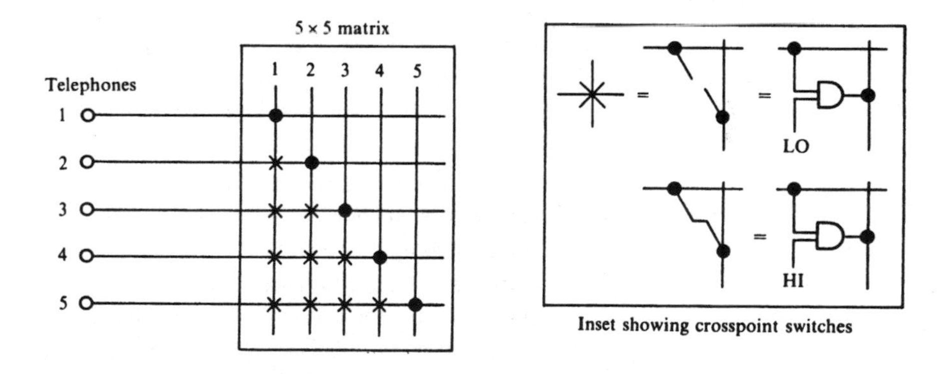

Figure 2.27. A cross-point matrix for interconnecting five telephones.

The number of cross-point switches needed to enable n telephones to be interconnected is the same as the number of links worked out in Exercise 2.6, namely $\frac{1}{2}n(n-1)$. The cross-point matrix has simply taken in n telephone lines and formed all the direct links within the matrix. This implies that a large number of switches will be unused for most of the time. Economies are made by reducing the number of switches and accepting the possibility of blocking. We shall consider a few examples to illustrate how this might be done.

In the arrangement shown in Fig. 2.28(a), any telephone in the first group can be connected to any telephone in the second group by closing one appropriate switch. The scheme does not allow for telephones in the

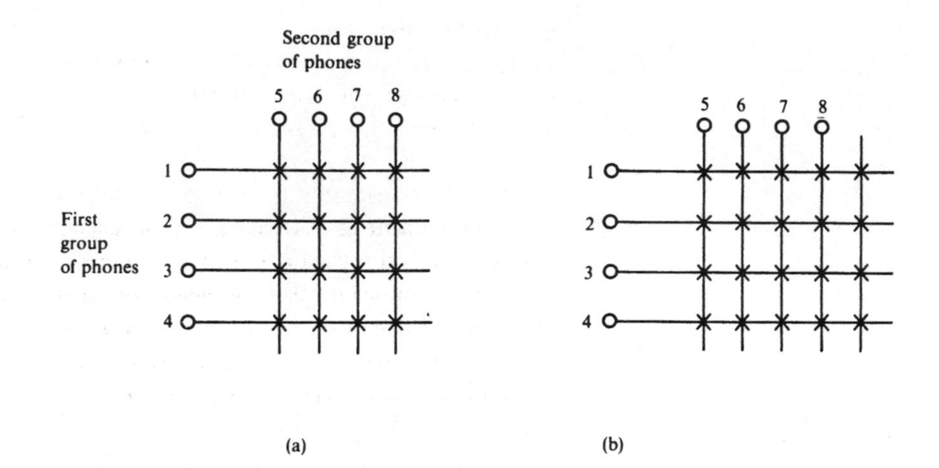

Figure 2.28. A cross-point matrix for interconnecting telephones.

same group to be connected. This can be done by adding more switches as shown in Fig. 2.28(b). With this arrangement, any telephone in the first group can be connected to any other telephone in that group by closing two appropriate switches in the right-hand column. This would allow for only two telephones in the group to be connected at the same time—the other two would be blocked. A similar *loop-back* arrangement could be used to connect the telephones in the second group.

Exercise 2.7
Check on page 49

Draw a cross-point arrangement, similar to that of Fig. 2.28, that would enable the telephones in the second group to be interconnected without blocking.

The scheme shown in Fig. 2.28(b) could be used to connect a group of subscribers on one local exchange with another group served by a remote local exchange. In order to economize on cabling between the two exchanges, an arrangement like that shown in Fig. 2.29 might be used. The arrangement shown in Fig. 2.29 enables any two subscribers served by local exchange 1 (LE1) to be connected with any two subscribers served by LE2. The matrix at LE1 *concentrates* four subscriber lines into two *trunk*

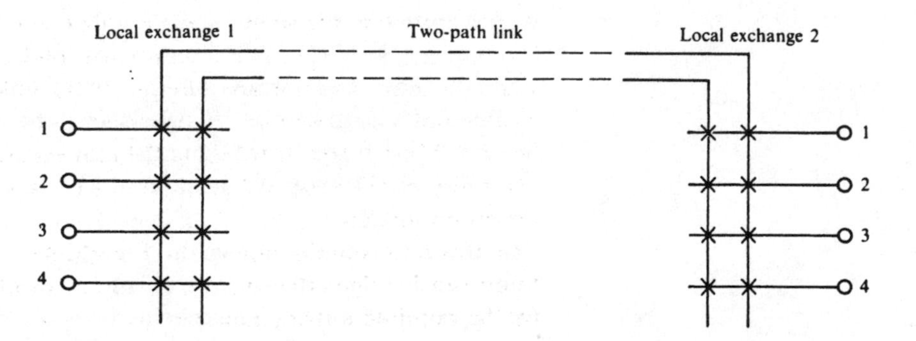

Figure 2.29. A simple connecting scheme for remote exchanges.

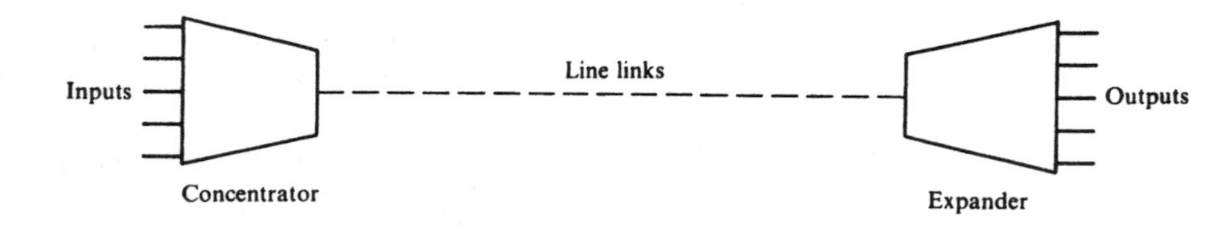

Figure 2.30. Concentration and expansion.

lines and the matrix at LE2 *expands* two *trunk* lines into four subscriber lines. The general principle involved is illustrated in Fig. 2.30.

Each telephone line between a subscriber and the local exchange accommodates one communication channel. Calls made to telephones served by other exchanges are carried on a number of trunk cables, each wire in the cable being allotted one channel by the switching system in the local exchange. Since there are less wires in the trunk cable than telephones linked to the exchange, blocking is possible. This multi-line trunking arrangement is space multiplexing.

Time-slot switching

The use of time-division multiplexing with pulse code modulation (PCM) for voice communication facilitates a *time-slot switching* technique, which is quite different from the *space* switching used in cross-point switching. British Telecommunications plc (BT) has digitized all the UK trunk routes of the *public switched telephone network* (PSTN) and is at present digitizing the exchanges. The PSTN will eventually operate digitally throughout, using a combination of time and space switching, controlled by processors under stored program control. This so-called System X will be described, in outline, in Chapter 10.

2.8 Private branch networks

The need to communicate within an industrial or commercial site is usually satisfied by the provision of a private branch network (PBN) switched either by an operator, using a private *manual* branch exchange (PMBX), or by employees operating keys attached to their telephone extension handsets (keyphones) connected to a private *automatic* branch exchange (PABX). The PABX links the PBN to the PSTN and routes calls to and from the extensions. Calls can be made free of charge within a PBN.

The older PBNs use analogue techniques. Developments in the automated office and personal computer (PC) applications have encouraged the installation of PBNs and PABXs that can handle data as well as speech signals. The modern PBN uses digital techniques and operates at 64 kb/s. Calls are routed either using a call transfer system with keyphones at each extension or by means of a central console under the control of an operator or a combination of these methods. Extensions may be connected to an outside line either manually or automatically by dialling 9 followed by the required outside number.

Digital operation enables a number of sophisticated facilities to be made available through the use of 'feature phones'. These are electronic telephones which, in addition to allowing such operations as abbreviated dialling and last number re-try, provide, in association with a PABX, such facilities as call diversion, queuing, barring, monitoring, conferencing and automatic call back. This last facility is used to re-try automatically a previously engaged extension by ringing it and the caller's extension when they are both free. The key that enables this facility is labelled CAMP (call again mutual party). Modern PABXs offer the above facilities and more, according to cost, thanks to the IC. The PABX may occupy pride of place in a company reception area where a receptionist may also act as a telephone operator by using a small console controlling, say, 24 outside lines, eight internal private lines and 80 extensions.

The 64 kb/s pulse rate makes the PBN quite suitable for many of the low-speed data exchange tasks carried out in the office such as the sharing of a high-quality laser printer. When large amounts of data need to be exchanged, as, for example, in PC links, the bit rate must be increased to several Mb/s. This means that a *local area network* (LAN) must be used instead of a PBN.

2.9 Local, wide and metropolitan area networks

A LAN is a wire or fibre network that is dedicated to providing data communication links within an office or workshop complex. It links a number of terminals (T) or PCs so that they can intercommunicate and share such expensive resources as software, databases, laser printers and plotters. A typical LAN arrangement is given in Fig. 2.31.

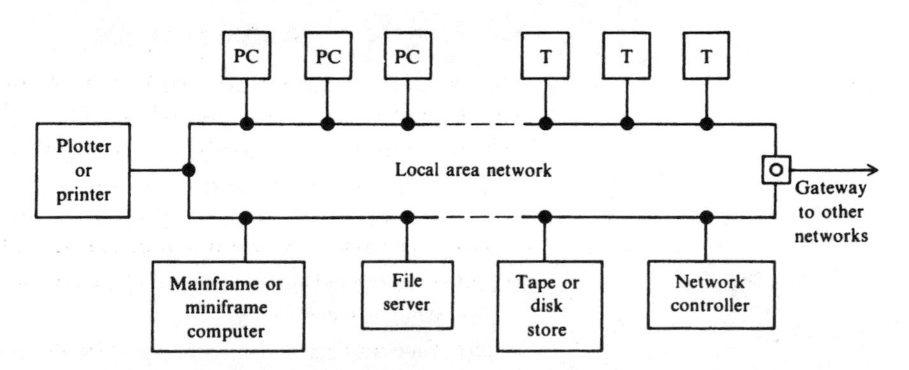

Figure 2.31. A typical LAN.

In a manufacturing organization LANs enable machines and robots to be controlled from a central computer.

A LAN must have arrangements to prevent more than one terminal using the network at the same time. Time sharing is not such a drawback as it may appear because the times needed to set up connections between dedicated and communal equipment and transferring data between them are very short so there is always plenty of spare time available until the traffic peaks. A *network controller* ensures efficient use of this time.

User terminals may have their own storage disks but if they want to share a network disk then arrangements have to be made to ensure that not more than one user is updating the same files on the disk. This may be done by allocating specific areas of the disk to each user. The network system files must be protected from corruption and a proper file back-up system must be in operation. All this is done by network management software housed in a *file server*.

Most LANs may be classified as star, ring or bus networks. The choice between which of these three topologies is used and which type of cable is used depends on the application. The way in which data is exchanged via various kinds of LAN will be described in Chapter 11.

The cabling and topology used for LANs restricts their use to linking data equipment within the same factory or office complex. When distances increase it becomes more economic to exchange data by using either the PSTN or leased private lines if the traffic between specific locations is very high. With the help of suitable interface hardware and software, data equipment of all kinds can exchange data by telephone. If a large amount of computer data is to be handled, then leased optical fibre connections would be the best option. Any system that transmits data in some way nationally or internationally is known as a *wide area network* (WAN). The PSTN can be used to transmit data over a wide area using modulator/demodulator devices called *modems*, which convert data signals into a form suitable for transmission over low-frequency cables.

Telex and Fax are examples of WAN services. Telex uses its own dedicated network to transmit printed information world-wide, whereas Fax uses the PSTN. BT's Prestel information service is an example of a WAN service in which the PSTN is used to link mainframe, mini- and microcomputers all over the UK. The Prestel database is held on a network of computers located in different parts of the country so that subscribers may access Prestel at local telephone charge rates. The mainframe computers of companies such as travel tour operators can be accessed via Prestel through a link called Gateway. The establishment of WAN services internationally depends on the use of under-sea cables and satellite microwave links.

Certain speech and data communications services are only economic if they are provided in areas of high demand. These are obviously areas of high population or business density and such services are carried by so-called *metropolitan area networks* (MANs). These networks fall, in terms of their geographical extent, between LANs and WANs and may be designed primarily for either speech or data communications. The financial market in the City of London, for example, is served by a MAN with glass-fibre links.

Summary

In this chapter we have seen how information is conveyed by electromagnetic signals carried along various kinds of link such as twin wires, coaxial cables, waveguides, glass fibres and radio waves in free space. Information flow may be either duplex, half-duplex or simplex. The links may carry one or many telecommunication channels—the latter requiring multiplexing. In FDM the link bandwidth is split up into frequency channels and communication is achieved by modulating a carrier frequency, equal to the centre of a channel, with the information signal. The signals are carried simultaneously by the link. TDM uses the whole of the link bandwidth on a time-shared basis.

Wire cables can use electrical signals of the same frequency as the information, giving baseband operation. In optical and radio links the transmission frequencies are much higher than the information frequencies, so modulation is essential if many channels are to be sent over such links.

Telecommunication between individuals requires a dedicated line but it is neither necessary nor possible to have everyone permanently directly connected. Instead, subscribers are connected, on demand, by an automatic telephone exchange using selectors, cross-point switches or time-slot interchanging.

Telephone communication within an industrial or commercial site usually takes place over a private branch network, which is switched either by an operator, using a private manual branch exchange, or by employees using keyphones. Local data communication uses a LAN, which, if connected to the PSTN or to private lines, can be extended to cover a wide area. Where local demand is high, a metropolitan area network would be installed. The LAN is to computer communications what the PBN is to human communications, although PBNs can be adapted for low-speed data exchange. A similar comparison may be made, on a larger scale, between a WAN and a PSTN.

► Checks on exercises

2.1

1. The capacitance per unit length of the insulated copper pair will be $10^{-9}/(36 \ln 6) = 16$ pF/m. The approximate inductance per unit length will be $4 \times 10^{-7} \ln 6 = 0.7$ μH/m.
2. The characteristic impedance of the twin cable if R and G are negligible will be $\sqrt{(L/C)} = 216\ \Omega$.

2.2

1. The capacitance per unit length of the coaxial cable will be $10^{-9}/(18 \ln 10) = 24$ pF/m and the approximate inductance per unit length will be $2 \times 10^{-7} \ln 10$ H/m. $Z_0 = \sqrt{(L/C)} = 138\ \Omega$. The velocity $= 1/\sqrt{(LC)} = 3 \times 10^8$ m/s.
2. At zero frequency, $Z_0 = \sqrt{(R/G)} = 4472\ \Omega$ and $\alpha^2 = RG = 80 \times 10^{-6}$, so the attenuation/km $= 8.7\alpha = 22$ dB.

2.3 The dipole length (l) is given for frequency f by $\lambda/2 = 3 \times 10^8/2f$ so $l = 15$ km at 10 kHz, 0.5 m at 300 MHz, 15 mm at 10 GHz and 50 μm at 3 THz. The only practicable dipole values are those of 0.5 m and 15 mm in the UHF and SHF bands, although monopoles of half these lengths would work well. It would clearly not be practicable to transmit the tones of commercial speech, which lie in the range 300 to 3400 Hz—some means must be found of converting this speech baseband into higher-frequency bands for radio telecommunication. These means are described in later chapters covering the various forms of modulation.

2.4 C, at 16 km from A, is twice as far from A as is B. Since the signal power falls off as the square of distance, the signal at the aerial at C is a quarter of that at B. To compensate for this reduction in signal power, the Yagi must have a gain of at least 4. In practice, there will be further reduction in the received signal power due to atmospheric water vapour absorption and scattering, so the Yagi gain will have to be greater than 4.

2.5 The ratio of the highest to the lowest frequency of commercial speech in the band 300 Hz to 3.4 kHz is $3400/300 = 11.3$. After translation through a range of 100 MHz the ratio will be $100\,003\,400/100\,000\,300 = 1.000\,031$. This narrowing of the frequency ratio is an example of an important effect known as *narrowbanding.*

An aerial transmitting at baseband has to cope with a bandwidth of over a decade and will be unable to transmit equally efficiently over the whole of such a large range of frequency. A VHF aerial, however, will transmit the whole speech band with practically equal efficiency because the fractional difference in frequency over the band is negligible.

2.6 The number of links needed to interconnect every telephone in a network consisting of n telephones is given by the number of combinations of pairings of n items. This will be ${}^nC_2 = n!/2(n-2)! = \frac{1}{2}n(n-1)$. In a local area with about 10 000 subscribers, this would mean about 50 million links.

2.7 A cross-point arrangement that would enable the telephones in the second group to be interconnected without blocking might be as shown in Fig. 2.32. In this scheme the use of two loop-back matrices overcomes blocking.

Self-assessment questions

2.1 Compare the telecommunications terms: signal, link and channel.

2.2 What are the main forms of energy loss at very high frequency in (a) twin cable and (b) coaxial cable?

2.3 What are the functions of a line repeater?

2.4 What are the advantages of optic fibres over waveguides?

2.5 Approximately how many telephone channels could be accommodated in a glass fibre operating at 300 THz?

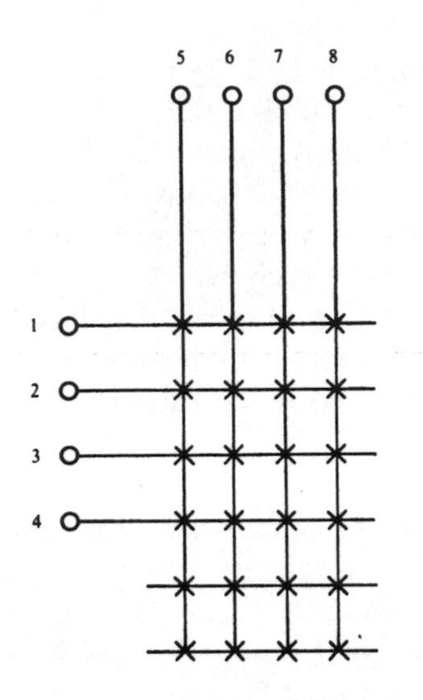

Figure 2.32. A scheme for second group connection without blocking.

2.6 Distinguish between space, frequency and time division multiplexing.
2.7 Define the bandwidth and gain of an aerial.
2.8 State two limitations imposed by the aerial on broadcasting at baseband.
2.9 Give two advantages of using microwave links.
2.10 Distinguish between duplex, half-duplex and simplex communication.
2.11 Why is network switching used?
2.12 Draw a diagram, showing the cross-points, of a concentrator that would enable any three of five callers to access another exchange.
2.13 Compare the functions of a PBN with those of a LAN.

3

Signal representation and channel frequency response

Mathematical modelling

Objectives

When you have completed this chapter you will be able to:

- Describe how a tone may be represented in the time and frequency domains
- Synthesize non-sinusoidal periodic waveforms with sinusoidal waveforms
- Recognize the general spectral content of a periodic waveform
- Give various mathematical expressions for a Fourier series
- Derive the Fourier series for a periodic pulse train
- Sketch spectra of pulse trains with various pulse widths and periods
- Derive a Fourier integral for a single pulse
- Sketch spectrum diagrams for a single pulse
- Describe magnitude and phase distortion in a simple low-pass channel
- Describe the magnitude and phase response of a simple low-pass channels
- Define cut-off frequency and break frequency
- Describe the ideal responses of high-pass and band-pass channels
- Define channel bandwidth

The signals that carry information can be described in terms of their current or voltage variation with respect to time. In practice, however, the resolution of signals into sinusoidal components is most useful. A sine wave (tone) has the property that, when it is input to a linear system, the output is another sine wave of the same frequency. A sine wave is thus an excellent standard signal for describing the performance of a system. The fact that such a signal is easily generated makes it an ideal practical test signal.

Many signals can be approximated by series of sine or cosine functions of time or frequency. Frequency is often preferred because the analysis is easier.

Periodic signals are made up of discrete sinusoidal components that are integer multiples of the signal frequency and may be expressed mathematically as a Fourier series. These components may be represented pictorially as a set of discrete spectral lines. Non-periodic signals have a continuous spectrum and may be expressed as a Fourier integral. In this chapter we consider in detail two signals, both of great importance in telecommunications: the periodic pulse train and the single pulse.

When complex signals pass through a channel, they will be distorted if the channel alters the relative amplitude or phase of the sinusoidal components.

3.1 The representation of a sine or cosine signal

A telecommunications signal is inherently an electrical change that occurs in an interval of time—it is an occurrence or an event. Many of these signals are periodic, as was explained in Chapter 1. A type of signal that forms the basis of many other signals is the tone, which has the sine form shown in Fig. 3.1.

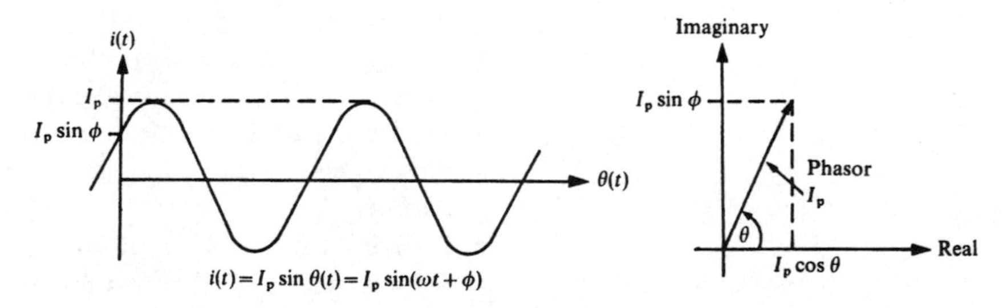

Figure 3.1. Two representations of a sinusoidal signal.

Although the current signal occurs in time, it is often mathematically more convenient to represent it in terms of its frequency. Both the sine and cosine forms of a signal may be expressed as in the Argand diagram of Fig. 3.1. The current of amplitude I_p is represented by a *phasor* of length I_p, which rotates anticlockwise in the complex plane, starting from an initial phase ϕ. The instantaneous angular velocity $\omega(t) = d\theta/dt$ is usually called the *instantaneous angular frequency*. The idea can be illustrated by Fig. 3.2.

A current phasor rotates in the complex plane and may be specified either in polar form by its magnitude (I_p) and phase angle (θ) or in cartesian form by its real and imaginary components $I_p \cos\theta$ and $I_p \sin\theta$, respectively. Alternative expressions exist as follows:

$$I(j\omega t) = I_p e^{j\omega t} = I_p[\cos(\omega t) + j\sin(\omega t)]$$

or

$$I(-j\omega t) = I_p e^{-j\omega t} = I_p[\cos(\omega t) - j\sin(\omega t)]$$

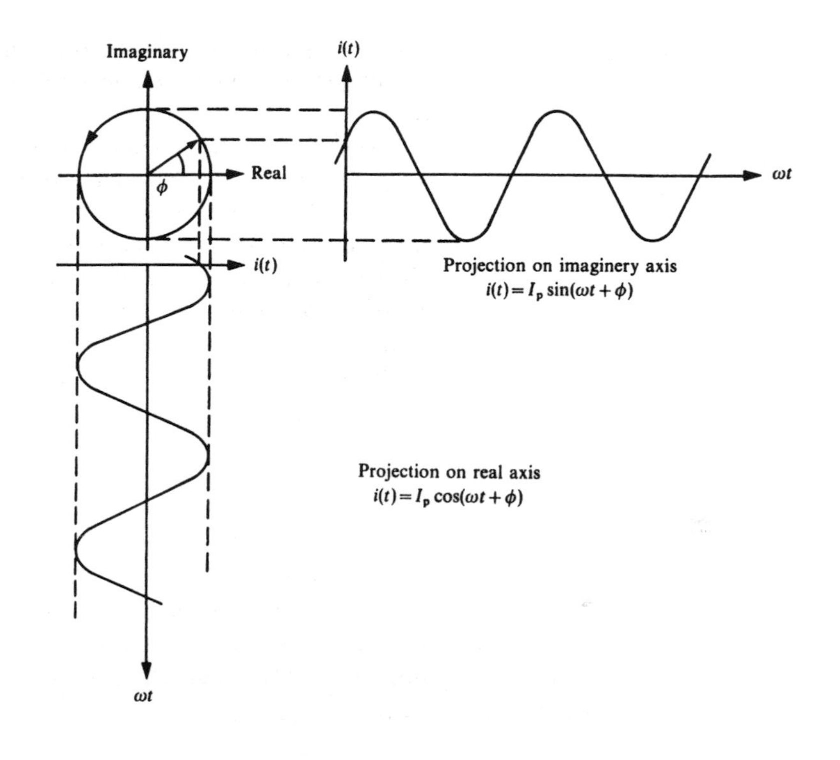

Figure 3.2. Phasor representation of sine/cosine signals.

so

$$\cos(\omega t) = \tfrac{1}{2}(e^{j\omega t} + e^{-j\omega t}) \quad \text{and} \quad j\sin(\omega t) = \tfrac{1}{2}(e^{j\omega t} - e^{-j\omega t})$$

Notice that the two complex-number representations of current give both sine and cosine forms simultaneously and that this leads to expressions for $\cos(\omega t)$ and $\sin(\omega t)$ involving both positive and negative frequency. We saw that a positive-frequency cosine term could be thought of as a phasor rotating anticlockwise with an angular velocity $\omega = d\theta/dt$. It follows then that a negative-frequency cosine term can be thought of as a phasor rotating clockwise with the same angular frequency. This concept gives two ways of visualizing the cosine (or sine) phasors as two contra-rotating phasors shown in Fig. 3.3. The two contra-rotating phasors of

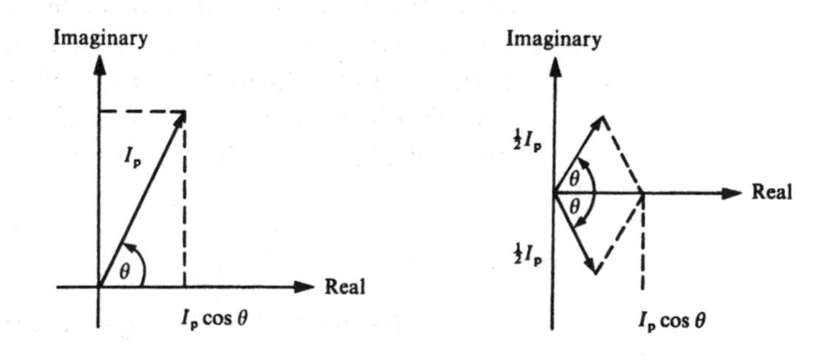

Figure 3.3. Two complex-plane representations of a cosine signal.

amplitude $\frac{1}{2}I_p$ have a resultant along the real axis of $I_p \cos\theta$ and are therefore equivalent to the single clockwise-rotating phasor of amplitude I_p.

Note that lower-case letters are used to denote time-dependent (known as instantaneous) values of current (or voltage) and upper-case letters for phasor, peak and r.m.s. values. It should be appreciated that the phasor representation of a tone signal applies to the *steady state* of a system, that is, the state of the circuits when they have settled down. Phasors take no account of the *transient* effects that occur on switching. In the interests of simplicity, the steady state will be assumed throughout this book. It is a reasonable assumption because, in practice, the transients will have died out before we start looking at a system output.

Exercise 3.1
Check on page 72

Draw two complex-plane representations of a sine signal similar to Fig. 3.3.

An alternative to the representation of tone signals as phasors shown in Fig. 3.3 is to use a frequency spectrum diagram as shown in Fig. 3.4. Note that the frequency spectrum display gives only amplitude and frequency information The phase angle does not appear.

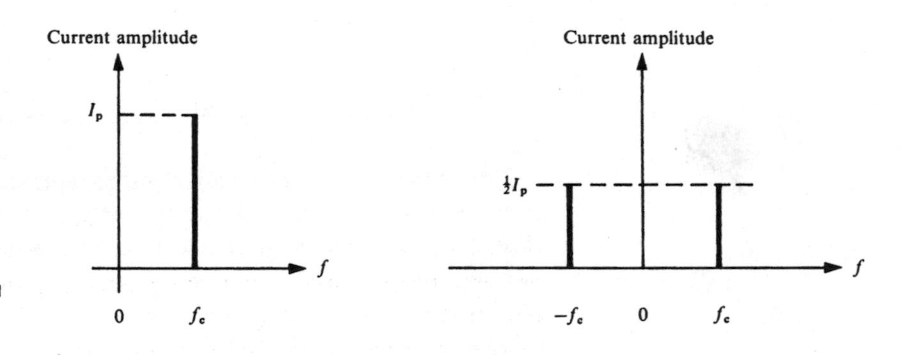

Figure 3.4. Two frequency spectra representations of a cosine signal.

3.2 The make-up of non-sinusoidal waveforms

Many periodic waveforms, such as the rectangular and ramp waveforms produced by a multivibrator, are not sinusoidal. Such *complex waves* may be shown to consist of a series of sinusoidal components. Notice that we use the word 'complex' in this context to mean 'complicated', not complex in the mathematical sense mentioned in Sec. 3.1.

Any periodic signal can be made up, or *synthesized*, from a number of sinusoidal components. Figure 3.5 gives an example using two components given by $v_1 = 10\sin(\omega t)$ and $v_2 = 5\sin(2\omega t)$. In Fig. 3.5, the two components, shown by broken curves, have their amplitudes at the same instant added together to give the synthesized waveform shown as a full curve.

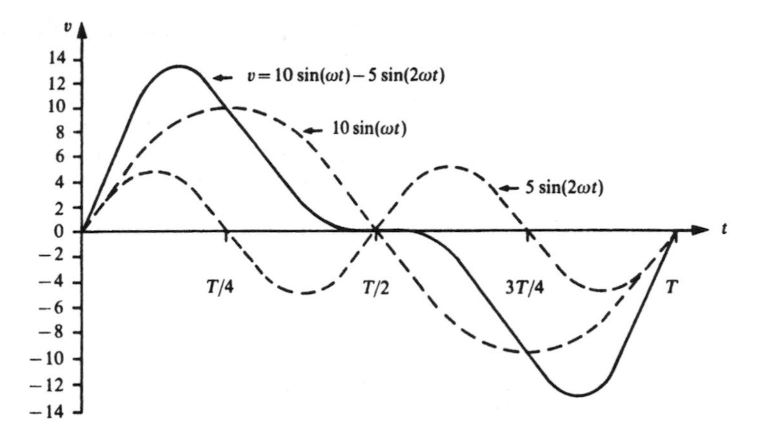

Figure 3.5. The synthesis of two sinusoids.

Exercise 3.2
Check on page 72

A voltage is made up of three sinusoidal components as follows:

$$v = 10\sin(\omega t) + 5\sin(2\omega t) + 2.5\sin(3\omega t)$$

Draw a graph for each component, with the help of Fig. 3.5, and then sketch in the waveform for v.

The synthesized waveforms shown in Figs 3.5 and 3.26 are periodic and of the same frequency as the lowest frequency component. This component is known as the *fundamental* or *first harmonic*. The other components are integer multiples of the first harmonic and are called *upper harmonics*. The voltage waveform of Fig. 3.5 contains the first and second harmonics only; that synthesized in Exercise 3.2 contains the first, second and third harmonics. Recall from Chapter 1 that, in sound, harmonics provide timbre—a property that distinguishes different musical instruments and voices irrespective of pitch (which is determined by the fundamental frequency).

In comparing Figs 3.5 and 3.26, it can be seen that the addition of the third harmonic has made the synthesized waveform more like a saw-tooth. As more harmonics of decreasing amplitude are added, this likeness to a saw-tooth shape becomes greater, as is shown in Fig. 3.6, where the second to fifth harmonics are added to the fundamental.

Exercise 3.3
Check on page 72

Complete a graph for each component in the follwing series and sketch in the synthesized waveform:

$$v = 10\sin(\omega t) + 5\sin(3\omega t) + 2.5\sin(5\omega t)$$

Exercise 3.3 suggests that, if the number of odd harmonics is increased, square-wave approximations are obtained as exemplified in Fig. 3.7.

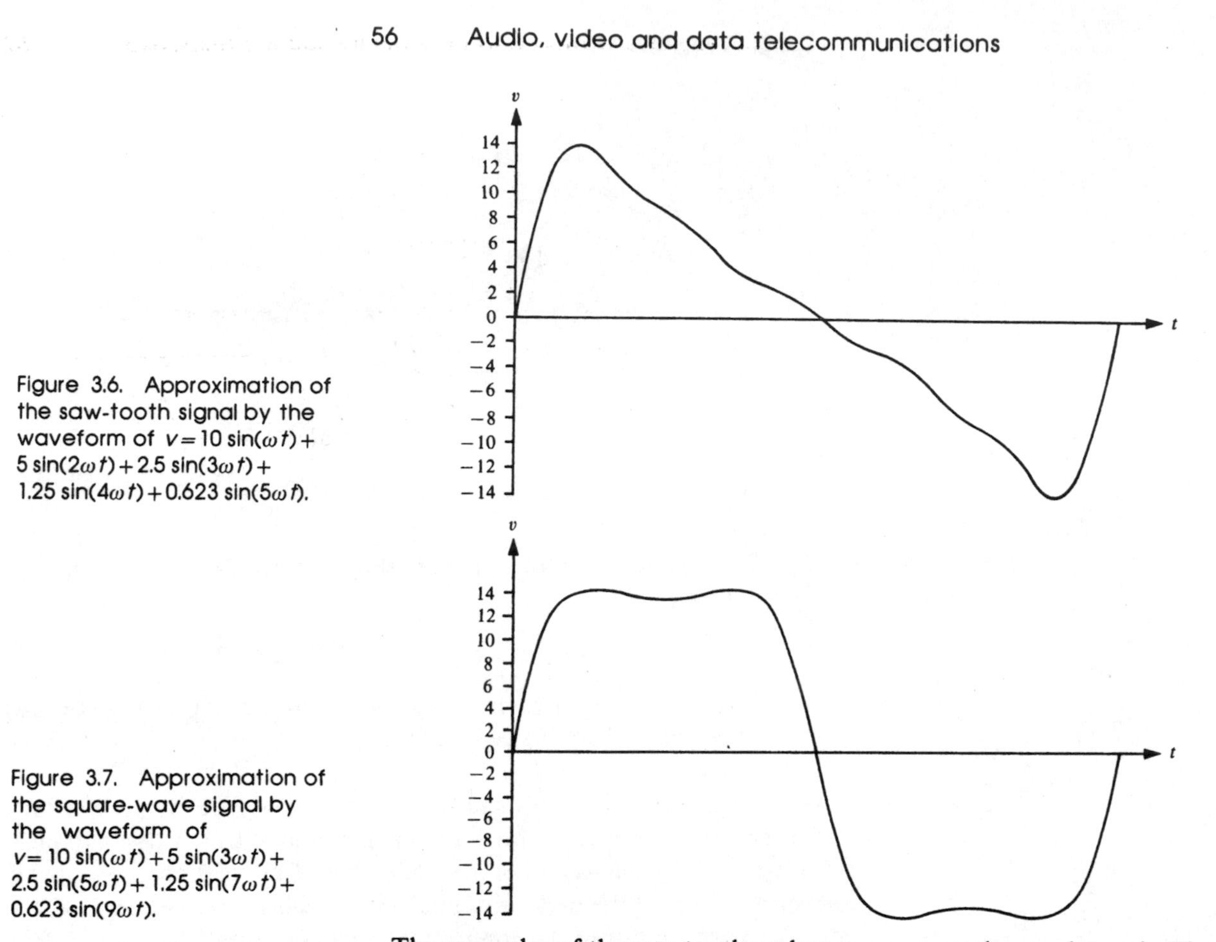

Figure 3.6. Approximation of the saw-tooth signal by the waveform of $v = 10\sin(\omega t) + 5\sin(2\omega t) + 2.5\sin(3\omega t) + 1.25\sin(4\omega t) + 0.623\sin(5\omega t)$.

Figure 3.7. Approximation of the square-wave signal by the waveform of $v = 10\sin(\omega t) + 5\sin(3\omega t) + 2.5\sin(5\omega t) + 1.25\sin(7\omega t) + 0.623\sin(9\omega t)$.

The examples of the saw-tooth and square-wave voltages shown in Figs 3.6 and 3.7 imply that periodic signals can be expressed as a series of sinusoids thus:

$$v = V_1\sin(\omega t + \phi_1) + V_2\sin(2\omega t + \phi_2) + V_3\sin(3\omega t + \phi_3) + \cdots$$

$$= \sum_{n=1}^{\infty} V_n \sin(n\omega t + \phi_n)$$

where V_n, $n\omega$ and ϕ_n are the harmonic amplitudes, frequencies and phases.

Waveforms have been considered that contain either all the harmonics or odd harmonics only. Figure 3.8 shows two examples of the effect of combining a fundamental with its even harmonics $v_1 = 10\sin(\omega t) + 5\sin(2\omega t)$ and $v_2 = 10\sin(\omega t) + 5\sin(2\omega t - \pi/2)$ in Figs 3.8(a) and 3.8(b) respectively.

The shapes of the synthesized waveforms that have been considered give a clue to the harmonics in a signal. A few general conclusions can be drawn, as follows:

1. When odd harmonics are added to a first harmonic, the positive and negative half-cycles of the synthesized waveform are identical.

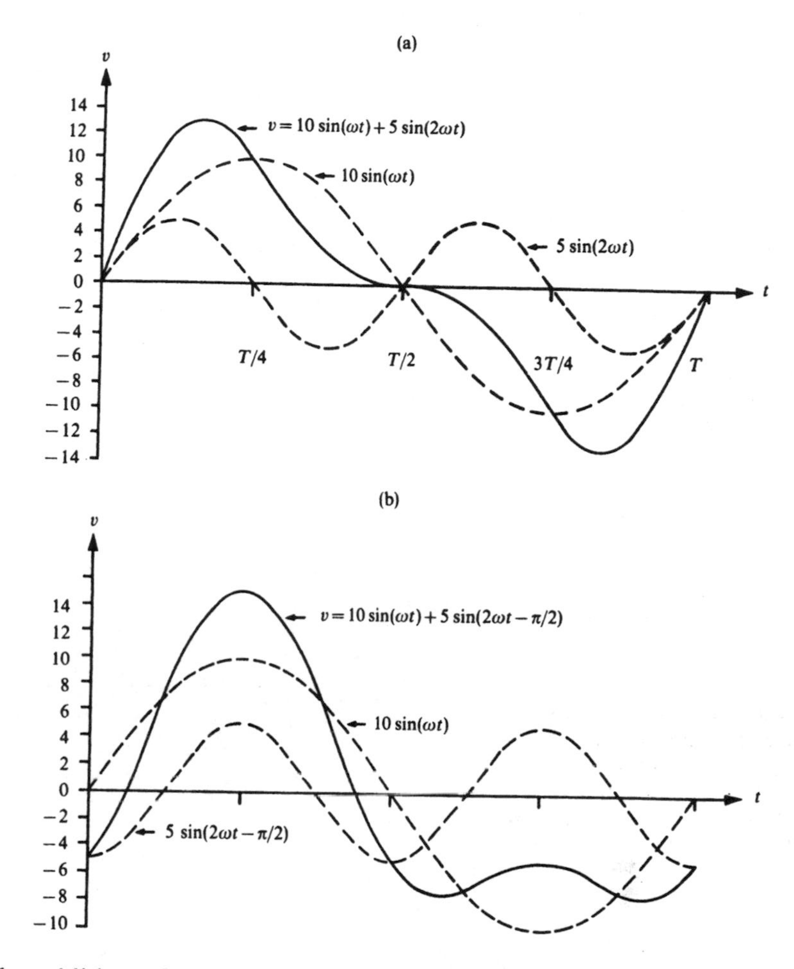

Figure 3.8. Synthesis of first and second harmonic waveforms: (a) fundamental plus second harmonic; (b) fundamental plus second harmonic with $\pi/2$ phase lag.

2. The addition of even harmonics only to a first harmonic makes the positive and negative half-cycles of the synthesized waveform symmetrical if the harmonics start off in phase or anti-phase. If the harmonics are otherwise initially out of phase, the synthesized waveform is asymmetrical.
3. If both even and odd harmonics are added to a first harmonic, the positive and negative half-cycles of the synthesized waveform are symmetrical when the harmonics are in phase or anti-phase to begin with. If the harmonics are otherwise out of phase initially, the synthesized waveform is asymmetrical.

A waveform may change its shape considerably if the relative phases of the harmonics are changed. To illustrate this, consider the synthesis of the waveforms $v_1 = 10\sin(\omega t) + 5\sin(3\omega t)$ and $v_2 = 10\sin(\omega t) + 5\sin(3\omega t - \pi/2)$, shown in Fig. 3.9. This shows that change in the relative phase may have a more drastic effect on the synthesized signal than change in the relative amplitude.

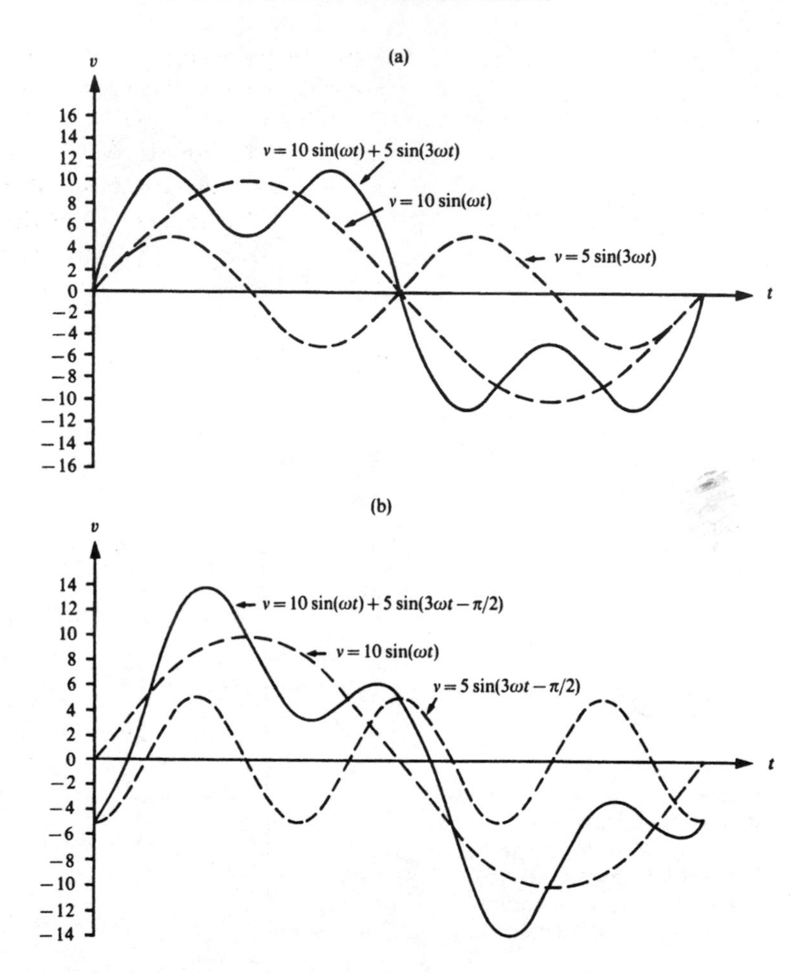

Figure 3.9. Waveform synthesis showing the effect of relative phase change: (a) $v = 10\sin(\omega t) + 5\sin(3\omega t)$; (b) $v = 10\sin(\omega t) + 5\sin(3\omega t - \pi/2)$.

3.3 Amplitude spectrum diagrams

The components that make up a complex wave can be displayed on a graph of the kind shown in Fig. 3.10, known as an amplitude spectrum diagram. Each value H_n represents the amplitude of one of the harmonics of a voltage or current waveform at frequency f_n. The diagram ignores

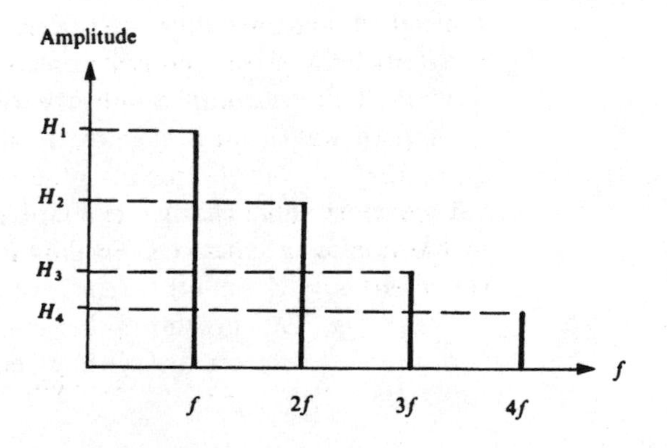

Figure 3.10. An amplitude spectrum diagram.

phase. In practice, such diagrams can be displayed on the screen of a *spectrum analyser*. This instrument is basically a *cathode ray oscilloscope* (CRO) in which the horizontal deflection timebase voltage is replaced by a voltage that is proportional to frequency. The vertical deflection voltage is produced by a rectified harmonic that has been obtained from the complex input signal by frequency-selective circuitry. We have seen how a general knowledge of the harmonics present in a signal can be inferred from the shape of the signal waveform that can be viewed on a CRO. A spectrum analyser gives more precise information about the harmonics present and therefore provides a useful alternative view of a signal as shown in Fig. 3.11.

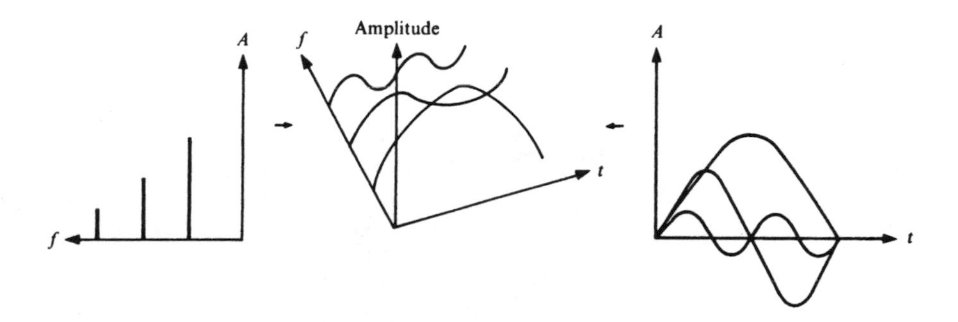

Figure 3.11. Time and frequency displays.

Exercise 3.4
Check on page 73

Draw a spectrum diagram for the waveform:

$$v = 5\sin(250\pi t) + 2.5\sin(500\pi t) + 1.25\sin(1000\pi t)$$

3.4 Signal analysis

So far in this chapter we have built up complex waveforms using harmonic components (a process of signal synthesis) and touched on how a signal may be analysed using a spectrum analyser. We now consider briefly the theoretical analysis of signals in terms of frequency in order to prepare for the signal analysis of later chapters.

A periodic signal consists of an integer number (n) of harmonic voltages or currents of amplitude H_n, frequency f_n and phase, ϕ_n. That is, a periodic signal is given by the summation of terms of the form $H_n\cos(2\pi f_n + \phi_n)$. The mean value of the signal over its period ($T = 1/f$) will be some value that we can conveniently call H_0. This is the d.c. level in the case of a current signal. A neat expression for a periodic voltage signal $v(t)$ is

$$v(t) = H_0 + \sum_{n=1}^{N} H_n \cos[2\pi(nt/T) + \phi_n] \qquad (3.1)$$

where n is the number of harmonics needed to make up the signal $v(t)$. We have seen that, for good square-wave and saw-tooth approximations, n tends to infinity. This is true for waveforms in general, although perfectly acceptable signals are obtained with finite bandwidths. Series of the kind specified by Eq. (3.1) were originated by the French mathematician Fourier (1768–1830). Alternative representations of Fourier series can be derived from Eq. (3.1).

One alternative uses the identity $\cos(A+B)=\cos A\cos B-\sin A\sin B$ to give

$$\begin{aligned}H_n\cos[2\pi(nt/T)+\phi_n]&=H_n\{\cos[2\pi(nt/T)]\cos\phi_n\\&\quad-\sin[2\pi(nt/T)]\sin\phi_n\}\\&=a_n\cos[2\pi(nt/T)]\\&\quad+b_n\sin[2\pi(nt/T)]\end{aligned}$$

where $a_n=H_n\cos\phi_n$ and $b_n=-H_n\sin\phi_n$
Thus Eq. (3.1) can be rewritten as

$$v(t)=H_0+\sum_{n=1}^{\infty}a_n\cos[2\pi(nt/T)]+\sum_{n=1}^{\infty}b_n\sin[2\pi(nt/T)] \tag{3.2}$$

Another alternative uses the relationship $\cos A=\frac{1}{2}(e^{jA}+e^{-jA})$ to give

$$\begin{aligned}H_n\cos[2\pi(nt/T)+\phi_n]&=\tfrac{1}{2}H_n\{\exp[^{j(2\pi nt/T+\phi_n)}]\\&\quad+\exp[^{-j(2\pi nt/T+\phi_n)}]\}\\&=c_n e^{j2\pi nt/T}+c_{-n}e^{-j2\pi nt/T}\end{aligned}$$

where $c_n=\frac{1}{2}H_n e^{j\phi_n}$, $c_{-n}=\frac{1}{2}H_n e^{-j\phi_n}$ and $c_0=\frac{1}{2}H_0$. Hence, Eq. (3.1) can be rewritten as

$$\begin{aligned}v(t)&=c_0+\sum_{n=1}^{\infty}c_n e^{j2\pi nt/T}+c_{-n}e^{-j2\pi nt/T}\\&=\sum_{n=-\infty}^{\infty}c_n e^{j2\pi nt/T}\\&=\sum_{n=-\infty}^{\infty}c_n[\cos(2\pi fnt)+j\sin(2\pi fnt)]\end{aligned} \tag{3.3}$$

The coefficient $c_0=\frac{1}{2}H_0$ is the mean value of the signal, so it is given by

$$c_0=(1/T)\int_{-T/2}^{T/2}v(t)\,dt$$

In general, the other coefficients (c_n) represent the harmonic amplitudes $\frac{1}{2}H_n$ and are given by

$$c_n=(1/T)\int_{-T/2}^{T/2} v(t)\,e^{-j2\pi nt/T}\,dt \tag{3.4}$$

Equation (3.3) gives a very compact expression for a Fourier series by using both positive and negative frequency (n takes on positive and negative values). The fact that we cannot ascribe a physical meaning to the concept of negative frequency does not preclude its use in mathematical descriptions. The concept may be pictured with a spectrum diagram as in Fig. 3.12. In Fig. 3.12, we get a spectral line at $f=n/T$. At each line $t=T$ so, in Eq. (3.3), the cosine term has magnitude unity and the sine term is zero.

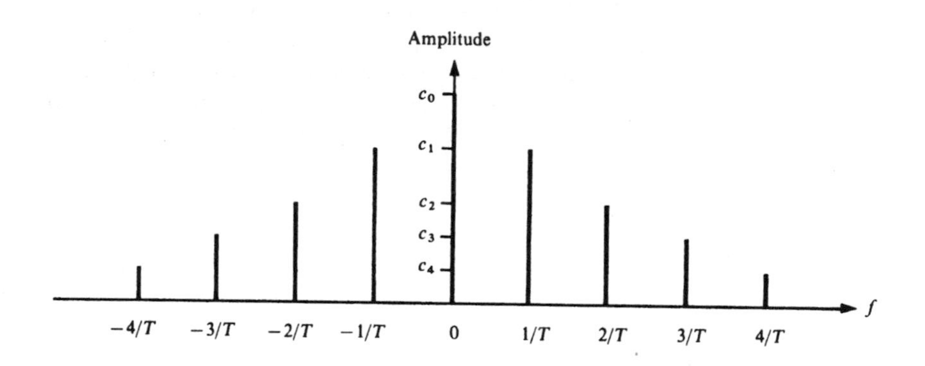

Figure 3.12. A double-sided spectrum diagram.

Exercise 3.5
Check on page 73

Sketch a double-sided spectrum diagram for the waveform:

$$v=10\sin(\omega t)+5\sin(3\omega t)+2.5\sin(5\omega t)$$

where $\omega=2\pi/T$.

3.5 The spectrum of a periodic pulse sequence

The periodic pulse is a very common form of signal in data telecommunications, so its spectral make-up is of great interest. Consider a train of rectangular voltage pulses of amplitude V, width W and period T as shown in Fig. 3.13. The signal of Fig. 3.13 may be described over one period by

$$v(t)=\begin{cases} V & \text{when } -\frac{1}{2}T\leqslant t\leqslant\frac{1}{2}T \\ 0 & \text{over the rest of the period}\end{cases}$$

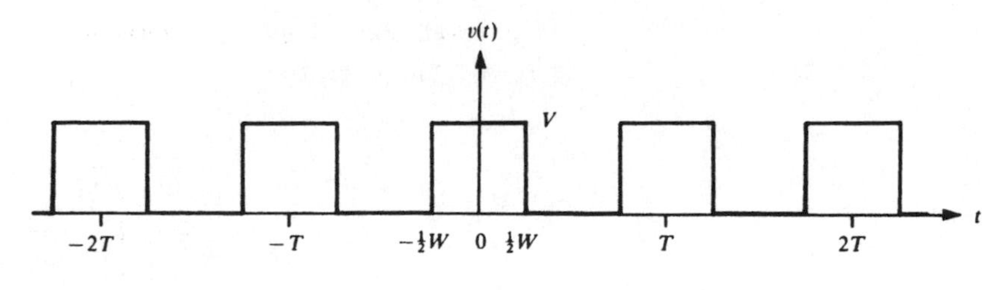

Figure 3.13. A pulse train.

Hence

$$c_n = (1/T) \int_{-W/2}^{W/2} V\mathrm{e}^{-\mathrm{j}2\pi nt/T}\,\mathrm{d}t = \frac{VW}{T}\frac{\sin(\pi nW/T)}{\pi nW/T}$$

or

$$c_n = (VW/T)\,\mathrm{sina}(\pi nW/T)$$

The sina function (defined generally as sin A divided by A or 'sine over argument') varies with the argument (A) as shown in Fig. 3.14. $(\sin A)/A \to 1$ as $A \to 0$ so the sina function will have a maximum of 1 at $A=0$. Whenever $A = n\pi$ (n integer), $\sin A = 0$ so $\mathrm{sina}\,A = 0$. Outside the range $-\pi < \beta < \pi$, all other maxima or minima will occur when A is an odd number of $\pi/2$.

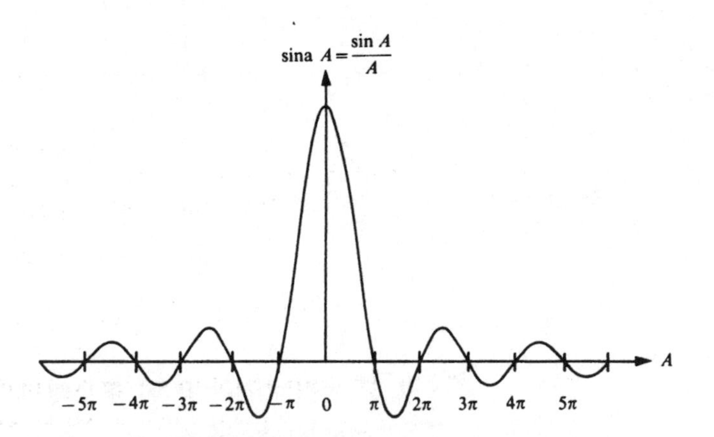

Figure 3.14. The sina function.

Having established the shape of the sina function, we can obtain a spectrum diagram of a periodic pulse train by plotting c_n, for various values of period (T) with constant pulse width (W). This gives the plots of Fig. 3.15.

The spectrum of a typical digital message (see Fig. 1.19 for example) will not be exactly repetitive, so it will be made up of spectra of the type shown in Fig. 3.15, with clock frequency ($1/T$) harmonics.

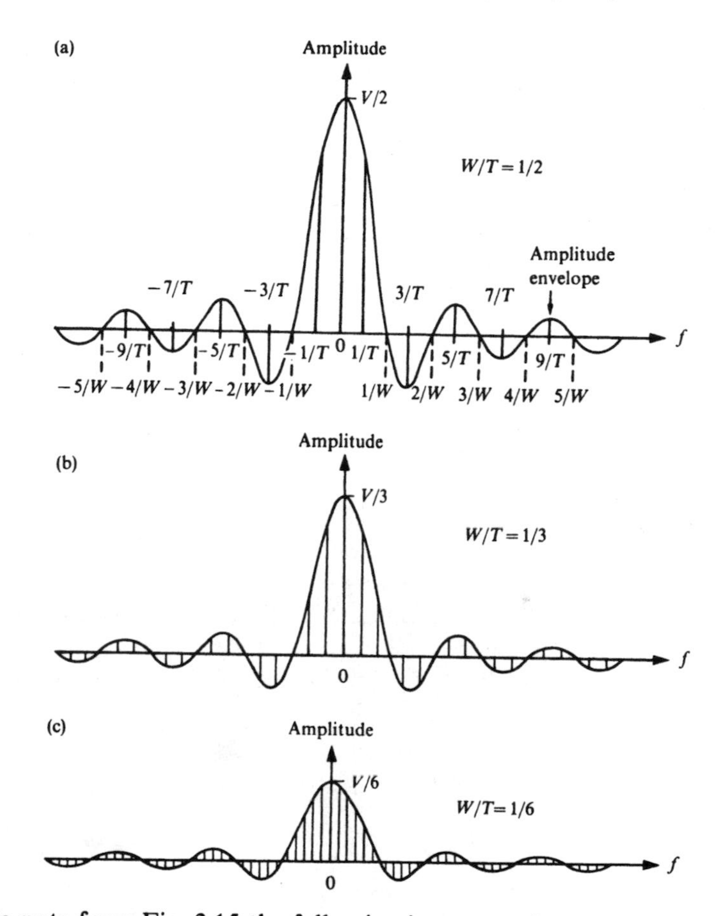

Figure 3.15. The double-sided spectra of pulse trains.

We note from Fig. 3.15 the following important features.

1. The spacing of the spectral lines is $1/T$.
2. The spectral line amplitudes are zero at integer multiples of $1/W$.
3. The size of the amplitude envelope decreases as the duty cycle W/T decreases.

Exercise 3.6
Check on page 73

1. Calculate the normalized values of the first minimum and maximum for the spectra of Fig. 3.15 when $W/T=\frac{1}{2}$. Compare these values to those when $W/T=\frac{1}{4}$.
2. Sketch a single-sided spectrum diagram (similar to Fig. 3.15) when $W/T=\frac{1}{2}$.

Feature 3 above arises because the energy in the pulse train decreases as W reduces and/or T increases. We now consider what happens to the spectrum of a pulse train as W is decreased at constant T. We allow the pulse amplitude to increase as W decreases in order to maintain the same energy in the pulse train. The effect is illustrated in Fig. 3.16. As the pulse width (W) shrinks and the pulse amplitude (A) increases to keep $AW=1$,

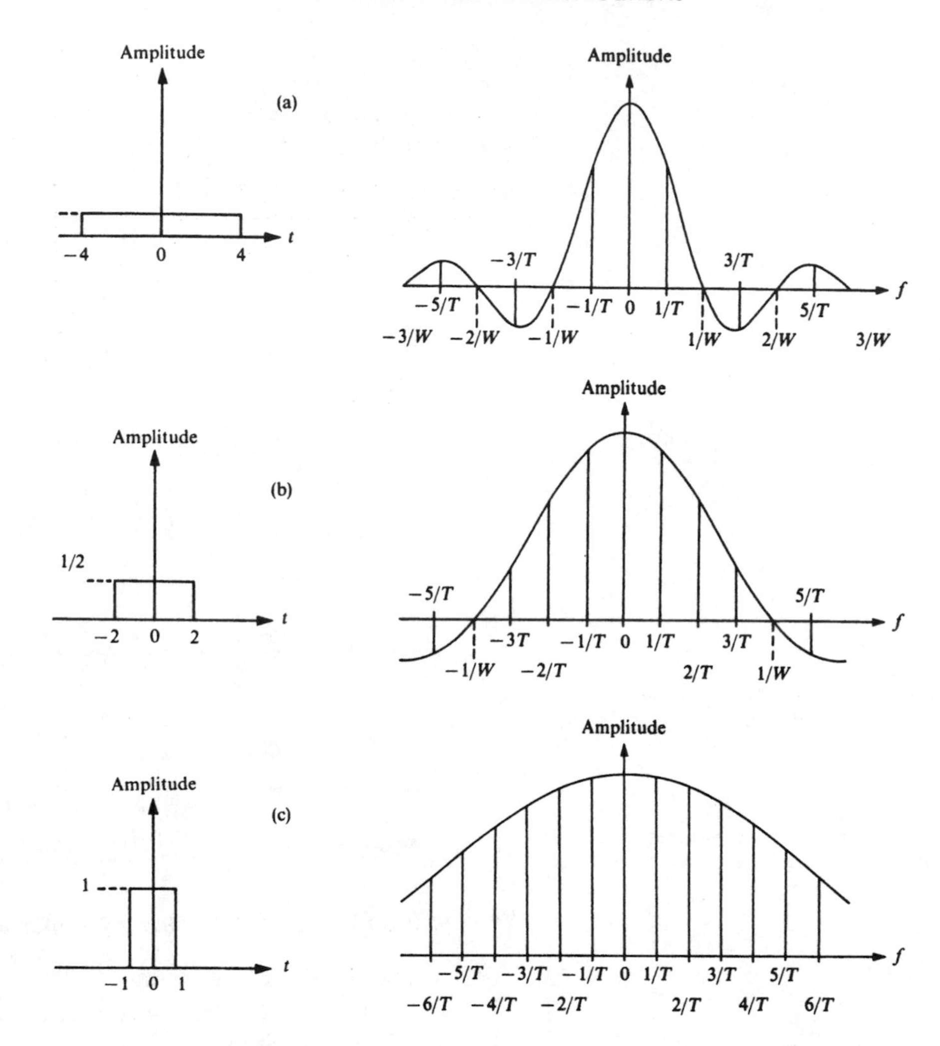

Figure 3.16. Effect of decreasing pulse width on constant-period pulse train.

we tend towards an *impulse train* and a uniform spectral amplitude. This effect is applied in Chapter 7 where *sampling* is described.

3.6 The spectrum of a single pulse

A single pulse may be obtained from a periodic pulse sequence by letting T tend to infinity. This may be illustrated by considering the effect, on a unit-amplitude pulse sequence, of increasing T and keeping W constant as illustrated in Fig. 3.17 for the time and frequency domains.

Since the spectral line spacings of Fig. 3.17 and $1/T$, this implies that as $T \rightarrow \infty$ the lines coalesce to form a *continuous* spectrum with an ever-shrinking sina function envelope. As more specral lines are added their amplitudes must decrease because the energy in the pulse signal remains finite. The limiting process cannot be illustrated, but we can see how this

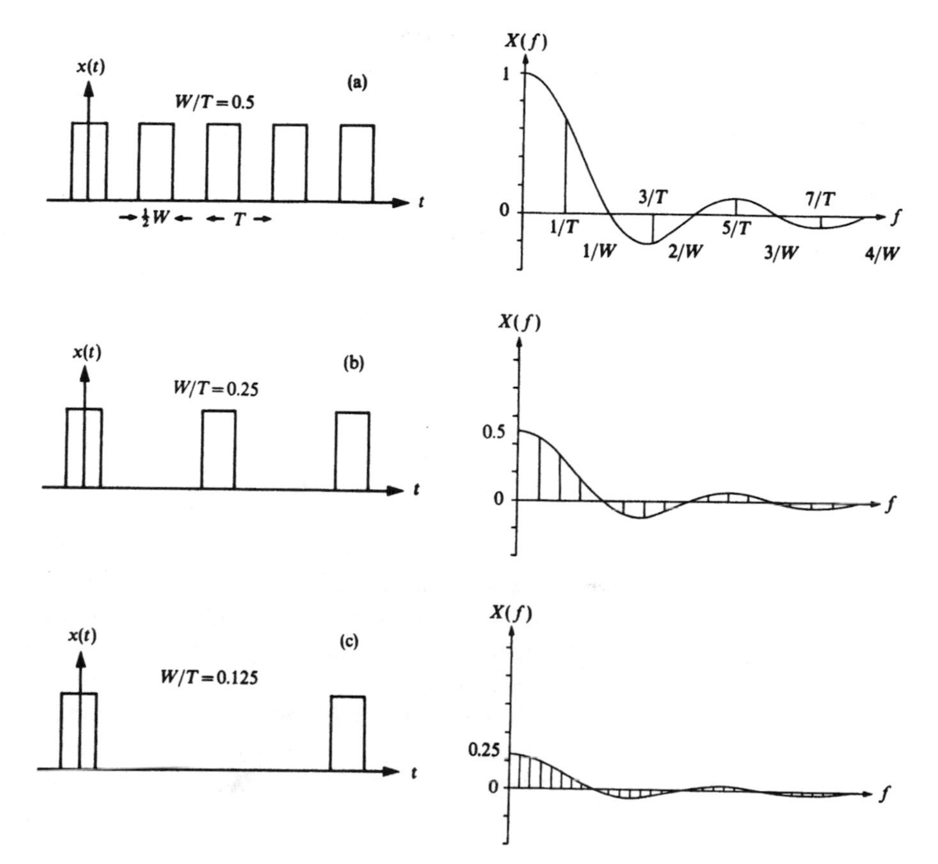

Figure 3.17. The single-sided spectra of pulse trains of increasing period.

idea works out when applied to the Fourier series as follows. Consider the series

$$x(t) = \sum_{n=-\infty}^{\infty} c_n \, e^{j2\pi nt/T}$$

where

$$c_n = (1/T) \int_{-T/2}^{T/2} x(t) \, e^{-j2\pi nt/T} \, dt$$

Defining $\delta f = 1/T$ and $f_n = n/T$ gives

$$c_n/\delta f = X(f_n) = \int_{-T/2}^{T/2} x(t) \exp(-j2\pi f_n t) \, dt$$

Thus

$$x(t)=\sum_{n=-\infty}^{\infty} X(f_n)\exp(\mathrm{j}2\pi f_n)\delta f$$

As $\delta f \to 0$, the discrete frequency f_n tends to a continuous frequency f and the discrete sum of the Fourier series for $x(t)$ becomes an integral equal to the area between the frequency axis and the curve defining the continuous function of frequency, namely $X(f)\mathrm{e}^{\mathrm{j}2\pi f}$. In the limit, therefore,

$$x(t)=\int_{-\infty}^{\infty} X(f)\,\mathrm{e}^{\mathrm{j}2\pi ft}\,\mathrm{d}f \tag{3.5}$$

We have thus converted a Fourier series into a Fourier integral by letting a periodic pulse sequence tend to a single pulse as the period tends to infinity. In Eq. (3.5) $X(f)$ is the Fourier transform of the signal function $x(t)$ and this equation links the time and frequency domains.

Most of the non-periodic signals encountered in telecommunications have a Fourier transform that enables their spectral envelope to be found. We will satisfy ourselves in this book with the observation that non-periodic signals have a continuous spectrum of some kind. Of particular interest is the *unit impulse* signal, which is a limiting case of the single pulse shown in Fig. 3.16 in which $W=\delta t \to 0$ and $A=1/\delta t \to \alpha$. The unit impulse is written functionally as $\delta(t)$ (delta function) and is defined by

$$\int_{-\infty}^{\infty} \delta(t)\,\mathrm{d}t=\begin{cases}1 & \text{at } t=0\\ 0 & \text{otherwise}\end{cases}$$

and, at time t'

$$\int_{-\infty}^{\infty} \delta(t-t')\,\mathrm{d}t=\begin{cases}1 & \text{at } t=t'\\ 0 & \text{otherwise}\end{cases} \tag{3.6}$$

Such a signal cannot exist practically but it turns out to be a useful figment of the imagination for advanced signal analysis where it is thought of as the derivative of a unit step voltage $H(t)$, that is $\delta(t)=\mathrm{d}H(t)/\mathrm{d}t$.

3.7 The effect of channel frequency response on waveshape

Having considered the spectra of various signals we next turn our attention to the way a communications channel might affect a signal. Telecommunication channels may be described in terms of their steady-state response to a specified input signal. The tone signal provides a simple standard input. The ratio $V_o(j\omega)/V_i(j\omega)$ gives the channel voltage *transfer function* $A(j\omega)$. A low-pass (LP) channel might

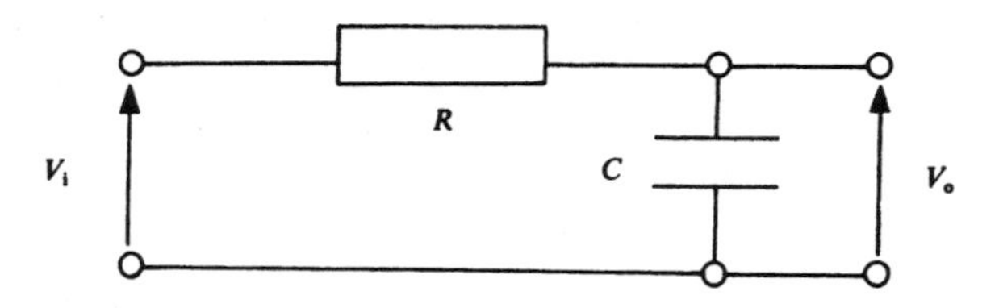

Figure 3.18. Simple model of a low-pass channel.

be simply modelled as shown in Fig. 3.18. For the LP model of Fig. 3.18

$$A(j\omega) = V_o/V_i = 1/(1 + j\omega/\omega_c)$$

so

$$|V_o/V_i|^2 = 1/[1 + (\omega/\omega_c)^2] = 1/[1 + (f/f_c)^2]$$

and

$$\phi = \tan^{-1}(\omega/\omega_c) = \tan^{-1}(f/f_c)$$

The amplitude ($|V_o/V_i| = |A|$) and phase (ϕ) frequency responses arising from these expressions are exemplified in Fig. 3.19. Note that the responses

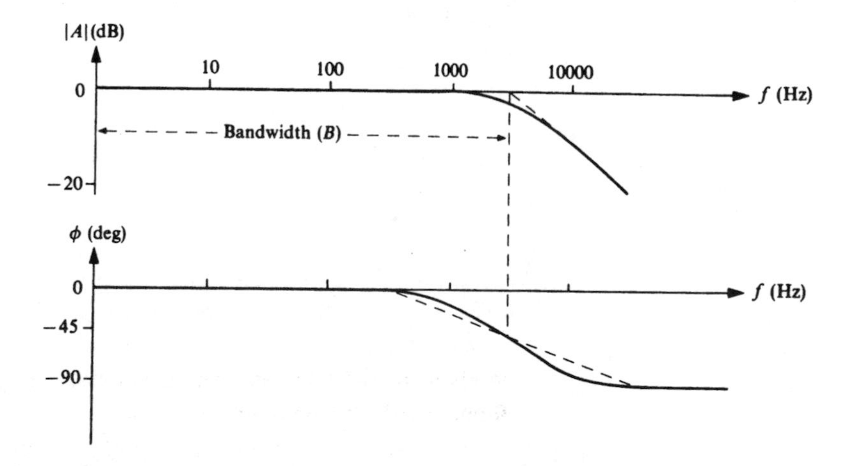

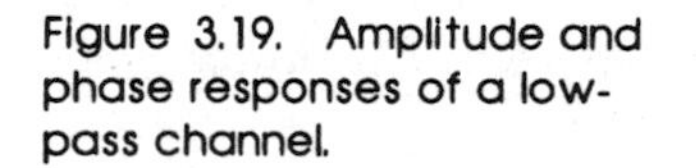
Figure 3.19. Amplitude and phase responses of a low-pass channel.

of Fig. 3.19 (known as Bode plots) can be approximated by straight lines. In the case of the magnitude response, these lines are the horizontal low-band response and the high-frequency roll-off at 20 dB/decade. The two lines 'break' at the cut-off or break frequency f_c where $|A| = -3$ dB.

The frequency at which the gain is '3 dB down' (the -3 dB frequency) is used to define the channel *bandwidth*. In the example of Fig. 3.18, it has been assumed that the channel passes all frequencies down to zero; that is, it will pass direct current (d.c.). Hence the bandwidth (B) is f_c in this case.

Exercise 3.7
Check on page 74

1. Referring to Fig. 3.19, state the cut-off frequency (f_c), the attenuation at this frequency and the slope per octave of the magnitude response as $f \to \infty$
2. If two circuits like that in Fig. 3.18 were cascaded with buffering (which ensured that one circuit did not load the other), sketch the cascade frequency responses.
3. Calculate the frequency at which the gain magnitude is 3 dB down.

If the capacitor and resistor of Fig. 3.18 were interchanged, the resulting circuit would have a first-order high-pass (HP) response as shown in Fig. 3.20. For the response of Fig. 3.20 it would not be possible to

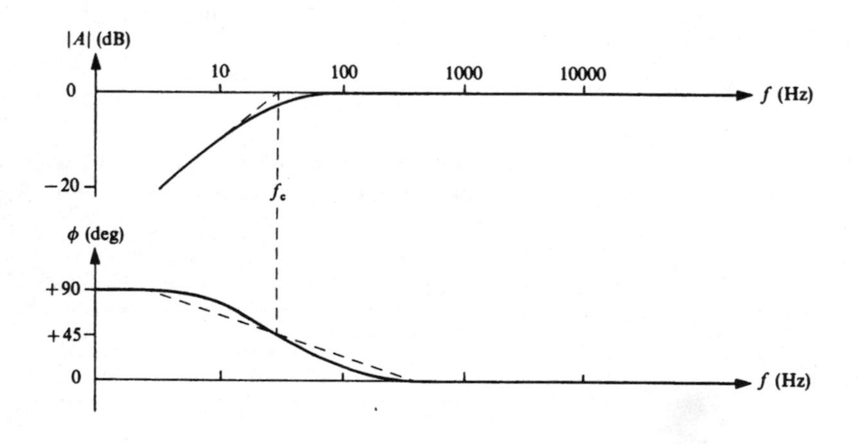

Figure 3.20. Amplitude and phase responses of a first-order high-pass channel.

define a finite bandwidth. In practice, channels will have high-frequency roll-off so the responses in Figs 3.19 and 3.20 will combine to give the magnitude response shown in Fig. 3.21, which has a finite bandwidth defined as the frequency range between the lower and upper −3 dB frequencies.

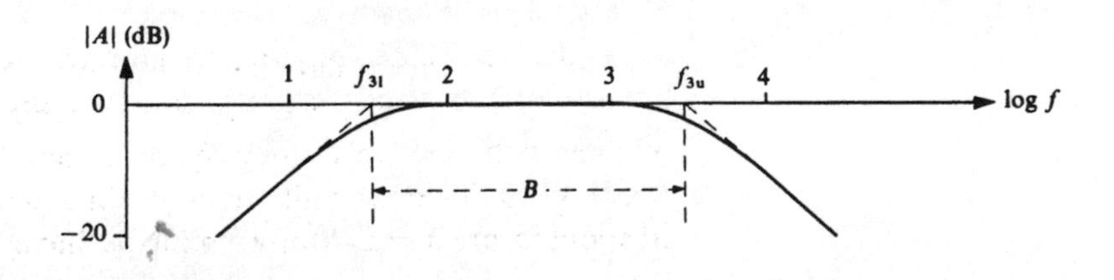

Figure 3.21. Amplitude and phase responses of a first-order band-pass channel.

Ideal channel frequency response

It might be thought that the ideal LP magnitude response would be one that was flat throughout the pass-band (0 to f_c) and zero above f_c as shown in Fig. 3.22. The response of Fig. 3.22 would certainly mean that the amplitude of the Fourier components would be unaltered as the signal passed through the channel, but the phase of these components in the region of f_c would be considerably altered. Such a response is perfectly satisfactory for audio signals because the ear is insensitive to phase changes in the signal harmonics.

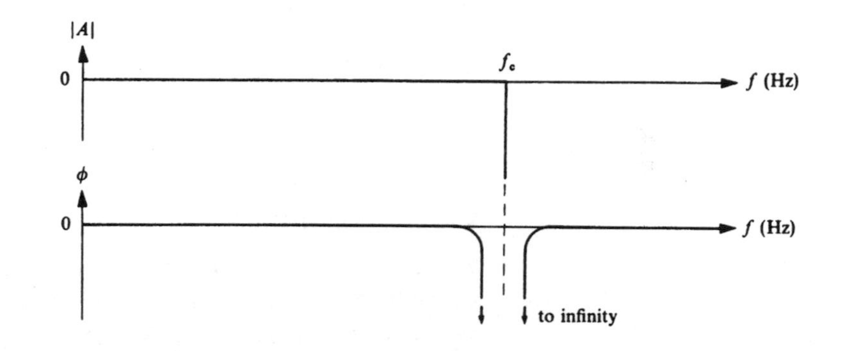

Figure 3.22. Ideal magnitude response of a low-pass channel with its phase response.

The effect of phase changes on the signal can be seen by considering an input signal $v_i = V_i \sin(\omega t)$. This will produce an output voltage $v_o = V_o \sin(\omega t + \phi)$. Writing this in the form $v_o = V_o \sin[\omega(t + \phi/\omega)]$ shows that the effect of the phase shift ϕ is to produce a *time delay* ($t_d = \phi/\omega$). In order to preserve the shape of a complex signal, the Fourier components must all suffer the same time delay. That is, t_d = constant (k), or $\phi = k\omega$. The eye is sensitive to phase variations between the Fourier components in a signal, so the ideal response of a video channel would be one that had a constant time delay characteristic, as shown in the two alternatives of Fig. 3.23.

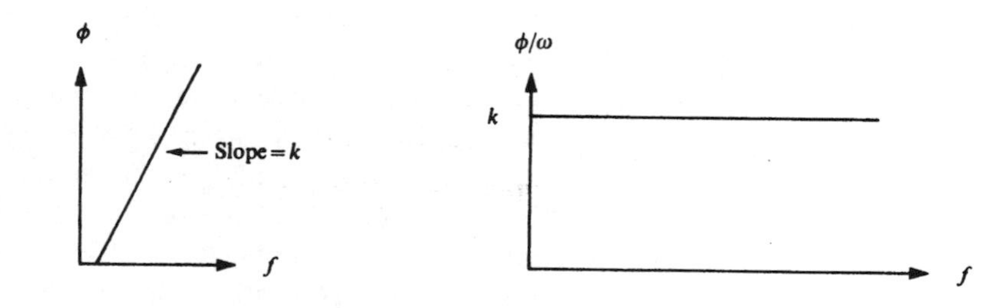

Figure 3.23. Ideal phase response of a low-pass channel.

A channel with a flat time delay response would pass data pulses with little distortion. Refer back to Fig. 3.8, which illustrates the effect of a phase change between the harmonics of a complex signal. It is not possible to have both ideal magnitude and time delay responses, so compromise responses are used. Many channels have a band-pass response (BP) in

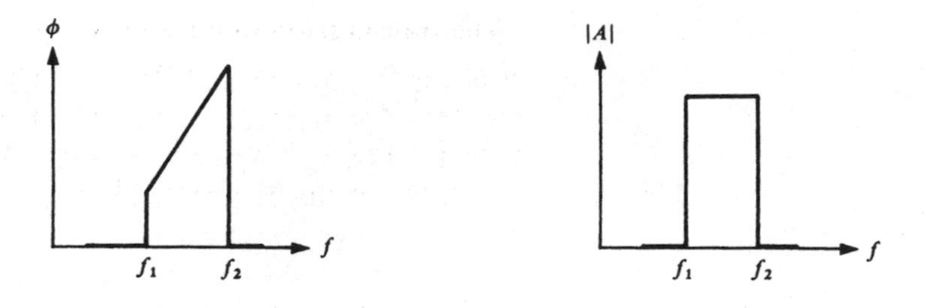

Figure 3.24. Ideal phase and magnitude responses of a band-pass channel.

practice, and in this case the ideal magnitude and phase responses would be as shown in Fig. 3.24.

It should be emphasized again that it is not practically possible to attain the ideal phase response without degrading the magnitude response and vice versa. You can aim to idealize one or the other, or accept some compromise.

Channel frequency and time domain responses

Frequency domain analysis, used in the previous subsection, may be done in the time domain, although with greater difficulty. Time domain analysis is beyond the scope of this book, so we conclude with a brief indication of how it could be done. One method is to think of a signal as an infinite number of successive samples, the amplitude of each sample being equal to the signal voltage $v_i(t)$ at the sampling instant. Each of these samples will produce an output and the total channel response is found by addition of each sample. Thus, a channel output is given by

$$v_o(t) = \int_{-\infty}^{\infty} v_i(t')\delta(t-t')\,dt' \qquad (3.7)$$

Equation (3.7) states that the output of a linear channel at time t is the sum of all instantaneous values of the input $v_i(t')$ each weighted by the impulse signal (sample) value at a particular instant (t'). Refer to Eq. (3.6) for the definition of impulse signal $\delta(t')$. The integration of Eq. (3.7) is referred to as the *convolution* integral of v_i and $\delta(t)$. Equation (3.7) is written $v_o(t) = v_i(t) * \delta(t)$, so that, if this equation is compared with the equation $V_o(j\omega) = V_i(j\omega)A(j\omega)$, we see that convolution in the time domain is the equivalent of multiplication in the complex frequency domain. The symbol $*$ denotes a type of multiplicative process, known as convolution, defined by

$$v_i(t) * \delta(t) = \int_{-\infty}^{\infty} v_i(t')\delta(t-t')\,dt'$$

The transmission of a step (V_i) through a low-pass (LP) channel of the kind shown in Fig. 3.18 provides a simple example of this idea. In this case the lower limit of the integral is 0 because there is no signal present when t' is less than zero, and the upper limit has a general finite value t for any practical signal. Hence, substituting $v_i(t') = V_i$ gives

$$v_o(t) = \int_0^t V_i \delta(t - t')\, dt'$$

The impulse response is $dH(t')/dt'$ where $H(t')$ is the unit step response of the LP channel, which is $1 - e^{-a(t-t')}$, where $a = 1/CR$. So, $\delta(t - t') = a\, e^{-a(t-t')}$. Thus

$$v_o(t) = \int_0^t V_i a\, e^{-a(t-t')}\, dt'$$

$$= V_i a\, e^{-at} \int_0^t e^{at'}\, dt'$$

$$= V_i a\, e^{-at} |e^{at'}/a|_0^t$$

$$= V_i e^{-at}(e^{at} - 1)$$

$$= V_i(1 - e^{-at})$$

The idea of sampling will be explained further in Chapter 7. It should, however, be appreciated that convolution can be applied to any types of signal—one of them does not have to be an impulse signal.

Summary

We have seen how periodic signals can be made up of a number of harmonics. The Fourier series provides a mathematical description of a periodic signal. The series can take on several forms; one with cosine components only, another with a mix of sine and cosine components and an exponential form. The signal components of a periodic pulse train were represented with single- and double-sided discrete spectrum diagrams. A single pulse was represented as a limiting case of a pulse train with a period tending to infinity and was expressed mathematically as a Fourier integral and pictorially as a continuous spectrum.

Channels have a non-uniform response to signal frequency components, which causes distortion of the transmitted signal waveform.

The distortion arises from changes in the relative magnitude and phase of the signal components. The ideal (non-distorting) response of a video channel should have a constant time delay characteristic, whereas the ideal audio channel should have a constant magnitude frequency response.

Checks on exercises

3.1 Complex-plane representations of a sine signal similar to Fig. 3.3 are shown in Fig. 3.25.

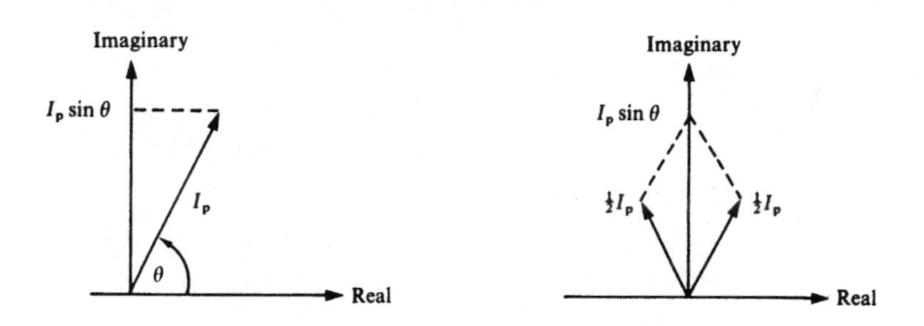

Figure 3.25. Two complex-plane representations of a sine signal.

3.2 The graph for the complex signal:

$$v = 10\sin(\omega t) + 5\sin(2\omega t) + 2.5\sin(3\omega t)$$

is given in Fig. 3.26.

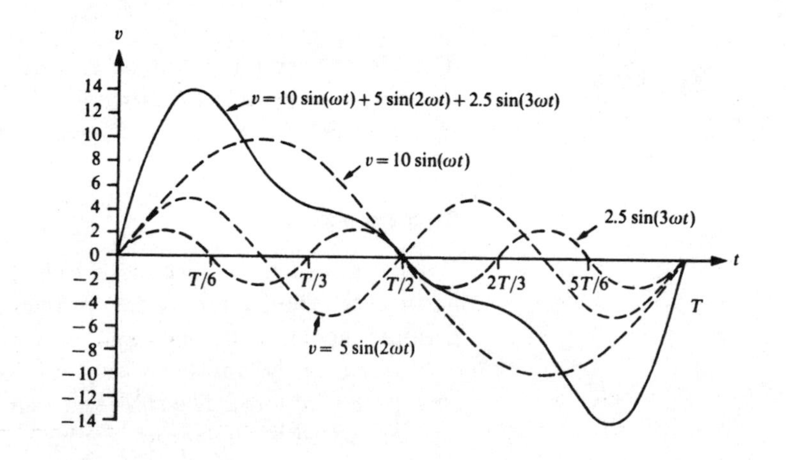

Figure 3.26. The synthesis of three sinusoids.

3.3 The graph for the signal:

$$v = 10\sin(\omega t) + 5\sin(3\omega t) + 2.5\sin(5\omega t)$$

is as shown in Fig. 3.27.

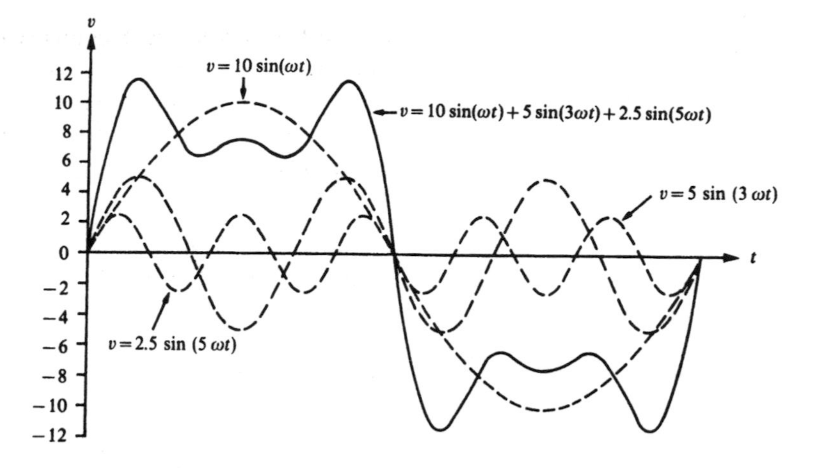

Figure 3.27. The synthesis of two odd harmonics.

3.4 The spectrum diagram for

$$v = 5\sin(250\pi t) + 2.5\sin(500\pi t) + 1.25\sin(1000\pi t)$$

is given in Fig. 3.28.

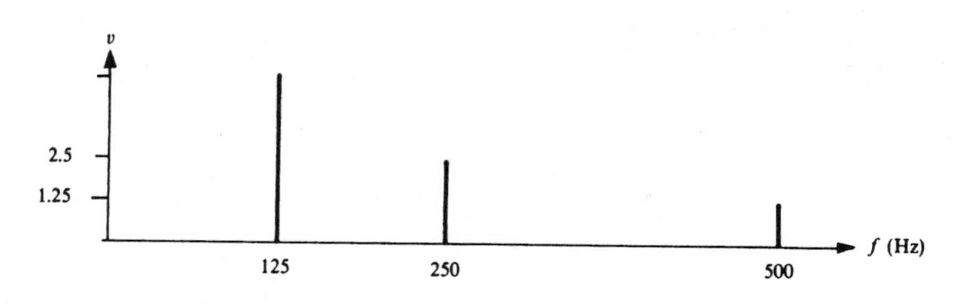

Figure 3.28. A spectrum diagram for three harmonics.

3.5 The double-sided spectrum diagram for

$$v = 10\sin(\omega t) + 5\sin(3\omega t) + 2.5\sin(5\omega t)$$

is given in Fig. 3.29.

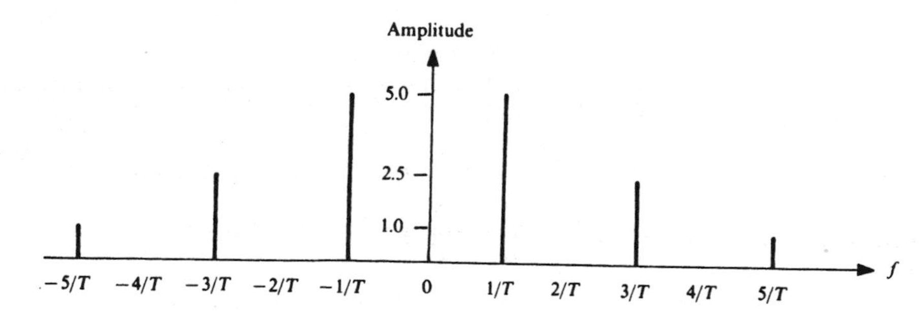

Figure 3.29. A two-sided spectrum diagram with odd harmonics.

3.6 1. When $W/T = \frac{1}{2}$, the first minimum will occur at the third harmonic, giving

$$c_3 = (W/T)\,\text{sina}(\pi n W/T) = [\tfrac{1}{2}\sin(3\pi/2)]/(3\pi/2) = -0.1060$$

The first maximum will occur at the fifth harmonic and will have the value

$$c_5=(W/T)\,\text{sina}(\pi nW/T)=[\tfrac{1}{2}\sin(5\pi/2)]/(5\pi/2)=0.0640$$

When $W/T=\frac{1}{4}$, these values will be halved.

2. The single-sided spectrum of a pulse train with duty cycle of $\frac{1}{2}$ is shown in Fig. 3.30. Note that the amplitude envelope of the single-sided spectrum representation is double the size of the double-sided spectrum representation of the same signal.

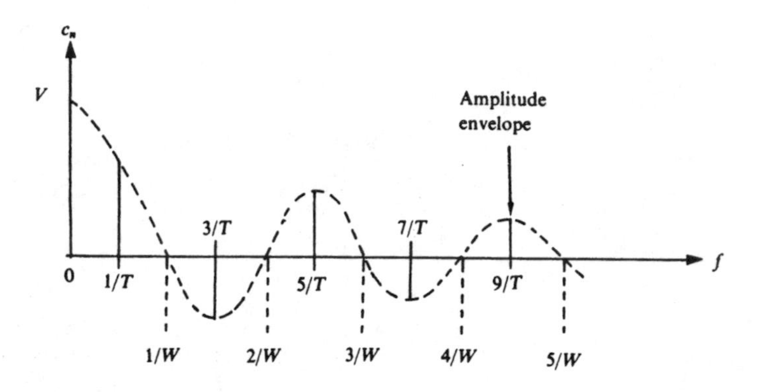

Figure 3.30. The single-sided spectrum of a pulse train with duty cycle of $\frac{1}{2}$.

3.7

1. The frequency axis has a log scale so mid-way between 1 and 10 kHz represents a cut-off frequency (f_c) given by $\log f_c=5000$. So $f_c=3162$ Hz. Now

$$|V_o/V_i|^2=1/[1+(f/f_c)^2]$$

so when $f=f_c$, $|V_o/V_i|=1\sqrt{2}=-3$ dB. At high frequency the magnitude response halves when the frequency is doubled. That is a drop of $20\log\frac{1}{2}$ per octave $=-6$ dB/octave.

2. If two of the circuits in Fig. 3.18 were cascaded with buffering, the cascade frequency responses would simply be the product of the two circuit responses. This would be a sum on a log scale. Hence the required sketch would be as in Fig. 3.31. Note that, for the LP cascade

$$|V_o/V_i|=1/[1+(f/f_c)^2]$$

f_c is still a break frequency but it is now a -6 dB instead of a -3 dB frequency.

3. The -3 dB frequency (f_3) is given by

$$|V_o/V_i|=1/[1+(f_3/f_c)^2]=1/\sqrt{2}$$

So $f_3=f_c\sqrt{(2^{1/2}-1)}=0.64f_c=2035$ Hz. For n non-interacting LP sections $f_3=f_c\sqrt{(2^{1/n}-1)}$ and the asymptotic slope is $20n$ dB/decade.

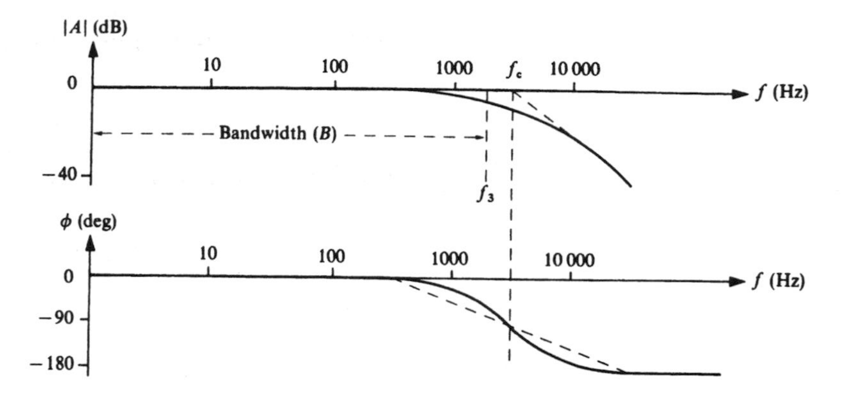

Figure 3.31. Amplitude and phase responses of a second-order low-pass channel.

Thus, the cascading of n first-order sections results in a progressively more rapid roll-off as n increases, but the bandwidth shrinks and the phase lag increases.

Self-assessment questions

3.1 State why the ability to express signals in terms of sinusoidal components is so useful in telecommunications.

3.2 Describe what is meant by a phasor and define instantaneous angular frequency.

3.3 Show that a single phasor may be expressed as the sum of two phasors with frequencies of opposite sign.

3.4 What combinations of harmonics lead to waveforms that are symmetrical about the time axis?

3.5 What is a spectrum diagram?

3.6 Write out three alternative forms of Fourier series.

3.7 Distinguish between single- and double-sided spectrum diagrams.

3.8 Define and sketch the sina function.

3.9 Sketch the spectrum diagrams for a periodic pulse train when:
(a) the pulse width is 1 ms and the period is 4 ms;
(b) the pulse width is 2 ms and the period is 4 ms;
(c) the pulse width is 2 ms and the period is 8 ms.

3.10 Give an expression for the Fourier integral of a signal $x(t)$ and sketch the double-sided spectrum diagram for a single pulse of width 1 μs.

3.11 Define the terms: −3 dB frequency, cut-off frequency and break frequency.

3.12 State how the bandwidth of a channel is defined.

3.13 State two simple criteria for an ideal channel frequency response.

4

Sinusoidal carrier modulation

Altering carrier frequencies to suit a channel

Objectives

When you have completed this chapter you will be able to:

- Describe the principle of amplitude and angle modulation
- Describe the following forms of amplitude modulation:
  double-sideband
  double-sideband suppressed-carrier
  single-sideband
  vestigial-sideband
- Describe how these types of amplitude modulation may be achieved
- Compare the practical advantages of each type
- Define phase modulation and frequency modulation
- Compare phase and frequency modulation
- Describe how frequency modulation may be accomplished
- State the advantages and disadvantages of frequency over amplitude modulation
- Describe how data may be carried by amplitude or frequency modulation

Baseband signals cannot be transmitted efficiently as radio waves because the antenna required would be too long and would have insufficient bandwidth. A baseband signal must be 'carried' by a radio wave at a frequency best suited to the propagation medium. Modulation is the process that combines the baseband and carrier signals so that the baseband is shifted to a higher frequency range. The process is used in frequency-division multiplexing (FDM).

The aim of this chapter is to describe how a sinusoidal signal may be used to carry information by having either its amplitude, phase or frequency varied by the baseband signal. The inherent continuity of the sinusoidal carrier has led to this technique being called continuous-wave modulation. A radio receiver recovers information by demodulating a carrier signal. This process is dealt with in Chapter 5.

In this chapter, modulation will be studied, first, by looking at the effect of a step change in the modulating signal on the carrier amplitude, phase and frequency in turn. Then a tone modulating signal will be considered and the principles, thus highlighted, applied to other modulating signals.

4.1 Radio systems

The basic processes that are fundamental to all radio communication are exemplified, for sound radio, in Fig. 4.1. The microphone (MIC) converts sound into an analogous electrical signal covering the audio-frequency range (the baseband). Amplification and modulation make the audio signal suitable for efficient radio transmission with an aerial of reasonable size. Recovery of the audio signal is the reverse of this process.

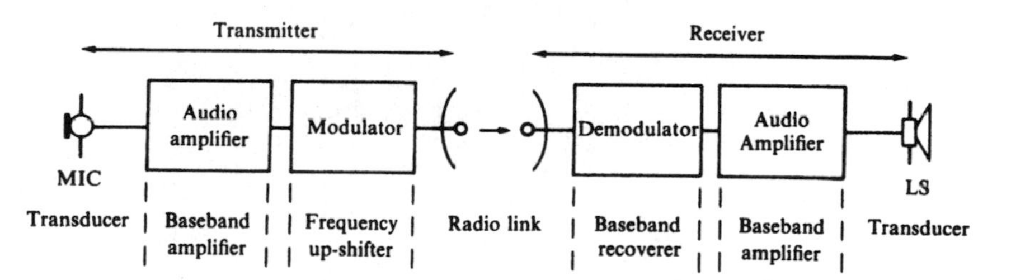

Figure 4.1. A one-way radio communications system.

Modulation of a continuous wave (CW) can be used to telecommunicate all types of signals: audio, video and data. CW modulation causes frequency translation from baseband to a chosen carrier band and thus allows FDM to be used, so increasing the number of channels available in the link. This is a very important feature of telecommunications made possible by modulation.

An electrical signal may be expressed in the form $i = I \cos \theta$ where i and θ are the instantaneous values of current and phase angle, respectively, and I is the current amplitude. A similar expression, $v = V \cos \theta$, may be used for signal voltage. In an unmodulated signal, the amplitude is constant and the phase changes at a constant rate known as the angular frequency (ω). There are thus two features of an electric signal that may be varied or modulated in order to convey information: the *amplitude* (I or V) and the *angle* (θ).

4.2 Amplitude modulation

The concept of amplitude modulation (AM) may be illustrated as in Fig. 4.2, which shows how the amplitude of a sinusoidal signal voltage may

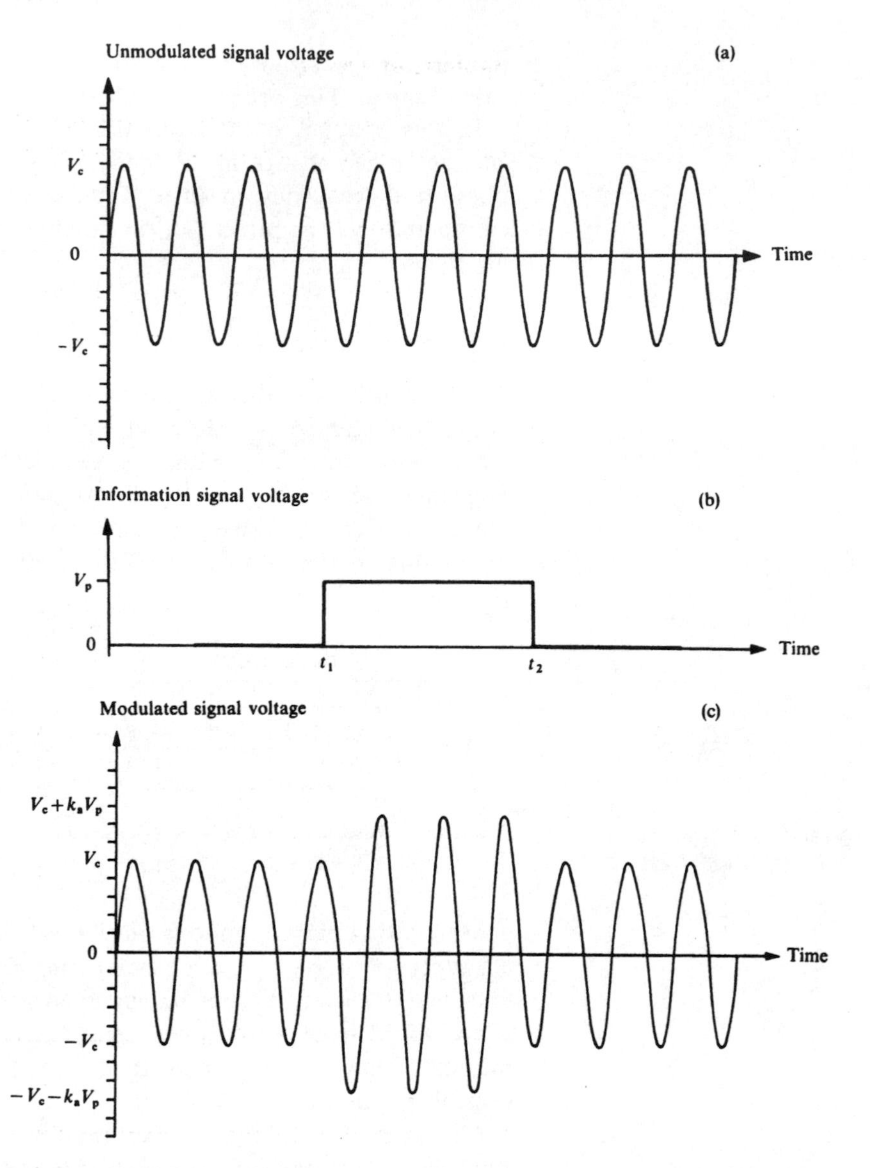

Figure 4.2. An example of amplitude modulation (AM).

vary according to information in the form of a pulse. This is sometimes referred to as *envelope* modulation because the carrier is enveloped by the shape of the modulating waveform. In this example, a pulse of amplitude V_p changes the magnitude of the carrier amplitude V_c by an amount V_d, which is the carrier *amplitude deviation*, $V_d = k_a V_p$, where k_a is the change in carrier voltage per volt of modulating signal.

Exercise 4.1
Check on page 102

Estimate the value of k_a in Fig. 4.2.

Draw the AM signal waveform if the pulse amplitude extends from $-V_p$ to $+V_p$ instead of from 0 to $+V_p$.

Comment on the effect of increasing V_p.

Analysis of AM signals

The expressions for a cosinusoidal signal voltage, $V_c\cos(\omega_c t)$, which has its amplitude (V_c) modulated by a pulse of the kind shown in Fig. 4.2, are

$$v_c=\begin{cases}(V_c+k_a V_p)\cos(\omega_c t) & \text{for } t_1<t<t_2\\ V_c\cos(\omega_c t) & \text{for } t<t_1 \quad \text{or } t>t_2\end{cases}$$

When the same carrier signal is amplitude-modulated by a tone, $v_t=V_t\cos(\omega_t)$, we have

$$\begin{aligned}v_c&=[V_c+k_a V_t\cos(\omega_t t)]\cos(\omega_c t)\\ &=V_c[1+m\cos(\omega_t t)]\cos(\omega_c t)\end{aligned}\tag{4.1}$$

where $m=k_a V_t/V_c$ is a useful parameter for analytical purposes, and is given by

$$m=\text{carrier amplitude deviation/carrier unmodulated amplitude}$$

$100m$ is the percentage modulation and m is known as the *modulation index*. In many modulators the carrier voltage deviation equals the modulating voltage (that is, $k_a=1$), so $m=V_t/V_c$ in this case.

Exercise 4.2
Check on page 103

Draw the modulated signal waveform, analogous to that of Fig. 4.2, if the pulse waveform is replaced by a single tone, whose amplitude gives a value of $m=0.2$.

A more general form of Eq. (4.1) is obtained when any modulating signal $v(t)$ is assumed. We then have

$$v_c=V_c[1+mv(t)]\cos(\omega_c t)$$

An important feature is revealed if Eq. (4.1) is expanded thus:

$$\begin{aligned}v_c&=V_c\cos(\omega_c t)+mV_c\cos(\omega_t t)\cos(\omega_c t)\\ &=V_c\cos(\omega_c t)+\tfrac{1}{2}mV_c\cos[(\omega_c-\omega_t)t]+\tfrac{1}{2}mV_c\cos[(\omega_c+\omega_t)t]\\ &=\text{carrier}+\text{lower side frequency}+\text{upper side frequency}\end{aligned}\tag{4.2}$$

So, for example, if a carrier signal of 100 kHz is modulated by the note A (440 Hz at international concert pitch), a modulated signal is generated, consisting of the original carrier and upper and lower side frequencies of 100 kHz + 440 Hz and 100 kHz − 440 Hz respectively. This effect can be illustrated, for both loud and soft notes, by means of the frequency spectra shown in Fig. 4.3.

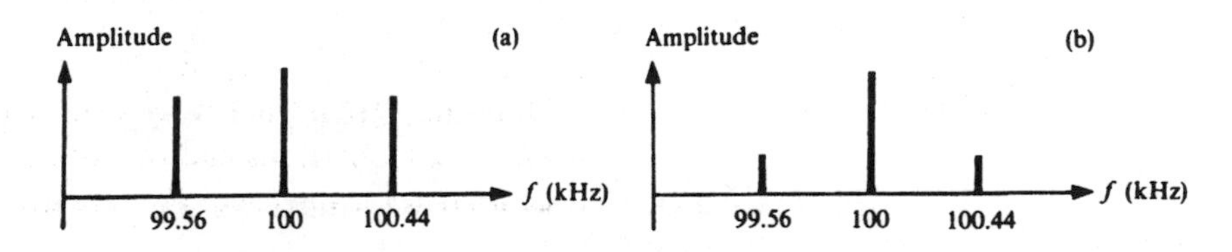

Figure 4.3. Frequency spectra of an AM signal modulated by a 440 Hz tone: (a) loud tone; (b) soft tone.

This type of modulation is often referred to as *double-sideband amplitude modulation* (DSB-AM). The range of frequency between the upper and lower side frequencies (880 Hz) is the minimum channel bandwidth required to convey the 440 Hz note. Whenever a tone modulates the amplitude of a carrier wave in this way, the resulting modulated wave has a bandwidth of twice the tone frequency. DSB-AM receivers must have at least this bandwidth. The maximum amount by which the sound signal is allowed to vary the amplitude of the carrier depends on the dynamic range with which the receiver can cope without distorting the signal. This technique is used in much commercial broadcasting because it simplifies the design of the receiver and so reduces its cost.

If the modulating signal happens to be music, then there will be a band of tones over a range of about 20 Hz to 20 kHz instead of a single tone. The side frequencies of Fig. 4.3 then become *sidebands.* This is the general case of which the single tone is a special example. Frequencies on either side of a carrier are generally referred to as sidebands, even though they may be single frequencies rather than frequency bands.

Exercise 4.3
Check on page 103

Draw spectrum diagrams for commercial-standard speech (300 Hz to 3.4 kHz) and a DSB-AM signal carrying such speech on a 3 MHz carrier, assuming that the speech signal has the same amplitude over the whole baseband. How many speech channels could be multiplexed in the 300–400 MHz band using 600 Hz guard bands?

DSB-AM may be accomplished by using a non-linear device. Since practically all electronic devices are non-linear, the choice is vast! We consider one that can be approximated by a square-law transfer characteristic: $v_o = av_i + bv_i^2$. If

$$v_i = V_c \cos(\omega_c t) + V_t \cos(\omega_t t)$$

then

$$\begin{aligned} v_o &= a[V_c \cos(\omega_c t) + V_t \cos(\omega_t t)] + b[V_c \cos(\omega_c t) + V_t \cos(\omega_t t)]^2 \\ &= aV_c \cos(\omega_c t) + aV_t \cos(\omega_t t) + \tfrac{1}{2} b\{V_c[1 + \cos(2\omega_c t)] \\ &\quad + V_t[1 + \cos(2\omega_t t)]\} + 2bV_c \cos(\omega_c t)\, V_t \cos(\omega_t t) \\ &= aV_c \cos(\omega_c t) + aV_t \cos(\omega_t t) + \tfrac{1}{2} b(V_c + V_t) + \tfrac{1}{2} bV_c \cos(2\omega_c t) \\ &\quad + \tfrac{1}{2} bV_t \cos(2\omega_t t) + 2bV_c \cos(\omega_c t)\, V_t \cos(\omega_t t) \end{aligned}$$

(refer to Appendix B). Since

$$2bV_c \cos(\omega_c t)\, V_t \cos(\omega_t t) = bV_c V_t\{\cos[(\omega_c - \omega_t)] + \cos[(\omega_c + \omega_t)t]\}$$

a band-pass filter of bandwidth $2\omega_t$, centred on ω_c, will give an output

(assuming that $2\omega_t \ll \omega_c$)

$$\begin{aligned} v_{o,\text{filter}} &= aV_c \cos(\omega_c t) + 2bV_c \cos(\omega_c t)\, V_t \cos(\omega_t t) \\ &= aV_c \cos(\omega_c t)[1 + (2b/a)\, V_t \cos(\omega_t t)] \\ &= aV_c \cos(\omega_c t)[1 + m \cos(\omega_t t)] \end{aligned} \tag{4.3}$$

Comparing Eq. (4.3) with Eq. (4.1) reveals that we get a DSB-AM signal at the filter output. An essential feature of this process is the multiplication of the terms $V_c \cos(\omega_c t)$ and $V_t \cos(\omega_t t)$. Any non-linear device will achieve this. One form of DSB-AM transmitter system is shown in Fig. 4.4.

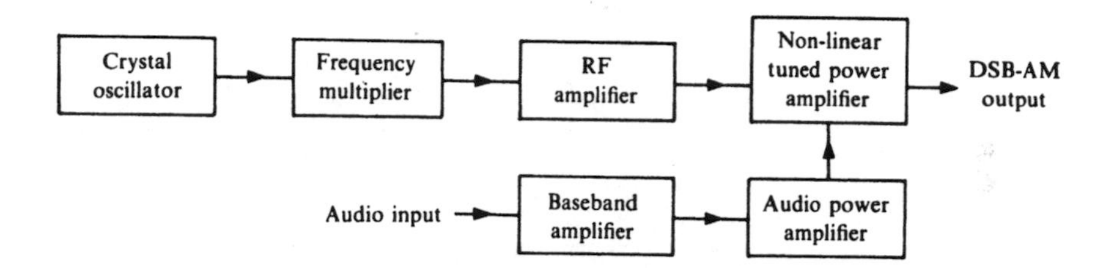

Figure 4.4. High-power amplitude modulation.

The tuned power amplifier performs the functions of modulation, BP filtering and signal power conversion. The crystal oscillator provides a stable frequency source from which the carrier frequency is derived by the frequency multiplier. An alternative system, in which a low-power, non-linear RF amplifier is fed by the audio power amplifier, produces a low-power modulated signal, which is then amplified by a linear amplifier. This gives a low-power modulation system.

Non-linear distortion

The action of a square-law device when a carrier and tone are input raises the question of what happens when two or more tones are present in the modulating signal. Suppose we have two tones $V_{t1} \cos(\omega_{t1} t)$ and $V_{t2} \cos(\omega_{t2} t)$. Each of these will combine with the carrier to produce sum, difference and product frequency components. However, tone 1 will also combine with tone 2 to produce sum and difference frequencies ($\omega_{t1} t + \omega_{t2} t$ and $\omega_{t1} t - \omega_{t2} t$) and harmonics ($2\omega_{t1} t$ and $2\omega_{t2} t$). Thus, not only will we get the desired modulation of the carrier by each tone, but also we will get undesired *intermodulation* between the tones. Non-linearity in any electronic device produces both harmonic and intermodulation distortion.

The presence of intermodulation components in the modulator output will mean that the demodulated signal will contain components that were not present in the modulating signal. Some of these components are harmonics, others are non-harmonic sum or difference terms. In audio applications the harmonics can be tolerated but the non-harmonics are objectionable. These components cannot be easily filtered out because many of them lie in the pass-band of the system. They must therefore be minimized at source. This implies that the amplitude of the tones must be curtailed so they are contained within a small and, therefore, almost linear range of the modulator—that is another way of saying that the modulation index ($m = V_t/V_c$) must not be too big.

The principle of restricting the amplitude or *dynamic range* of signals that are to be sent among non-linear channels in order to minimize distortion applies generally. The signal *compression* that takes place at the transmitter must be made good by *expansion* at the receiver. This combined operation of compression followed by expansion is the *companding* process shown in Fig. 4.5.

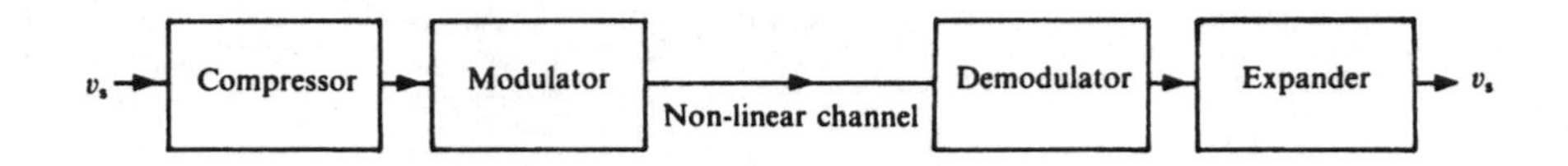

Figure 4.5. Companding.

Signal power

The amplitudes of the sidebands (and, hence, their power) depend on the amplitude of the modulating signal. In AM the power in the carrier is unaltered by the modulation process: the information power is carried by the sidebands. The power (p) developed in a load R by the signal of Eq. (4.1) would be given by

$$p = [\tfrac{1}{2}V^2 + \tfrac{1}{4}(mV_c)^2 + \tfrac{1}{4}(mV)^2]/R$$

Even with the maximum amount of modulation of a single tone ($m = 1$), half the power is wasted because it is in the carrier and not in the sidebands where the signal information is.

Double-sideband suppressed-carrier amplitude modulation (DSB-SC-AM)

Transmitter power can be saved by suppressing the carrier and transmitting the sidebands only. This idea seems preposterous at first sight because, if the carrier is suppressed, there will be no carrier amplitude to modulate! However, all that is being proposed is to send the sidebands only and combine them with a signal of carrier frequency at the receiver. The carrier frequency is known, so it does not need to be sent with the sidebands. In effect, envelope modulation would take place at the receiver, where the carrier would be generated, rather than at the transmitter. Power saving in this way would make the receiver more complicated—an example of an engineering trade-off. If the carrier is suppressed, Eq. (4.2) becomes

$$\begin{aligned} v_c &= \tfrac{1}{2}mV_c\cos[(\omega_c - \omega_t)t] + \tfrac{1}{2}mV_c\cos[(\omega_c + \omega_t)t] \\ &= mV_c\cos(\omega_t t)\cos(\omega_c t) \end{aligned} \qquad (4.4)$$

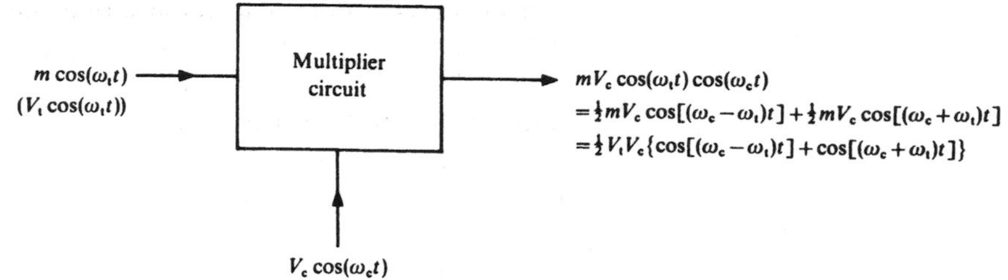

Figure 4.6. Double-sideband suppressed-carrier amplitude modulation (DSB-SC-AM).

Equation (4.4) suggests that DSB-SC-AM can be accomplished as shown in Fig. 4.6.

The multiplying operation shown in Fig. 4.6 is known as *mixing* or *heterodyning*, and the upper and lower sidebands are the sum and difference frequencies $\omega_c+\omega_t$ and $\omega_c-\omega_t$. In practice the multiplier circuit would be a non-linear electronic device such as a diode or a transistor, which, as we saw in the case of the DSB-AM square-law modulator, produces a carrier signal, second harmonics and constant voltages as well as the product of the carrier and tone frequencies, which have to be filtered. Filtering may be avoided by using a *balanced modulator* to suppress the carrier in the manner shown in Fig. 4.7.

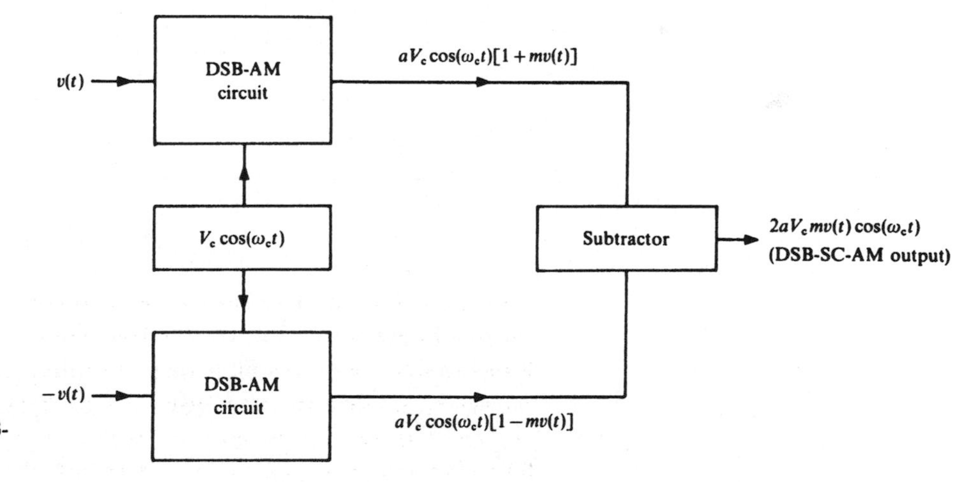

Figure 4.7. A balanced modulator for producing DSB-SC-AM.

Single-sideband amplitude modulation (SSB-AM)

The bandwidth needed for DSB-SC-AM is twice the baseband. Since the information content of the upper and lower sidebands is the same, it would seem that a 50% bandwidth saving could be achieved by suppressing one of them. Inevitably, there is a price to pay for this: the receiver is going to be more complicated, as we will now show.

If the carrier and the lower sideband are suppressed, Eq. (4.2) becomes

$$v_c = \tfrac{1}{2} m V_c \cos[(\omega_c + \omega_t)t]$$

which represents the SSB-AM transmission of a tone of frequency ω_t.

Exercise 4.4
Check on page 103

Draw spectra as in Exercise 4.3 for DSB-SC-AM and SSB-AM transmissions.

Conceptually, the simplest way to accomplish SSB-AM would be to add a low- or high-pass filter to the multiplier output of Fig. 4.4. which filters out one of the sidebands. In practice, however, the closeness of the two sidebands makes it difficult to filter out one of them completely. The problem can be eased if one multiplication is done at low frequency to be followed, after filtering, by a second modulation, which shifts the SSB to the required higher frequency range. The principle is illustrated in Figs 4.8 and 4.9.

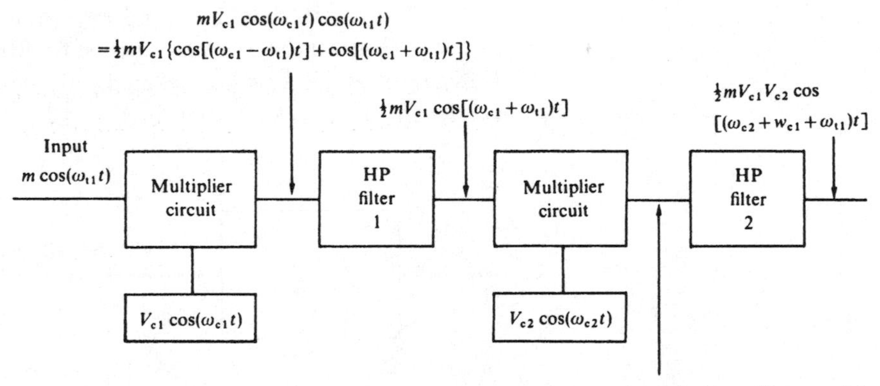

Figure 4.8. Two-stage SSB-AM.

Figure 4.9 shows two sidebands at 99 and 101 kHz, which are formed when a 1 kHz tone modulates a 100 kHz sub-carrier. The 101 kHz tone from the first high-pass filter then modulates a 3 MHz carrier, producing two sidebands separated by 3101 − 2899 = 202 kHz. This greater separation

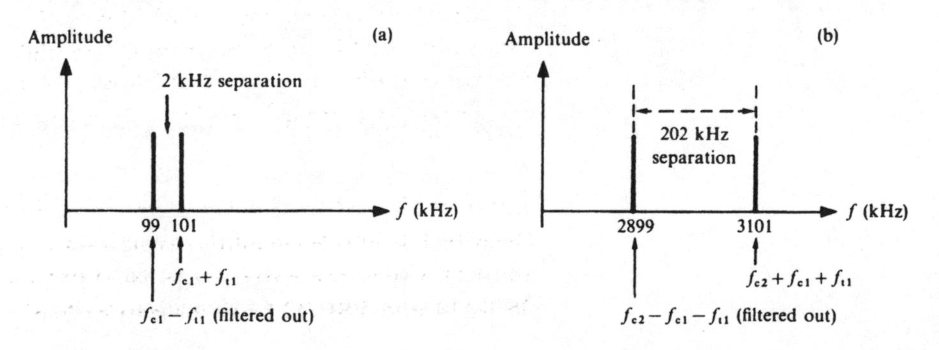

Figure 4.9. Spectra of two-stage SSB-AM: (a) first modulation; (b) second modulation.

expedites the filtering by the second high-pass filter. If the 1 kHz tone had modulated the 3 MHz carrier, then two sidebands of 2999 and 3001 kHz would have been produced—a separation of only 2 kHz, making it difficult to suppress the lower sideband completely.

Exercise 4.5
Check on page 104

Verify that the system shown in Fig. 4.10 produces SSB-AM.

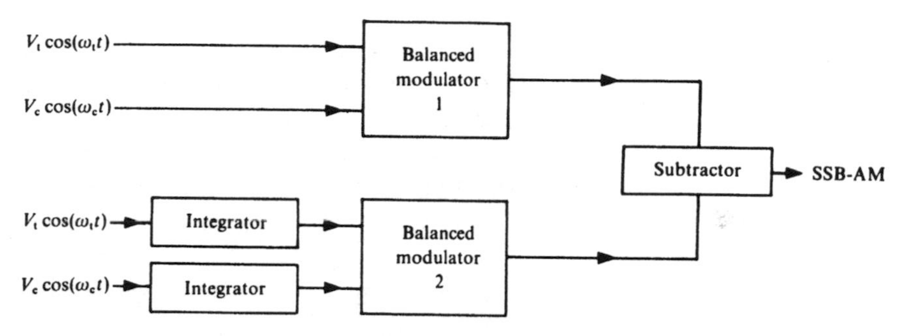

Figure 4.10. Another form of SSB modulator.

Radio signals may take different paths: the line of sight and a reflection path via the ionosphere. If the path difference is an odd number of half-wavelengths of a pair of sidebands, then destructive interference will occur. This is known as *selective fading*—an effect that cannot occur in SSB transmission.

The reduced bandwidth of SSB transmission has led to its use in FDM. Each signal lower sideband (LSB) is allotted a frequency band within the link bandwidth. So, for example, a coaxial cable with a bandwidth of 1 GHz, carrying speech of 3.4 kHz bandwidth, could accommodate 1 GHz/4 kHz = 250 000 channels, assuming guard bands of 600 Hz.

The price paid for the advantages of SSB-AM is the increased complexity of the receiver. In systems such as mobile radio, SSB-AM is used because each receiver has a transmitter and the relatively high cost of the receiver is acceptable because a saving can be made on the lower-power transmitter. In national sound broadcasting, where there are few transmitters and a huge number of receivers, it is economic to use DSB-AM. However, in a system such as 625-line TV, where the video transmission bandwidth is 11 MHz, a compromise modulation technique is chosen (to economize on bandwidth) in which one of the sidebands of the DSB-AM video signal is curtailed, leaving only a *vestige* of this sideband.

Vestigial-sideband amplitude modulation (VSB-AM)

The principle of this technique, as it applies to 625-line TV, is illustrated in Fig. 4.11, which compares the lower (LSB) and upper sidebands (USB) of DSB-AM and VSB-AM using a carrier frequency of 500 MHz.

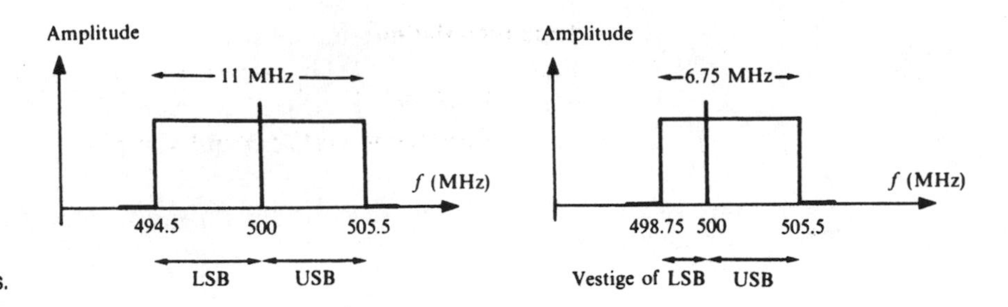

Figure 4.11. The spectra of DSB and VSB TV video signals.

The video baseband covers the range 0 to 5.5 MHz for 625-line pictures. The need to transmit a d.c. component precludes carrier suppression. A reduction of the LSB from 5.5 to 1.25 MHz produces a practical compromise between picture quality and bandwidth economy. The VSB modulated signal is obtained by filtering a DSB signal using a cascade of tuned amplifiers. Each amplifier is *stagger tuned* (that is, tuned to a slightly different centre frequency), so that the overall frequency response covers the range 498.75 to 505.5 MHz shown in Fig. 4.11.

The various AM systems we have considered all have the attractive features, to a greater or lesser extent, of bandwidth economy and transmitter and/or receiver simplicity. They all, unfortunately, suffer from the inherent disadvantage of susceptibility to noise. Frequency modulation (FM) is widely used in order to increase the received signal-to-noise ratio (SNR) at the expense of increased bandwidth. The next section will show that FM requires much more bandwidth than AM, so its use is generally restricted to audio signal applications such as high-quality stereo sound broadcasting at VHF and to the 625-line TV sound channel at UHF.

4.3 Angle modulation

We have seen that a voltage $v_c(t) = V_c \cos\theta = V_c \cos(\omega_c t + \phi)$ may carry signal information $v(t)$ by modulating the carrier signal amplitude (V_c) so that it becomes $V_c[1 + mv(t)]$. In AM the amplitude thus varies as a function of the modulating signal, while the carrier angle (θ) changes at a constant rate as the carrier goes through cycle after cycle. Thus, $\mathrm{d}\theta/\mathrm{d}t = \omega_c$, where ω_c is the constant angular velocity or angular frequency of the carrier. If $\mathrm{d}\theta/\mathrm{d}t$ instead of V_c is allowed to vary with time, then, at any instant, we have an instantaneous frequency $\omega_i(t)$. Variation of $\omega_i(t)$ by the modulating signal gives a form of angle modulation known as *frequency modulation* (FM).

Since $\theta = \omega_c t + \phi$, another obvious form of angle modulaton occurs when the phase (ϕ) is varied by the modulating signal. This results in *phase modulation* (PM). We will look briefly at PM and then consider FM, which is used much more widely, in greater detail.

Phase modulation

An unmodulated carrier signal will be given by

$$v_c = V_c \cos\theta = V_c \cos(\omega_c t + \phi_c)$$

If a binary pulse modulates ϕ_c, then ϕ_c will *deviate* from its initial value each time the pulse changes amplitude. This effect is shown in Fig. 4.12.

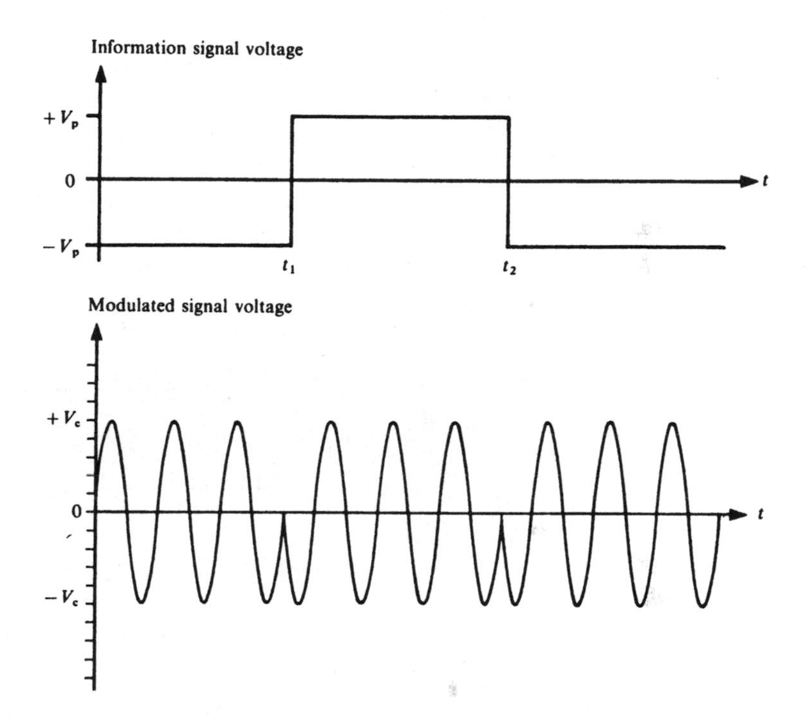

Figure 4.12. An example of PM.

If a voltage $v(t)$ modulates ϕ_c, then $\phi_c = \phi_0 + k_p v(t)$. The constant k_p is thus the carrier phase deviation in radians per volt of $v(t)$.

The expressions for the PM signal shown in Fig. 4.12 are

$$v_c = \begin{cases} V_c \cos(\omega_c t + \phi_0 + k_p V_p) & \text{for } t_1 < t < t_2 \\ V_c \cos(\omega_c t + \phi_0) & \text{for } t < t_1 \text{ or } t > t_2 \end{cases}$$

If a tone $v_t = V_t \cos(\omega_t t)$ modulates ϕ_c, then $\phi_c = \phi_0 + k_p V_t \cos(\omega_t t)$. The maximum carrier phase deviation ($d\phi_m$) will be $k_p V_t$ so

$$\phi_c = \phi_0 + d\phi_m \cos(\omega_t t)$$

The initial phase (ϕ_0) is arbitrary and, therefore, carries no information. It is the phase deviation that carries the information, so we can ignore the initial phase without affecting the results of our analysis. Thus, the effective expression for a continuous wave phase-modulated by a tone is

$$v_c = V_c \cos[\omega_c t + d\phi_m \cos(\omega_t t)] \tag{4.5}$$

Equation (4.5) is analogous to Eq. (4.1). The instantaneous amplitude deviation in Eq. (4.3) is $mV_c \cos(\omega_t t) = k_a V_t \cos(\omega_t t)$; the instantaneous phase deviation in Eq. (4.5) is $d\phi_m \cos(\omega_t t) = k_p V_t \cos(\omega_t t)$. This last expression indicates that the frequency of the phase deviation is the tone of frequency, while the extent of the phase deviation is proportional to the tone amplitude.

Frequency modulation

Exercise 4.6
Check on page 104

Sketch waveforms, similar to those of Fig. 4.12, to show the effect you would expect when a pulse modulates the frequency of a continuous carrier.

In FM a voltage $v(t)$ would modulate the carrier's instantaneous frequency (ω_i) so that it deviated by an amount $d\omega$ proportional to $v(t)$. Thus, at any instant, the carrier frequency would be given by

$$\omega_i = \omega_c + d\omega = \omega_c + k_f v(t)$$

where k_f is the carrier frequency deviation in radians per second per volt of $v(t)$.

In order to find an expression for a frequency-modulated signal carrying $v(t)$, we need to find an expression for θ and use this in $v_c = V_c \cos\theta$. Since

$$d\theta/dt = \omega_c + k_f v(t)$$

we have

$$\theta = \int [\omega_c + k_f v(t)]\, dt$$

$$= \omega_c t + k_f \int v(t)\, dt + \phi_0$$

Ignoring ϕ_0 as we did in PM, an FM signal carrying $v(t)$ is thus given by

$$v_c = V \cos\theta = V_c \cos\left[\omega_c t + k_f \int v(t)\, dt\right]$$

Comparison of AM, PM and FM

Having derived the equations obtained when a voltage $v(t)$ modulates the amplitude, phase or frequency of a CW, we now compare these by considering an unmodulated CW: $v_c = V_c \cos(\omega_c t + \phi_c)$. If we modulate with a voltage $v(t)$ the CW becomes

$$v_c = [V_c + k_a v(t)] \cos(\omega_c t) \qquad \text{with AM}$$
$$v_c = V_c \cos[\omega_c t + k_p v(t)] \qquad \text{with PM}$$
$$v_c = V_c \cos\left[\omega_c t + k_f \int v(t)\, dt\right] \qquad \text{with FM}$$

where k_a, k_p and k_f are the amplitude, phase and frequency deviations per volt of modulating signal.

4.4 Frequency modulation by a single tone

In FM, a tone $v_t = V_t \cos(\omega_t t)$ would modulate the carrier's instantaneous frequency ($d\theta/dt$), making it vary with $V_t \cos(\omega_t t)$ thus:

$$d\theta/dt = \omega_c + d\omega = \omega_c + k_f v_t$$

The maximum carrier frequency deviation ($d\omega_m$) will be $k_f V_t$, so

$$d\theta/dt = \omega_c + d\omega = \omega_c + d\omega_m \cos(\omega_t t)$$

giving

$$\theta = \int [\omega_c + d\omega_m \cos(\omega_t t)]\, dt$$
$$= \omega_c t + (d\omega_m/\omega_t) \sin(\omega_t t) + \phi_0$$

Ignoring ϕ_0, as we did in PM, an FM signal carrying a tone of frequency ω_t is thus given by

$$v_c = V_c \cos\theta$$
$$= V_c \cos[\omega_c t + \beta \sin(\omega_t t)] \tag{4.6}$$

where $\beta (= d\omega_m/\omega_t = df_m/f_t = k_f V_t/f_t)$ is a parameter known, in this context as the *frequency modulation index*. Comparing Eqs (4.5) and (4.6), we see that β represents the maximum phase deviation of an FM signal. In the UK, the value of $k_f V_t$ ($= df_m$) is chosen so that $df_m = \pm 75$ kHz. This frequency deviation corresponds to maximum amplitude in the modulating tone. The frequency with which the frequency deviation changes is the tone frequency.

The expressions derived for PM and FM suggest the system shown in Fig. 4.13 for achieving FM by using PM. The system shown in Fig. 4.13 forms the basis of a practical FM modulator (see Fig. 4.18) invented by

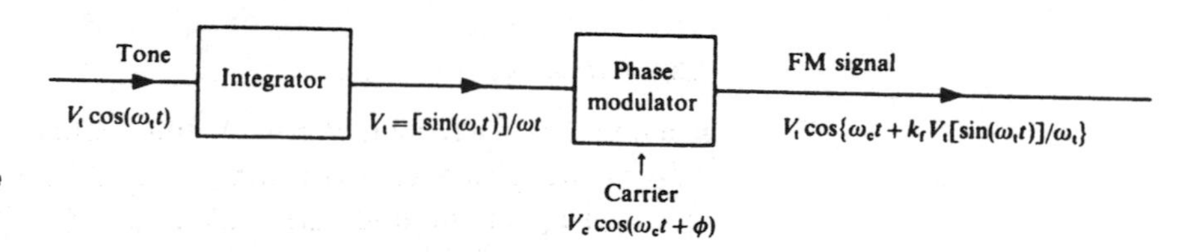

Figure 4.13. FM from a phase modulator.

Edwin Armstrong, a British engineer who pioneered FM. A particular point to note from Fig. 4.13 is that, since an integrator is a LP filter, FM may be regarded as PM with treble cut. Therefore, where bandwidth economy is important, FM will be preferred to PM. Another frequency modulator, based on a *voltage-controlled oscillator* (VCO), is shown in Fig. 4.14. A VCO has a frequency-determining component that can be varied

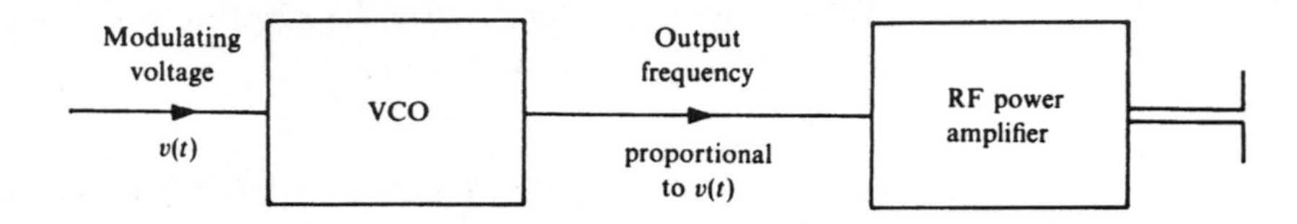

Figure 4.14. A basic frequency modulator.

by a voltage, such as a capacitor. This is described later with the help of Fig. 4.16.

The spectrum of a CW that has been frequency-modulated by a tone

The spectral components may be found, first, by expanding Eq. (4.6), normalized by setting $V_c=1$ for convenience, using $\cos(A+B)=\cos A\cos B-\sin A\sin B$. Thus

$$v_c=\cos(\omega_c t)\cos[\beta\sin(\omega_t t)]-\sin(\omega_c t)\sin[\beta\sin(\omega_t t)]$$

The second, and much larger, step is to expand these products further. Thus

$$\begin{aligned}v_c=&\cos(\omega_c t)\{1-[\beta\sin(\omega_t t)]^2/2!+[\beta\sin(\omega_t t)]^4/4!\\&-[\beta\sin(\omega_t t)]^6/6!+\cdots\}-\sin(\omega_c t)\{[\beta\sin(\omega_t t)]\\&-[\beta\sin(\omega_t t)]^3/3!+[\beta\sin(\omega_t t)]^5/5!-\cdots\}\end{aligned}$$

If $\beta\ll 1$,

$$\begin{aligned}v_c&=\cos(\omega_c t)-\sin(\omega_c t)\beta\sin(\omega_t t)\\&=\cos(\omega_c t)-\tfrac{1}{2}\beta\cos[(\omega_c-\omega_t)t]+\tfrac{1}{2}\beta\cos[(\omega_c+\omega_t)t]\end{aligned}\qquad(4.7)$$

Comparing Eqs (4.2) and (4.7) shows that FM, with a very small modulation index, is the same as DSB-AM with the lower sideband shifted in phase by 180°. The modulated signal has twice the bandwidth of the modulating tone.

As β $(=k_f V_t(\max)/\omega_t(\max))$ increases, the other terms of the second expansion become progressively more significant. For example, a slight increase in the tone amplitude or a decrease in its frequency will give a new approximation:

$$\begin{aligned}v_c=&\cos(\omega_c t)\{1-[\beta\sin(\omega_t t)]^2/2!\}-\sin(\omega_c t)\{[\beta\sin(\omega_t t)]\\&-[\beta\sin(\omega_t t)]^3/3!\}=(1+\tfrac{1}{4}\beta^2)\cos(\omega_c t)-\tfrac{1}{2}\beta\{\cos[(\omega_c-\omega_t)t]\\&-\cos[(\omega_c+\omega_t)t]\}+\tfrac{1}{8}\beta^2\{\cos[(\omega_c-2\omega_t)t]+\cos[(\omega_c+2\omega_t)t]\}\end{aligned}\qquad(4.8)$$

An additional lower and upper sideband have appeared of frequency $\omega_c - 2\omega_t$ and $\omega_c + 2\omega_t$ respectively, resulting in a doubling of the FM signal bandwidth.

It is now fairly easy to imagine the huge increase in the number of sideband pairs that occurs as β increases. Further expansion of Eq. (4.6) yields

$$\begin{aligned} v_c = J_0 \cos(\omega_c t) - J_1\{\cos[(\omega_c - \omega_t)t] - \cos[(\omega_c + \omega_t)t]\} \\ + J_2\{\cos[(\omega_c - 2\omega_t)t] + \cos[(\omega_c + 2\omega_t)t]\} \\ - J_3\{\cos[(\omega_c - 3\omega_t)t] + \cos[(\omega_c + 3\omega_t)t]\} + \cdots \\ + J_n\{\cos[(\omega_c - n\omega_t)t] + \cos[(\omega_c + n\omega_t)t]\} \end{aligned} \tag{4.9}$$

The terms (J_n) of this Fourier series are all functions of β. They are known as 'Bessel functions of the first kind and of order β'. Figure 4.15 shows the way in which the first 11 terms vary with β.

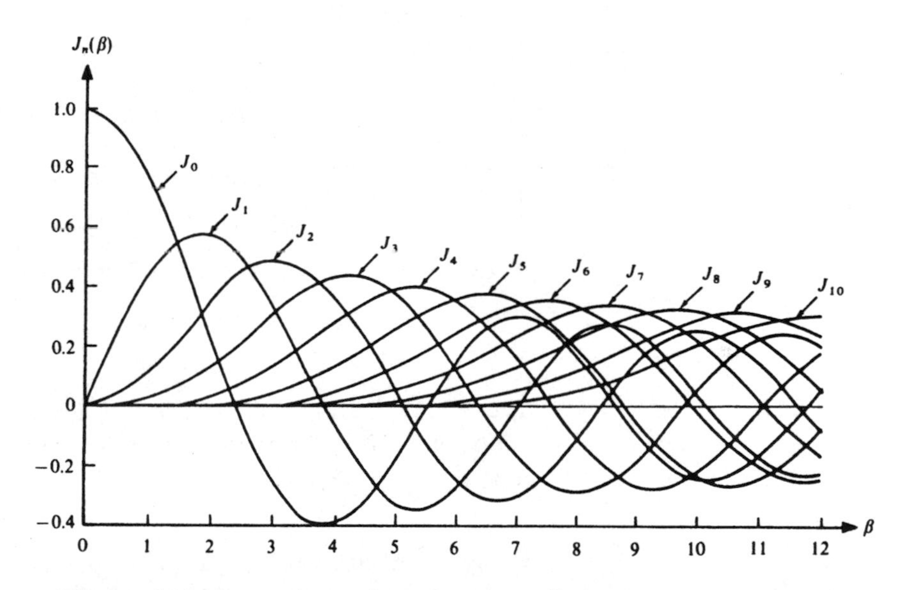

Figure 4.15. Bessel functions of the first kind for various values of β.

Figure 4.15 shows that, when β varies, all the frequency components vary in magnitude, including the carrier. This means that the FM carrier signal makes a contribution to the information content of the modulated signal. This is not the case in AM, suggesting that FM produces more information power than an AM signal of the same total power. Thus, an important feature of FM is its ability to perform better than AM in the presence of noise.

Exercise 4.7
Check on page 104

Use Fig. 4.15 to sketch the frequency spectra of an FM signal modulated by a tone, when $\beta = 0.75$ and then 1.5.

The number of significant values of J_n increases with large values of β (in practice a low tone frequency). This leads to the approximation that there will be $\beta+1$ pairs of sidebands each separated by ω_t in the spectrum of an FM signal, giving a bandwidth (B) of $2(\beta+1)\omega_t = 2(d\omega_m + \omega_t)$. This approximation is *Carson's rule.* If $\beta \ll 1$ (narrow-band FM) or $\beta \gg 1$ (wide-band FM), then B becomes $2\omega_t$ or $2d\omega_m$ respectively.

Exercise 4.8
Check on page 104

In the UK and many other countries, commercial FM broadcasting restricts the carrier frequency deviation to 75 kHz. Calculate the values of β for maximum amplitude tone frequencies of 25 Hz and 15 kHz.

Estimate the bandwidth required for the transmission of high-quality (Hi-Fi) sound covering this range and comment on the effects on sound quality of reducing the bandwidth of the FM signal.

The bandwidth requirement of PM may be found by bearing in mind that, for a modulating signal $v(t)$, the carrier voltages for PM and FM are, respectively,

$$V_c \cos[\omega_c t + k_p v(t)] \qquad \text{and} \qquad V_c \cos\left[\omega_c t + k_f \int v(t)\,dt\right]$$

If $v(t) = V_t \cos(\omega_t t)$ then the FM carrier voltage becomes

$$v_c = V_c \cos[\omega_c t + \beta \sin(\omega_t t)]$$

It follows then that the corresponding PM voltage will be

$$v_c = V_c \cos[\omega_c t + \beta\omega_t \cos(\omega_t t)]$$
$$= V_c \cos[\omega_c t + d\omega_m \cos(\omega_t t)]$$

The modulation index for PM is $d\omega_m$ ($= k_p V_t$). Unlike β ($= k_f V_t/\omega_t$), this index does not decrease as ω_t increases, but remains constant for a given tone amplitude (V_t).

Power in FM and PM signals and their signal-to-noise ratios

Since the amplitude of FM signals is constant, the total power must be constant for all values of β. Figure 4.15 indicates that, as β increases, the carrier amplitude decreases. This means that power is transferred from the carrier to the sidebands as β increases, so extra information is sent as this power transfer takes place. This suggests that the received signal-to-noise ratio (SNR) can be improved by increasing β rather than the transmitter power. This, in effect, is increasing signal bandwidth to improve the SNR. The fact that FM may be regarded as PM with treble cut suggests that PM leads to a greater SNR than FM. These factors are important when considering the best modulation method for a given application. PM is used in satellite communications because the large bandwidth improves the signal-to-noise ratio, thus allowing reduced satellite transmitter power and transmitter weight. Narrow-band FM allows many voice channels to be carried over terrestrial radio links where transmitter weight is not a problem.

4.5 Frequency modulators

FM may be produced directly by varying the frequency of a voltage-controlled oscillator (VCO) with a modulating voltage. The basis of the VCO is the capacitance of a reverse-biased pn junction, which can be varied by the bias voltage across the junction. This *varactor* (or varicap) diode forms part of the capacitance placed across an inductor to form a tuned circuit as in Fig. 4.16.

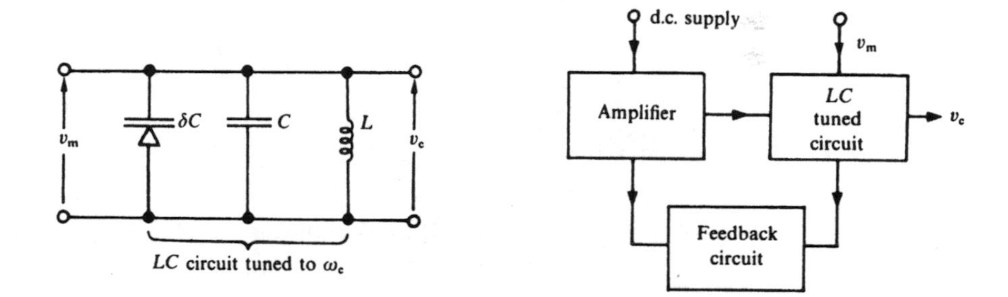

Figure 4.16. A varactor frequency modulator.

The circuit on Fig. 4.16 forms part of an electronic oscillator in which a small fraction of v_c is fed back to an amplifier feeding the circuit in order to sustain the oscillations. The frequency of v_c varies as the capacitance of the tuned circuit changes with the modulating signal v_m.

The total capacitance across the inductor (L) will be $C+\delta C$ where $\delta C = Kv_m$, K being the capacitance change per volt of modulating voltage. The unmodulated carrier frequency (ω_c) is $1/\sqrt{(LC)}$, so the instantaneous frequency when the signal v_m is applied is given by

$$\frac{d\theta}{dt}=\frac{1}{\sqrt{[L(C+\delta C)]}}=\frac{1}{\sqrt{(LC)}}\times\frac{1}{\sqrt{(1+\delta C/C)}}$$

If $\delta C \ll C$ then

$$d\theta/dt=\omega_c(1-\tfrac{1}{2}\delta C/C)=\omega_c(1-\tfrac{1}{2}Kv_m/C)$$

and

$$v_c=V_c\cos\theta=V_c\cos\left[\omega_c t-(\tfrac{1}{2}Kv_m\omega_c/C)\int v_m\,dt\right]$$

$$=V_c\cos\left[\omega_c t+k_f\int v(t)\,dt\right]$$

which shows that this form of modulator can produce FM, albeit narrow-band because the change in capacitance must be small.

The output produced by the modulator of Fig. 4.16 will also be amplitude-modulated by v_m but this will not affect the recovery of the FM signal components. In practice, FM receivers limit the amplitude of the carrier before demodulation so that all spurious AM is removed. This may, in hard limiting, result in a demodulator receiving a pulse-like carrier as shown in Fig. 4.17.

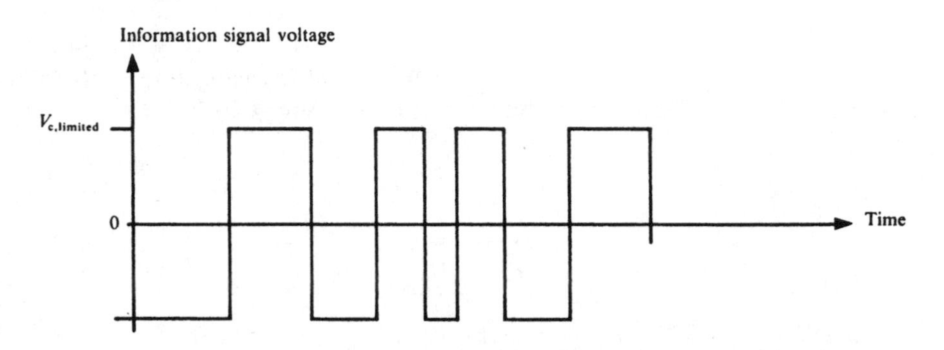

Figure 4.17. An example of hard amplitude-limiting in FM.

Hard amplitude-limiting results in severe phase distortion of the CW. This would make PM information difficult to recover. In a frequency-modulated pulse carrier, however, the information is contained in the frequency deviation of the zero crossings and is easily recovered.

In FM, the carrier frequency must not vary except as dictated by the modulating voltage. Any other frequency variations will appear as noise at the receiver. A crystal-controlled oscillator (XCO) modulator, suggested by Armstrong and shown in Fig. 4.18, replaces the inductor of the *LC* oscillator with a specially cut quartz crystal. The extremely low drift in quartz crystal parameters ensures carrier frequency stability.

The XCO forms part of a narrow-band phase modulator. To achieve FM the modulating signal simply has to be integrated before applying it to the phase modulator as in Fig. 4.13.

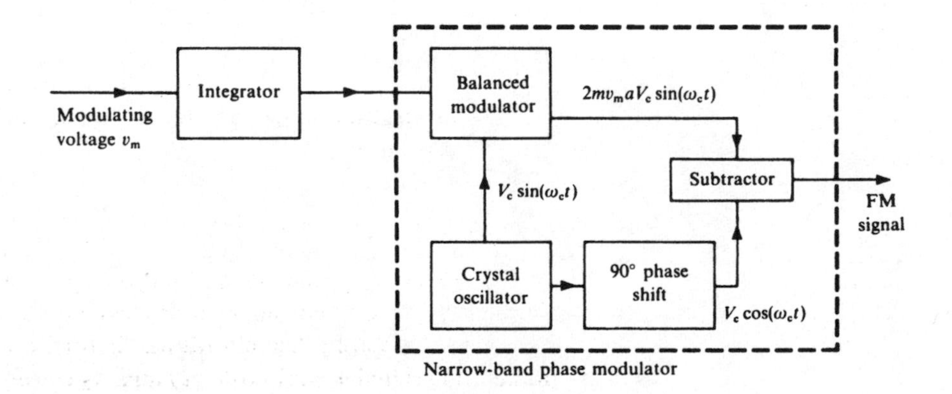

Figure 4.18. Crystal-controlled frequency modulator.

To show that narrow-band PM occurs in the phase modulator of Fig. 4.18, consider the expression for a PM signal modulated by v_m:

$$v_c = V_c \cos[\omega_c t + k_p v_m]$$
$$= V_c[\cos(\omega_c t)\cos(k_p v_m) - \sin(\omega_c t)\sin(k_p v_m)]$$

But if $k_p \ll 1$ then

$$v_c = V_c \cos(\omega_c t) - V_c k_p v_m \sin(\omega_c t) \tag{4.10}$$

Thus a narrow-band PM signal is a carrier signal minus a DSB-SC signal with a carrier signal in quadrature to the first carrier signal.

In Fig. 4.18, the output (v_o) of the phase modulator will be

$$v_o = V_c \cos(\omega_c t) - 2mv_m a V_c \sin(\omega_c t) \tag{4.11}$$

Comparing Eqs (4.10) and (4.11) verifies that narrow-band PM is obtained. The constraint imposed by $k_p \ll 1$ results in narrow-band FM in the Armstrong modulator. In order to increase the deviation, frequency multiplication may be used. Figure 4.19 exemplifies this technique for a VHF tone-modulated carrier. Notice that multiplication increases the carrier and deviation frequencies by the same factor, whereas mixing translates the FM spectrum only, without affecting the deviation or the modulation index β.

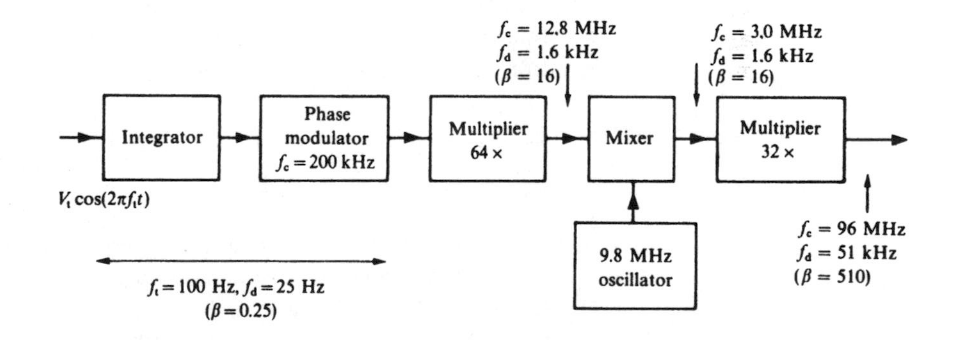

Figure 4.19. Frequency multiplication used to increase β.

Exercise 4.9
Check on page 105

Modify Fig. 4.19, by interchanging the 64 × multiplier and the mixer, and show the effect of using a 9.8 MHz/64 = 153.125 kHz oscillator frequency.

Multi-tone AM and FM

So far, we have considered modulation by a single tone in the interests of simplicity, but our considerations also apply to multi-tone modulation. However, there are a few additional factors. Suppose we have a two-tone modulating signal $v_t = A_1 \cos(\omega_1 t)$ and $A_2 \cos(\omega_2 t)$. Then AM would simply translate these baseband frequencies to $\omega_c \pm \omega_1$ and $\omega_c \pm \omega_2$. It can be seen from Eq. (4.9), however, that FM would produce many other intermodulation terms of the form $\omega_c \pm (m\omega_1 \pm n\omega_2)$ where m and n are integers.

The two-tone version of Eq. (4.6) will be

$$v_c = V_c \cos[\omega_c t + \beta_1 \sin(\omega_1 t) + \beta_2 \sin(\omega_2 t)]$$

Using the identity $\cos(A+B) = \cos A \cos B - \sin A \sin B$ gives

$$v_c = V_c \{\cos(\omega_c t)\cos[\beta_1 \sin(\omega_1 t) + \beta_2 \sin(\omega_2 t)] - \sin(\omega_c t) \times \sin[\beta_1 \sin(\omega_1 t) + \beta_2 \sin(\omega_2 t)]\}$$

If $\beta \ll 1$, v_c approximates to $V_c\{\cos(\omega_c t) - [\beta_1 \sin(\omega_1 t) + \beta_2 \sin(\omega_2 t)]\sin(\omega_c t)\}$, which shows that superposition almost applies. If β is not very much less than unity, then the two tones cannot be superimposed to obtain v_c. Since the principle of superposition does not apply, it means that wide-band FM is a non-linear process.

4.6 Modulation by data signals

Data communication can occur at baseband providing the channel bandwidth is adequate. In narrow-band channels a modulated carrier must be used. An important example of data transmission is the exchange of computer data over the PSTN. This network was designed to carry analogue speech signals over a bandwidth of 300 Hz to 3.4 kHz. In order to transmit pulses over the PSTN, it is necessary to convert them into a form similar to voice signals and then recover them at the receiver. The devices that do this are called *modems* because they use the processes of modulation and demodulation as shown in Fig. 4.20.

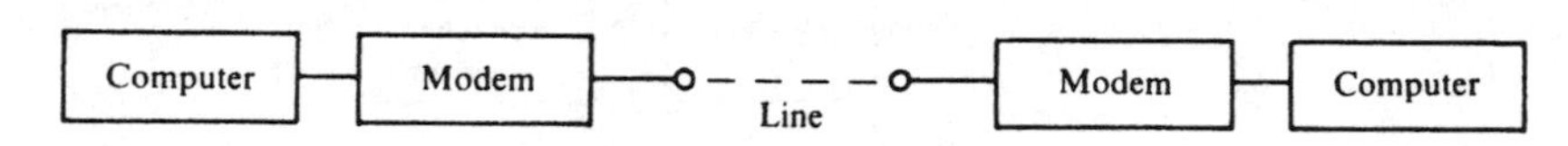

Figure 4.20. Modems used in duplex computer communications.

Consider first amplitude and frequency modulation of a CW by a pulse. In Fig. 4.21 the 1s are represented by a higher carrier frequency than the 0s to give a form of FM called *frequency shift keying* (FSK)—the carrier frequency being switched or *keyed* between two values of frequency. The

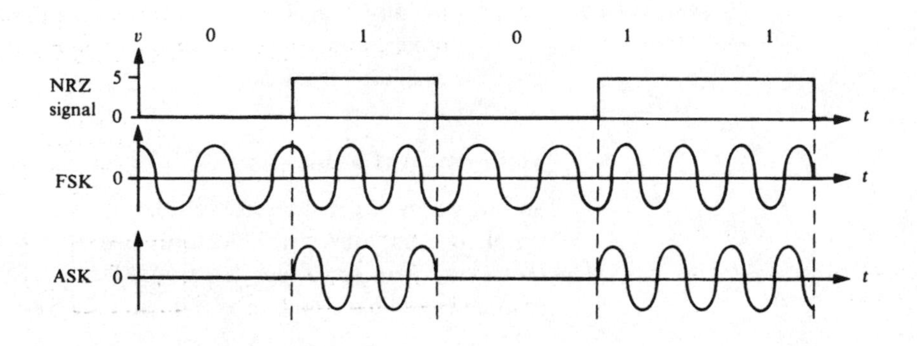

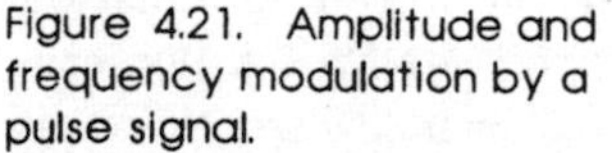
Figure 4.21. Amplitude and frequency modulation by a pulse signal.

receiving modem converts the two-frequency signal back into 0s and 1s. A variation of this technique is *amplitude shift keying* (ASK), in which the 1s are represented by a burst of carrier signal and the 0s by zero signal—it is DSB-AM with $m = 1$. A typical 300 b/s FSK system is outlined in Fig. 4.22.

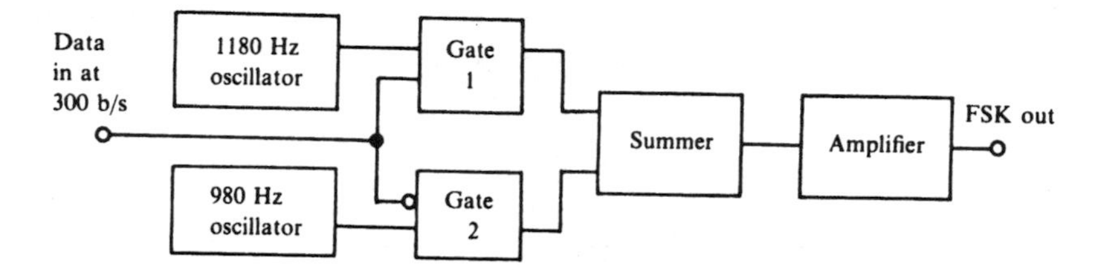

Figure 4.22. A 300 b/s modulator.

Data pulse 'high' enables gate 1, thus keying a 1180 Hz carrier. Data pulse 'low' will enable gate 2, keying a 980 Hz carrier. The carrier frequencies change at the bit rate. These carrier frequencies apply to one direction of transmission only. The modulator in the modem at the other end of the line would operate at 1650 and 1850 Hz. The modulator of Fig. 4.22 treats FSK as a combination of two ASK signals as shown in Fig. 4.23.

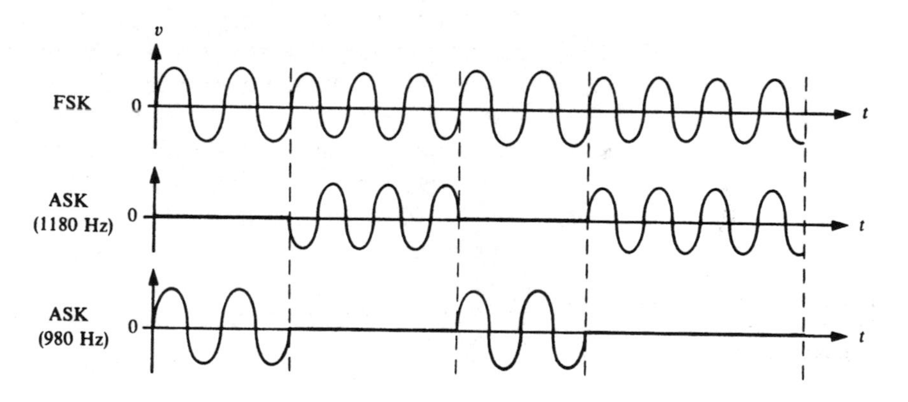

Figure 4.23. FSK as a combination of two ASK signals.

The data is recovered by the receiving modem as shown in Fig. 4.24. The process involves the separation of the two ASK signals making up the FSK signal with tuned circuits (band-pass filters) and their demodulation by *envelope detection*—a technique that is explained in Chapter 5.

A disadvantage of ASK is the difficulty of distinguishing between the 0 and 1 voltages in low SNR conditions. FSK operates symmetrically about

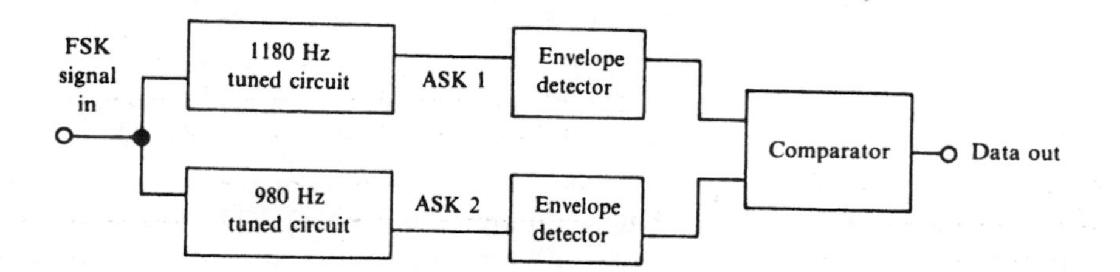

Figure 4.24. A 300 b/s demodulator.

a zero decision threshold whatever the signal strength. Since there is little difference in the complexity of ASK and FSK modems, FSK is preferred to ASK. However, as data rates increase, it becomes necessary to raise the carrier frequencies, otherwise the pulses would be represented by too few cycles to enable the demodulator to operate satisfactorily. Above a data rate of 600 b/s, *phase shift keying* (PSK), the digital equivalent of PM, is used in preference to FSK (Fig. 4.25).

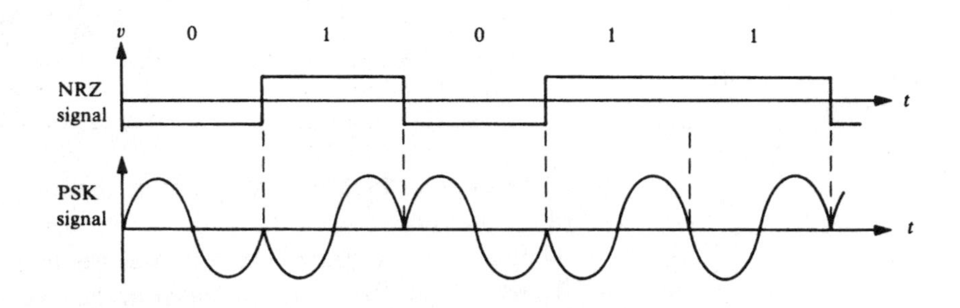

Figure 4.25. Phase shift keying.

The information is carried as a π phase shift on each transition of the data pulse, PSK is preferred to FSK at high bit rates because it is easier to detect abrupt changes in phase than it is to sense frequency changes. In effect, the carrier is being inverted each time the data changes. PSK signals require about half the transmitted power for a given error rate. But, unlike ASK and FSK signals, they must be demodulated coherently (that is, with an exact copy of the original carrier for phase reference), which makes for a more expensive receiver. This suggests that PSK is a form of DSB-SC-AM and that PSK signals will have the same bandwidth as their ASK and FSK counterparts. To overcome the need for frequency synchronization between modems, *differential phase shift keying* (DPSK) is used. Figure 4.26 shows how PSK and DPSK signals are related.

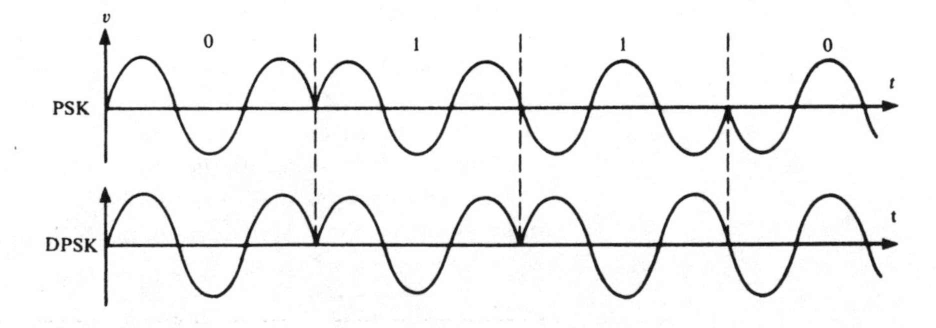

Figure 4.26. Waveforms for PSK and DPSK.

The idea in DPSK is to use the phase of the carrier in the previous bit interval as the reference for the phase in the next interval. This means transmitting a 1 by reversing the phase representing the previous bit and a 0 by maintaining this phase. This is done by using an exclusive-OR (XOR) gate, as explained in Fig. 4.27, to generate a modulating signal.

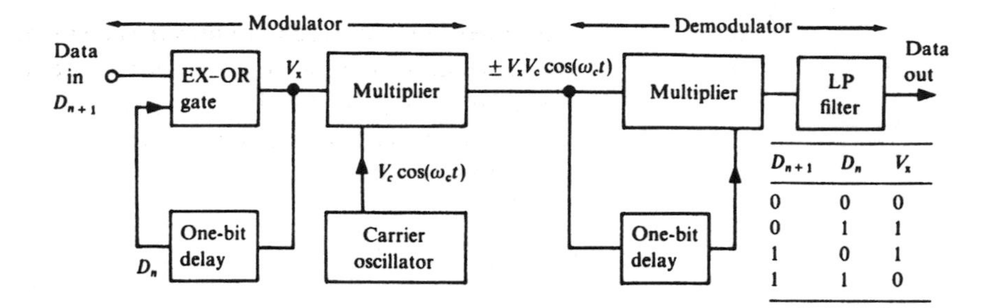

D_{n+1}	D_n	V_x
0	0	0
0	1	1
1	0	1
1	1	0

Figure 4.27. A DPSK modem arrangement.

The demodulator multiplier outputs $+[V_x V_c \cos(\omega_c t)]^2$ if the received pulse is the same as the previous pulse and $-[V_x V_c \cos(\omega_c t)]^2$ otherwise. These outputs may be written $\pm(V_x V_c)^2 \frac{1}{2}[1+\cos(2\omega_c t)]$, so the data can be recovered with a LP filter.

Variations on PSK and ASK

In PSK the phase is changed by 180° on each pulse transition. In effect, 0 is represented by one phase and 1 by the opposite phase. If phase changes of 90° were used, we would have four symbols instead of two to carry data. This is called four-phase or *quadrature PSK* (QPSK). PSK transmits two different symbols (0 or 1) and the bit rate is the same as the symbol rate, QPSK transmits four different symbols (00, 01, 10 or 11), and each symbol takes the time of two bits; that is, the bit rate is twice the *symbol* or *signalling* rate (see Chapter 9 for explanation of these terms).

If the carrier is amplitude-modulated while QPSK is taking place, we would increase the number of symbols still further by a process known as *quadrature amplitude PSK* (should be QAPSK but QASK is used). The various states of QPSK and QASK signals can be visualized with signal space diagrams as in Fig. 4.28.

The signal space diagram is a form of phasor diagram based on a carrier voltage given by

$$v = V_1 \cos(\omega t) + V_2 \sin(\omega t)$$

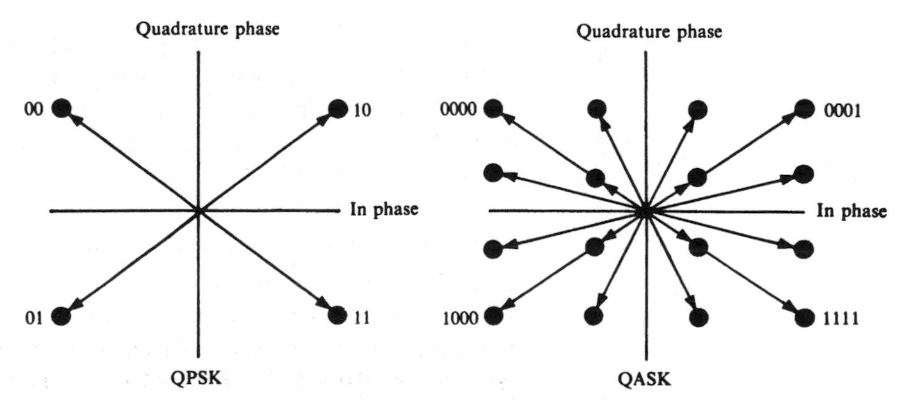

Figure 4.28. Signal space diagrams for QPSK and QASK signals.

In QPSK, V_1 and V_2 would be of magnitude, say, V, and ωt might take the values $\pm\pi/4$ and $\pm 3\pi/4$, giving four possible values of v. In QASK, V_1 and V_2 would have different values, with ωt taking on the values $\pm\pi/8$, $\pm\pi/4$, $\pm 3\pi/8$, $\pm 5\pi/8$, $\pm 3\pi/4$ and $\pm 7\pi/8$, giving 16 possible values of v as exemplified in Fig. 4.28.

Applications of data modulation techniques

ASK is seldom used because, like its AM counterpart, much of the signal power resides in the carrier, resulting in a low SNR. It also suffers from fading when used in radio links. FSK is used in low-frequency modems but it suffers from the same disadvantages just mentioned for ASK. FSK is not a true FM system—it is a combination of alternate AM signals—so it does not have the SNR advantages over ASK that its sinusoidal modulating signal counterpart has over AM. PSK and its variants are the preferred modulation methods because PSK has a higher noise immunity than FSK. Either QPSK or QASK is used to carry the pulse code modulated (PCM) signals used in satellite communication systems (see Chapter 11) because they have high noise immunity and modest bandwidth.

In line communications, the cheaper modems provide only a fixed bit rate up to 2400 b/s using FSK or PSK, while the more expensive offer a choice of rates up to 9600 b/s using quadrature amplitude modulation (QAM). There is also the choice of one- or two-directional operation—*simplex* and *duplex* respectively. In the former, the modem is connected to a two-wire telephone line and the information is sent once it is prepared on the computer. Full duplex operation, which is necessary to cope with computer 'conversation' at the top bit rate, requires four wires. This mode is restricted to private lines and many users find *half-duplex* adequate. Half-duplex operation allows two modems to 'converse' by sending information to each other, in rapid alternation, along the same twin wire.

Summary

The modulation of a relatively high-frequency continuous wave by a baseband signal facilitates telecommunications in two fundamental respects. First, the frequency translation that occurs from baseband to carrier frequency allows for the design of high-gain aerials of a practicable length which can work over the whole translated baseband range. Secondly, modulation enables FDM to be used, in both radio and line links, so increasing the number of channels in the link.

An analogue (CW) signal,

$$v_c = V_c \cos(\omega_c t + \phi_c)$$

may be used to carry information by having its amplitude, phase or frequency varied by the message signal. If the CW is modulated with a voltage $v(t)$, it can be written as

$$v_c = [V_c + k_a V_t \cos(\omega_t t)] \cos(\omega_c t)$$

where k_a is the change in carrier voltage per volt of modulating signal. Often, the carrier voltage deviation equals the modulating voltage (so $k_a = 1$). Then

$$v_c = V_c[1 + m \cos(\omega_t t)] \cos(\omega_c t)$$

where $m = V_t/V_c$ is the modulation index.

The spectrum of such a signal consists of a carrier plus an upper and lower sideband, so it is known as double-sideband AM. The bandwidth required is twice the highest modulating frequency and increases linearly with frequency. All the information is contained in the sidebands, so it is possible to save power, without loss of information, by suppressing the carrier. Further, since the same information is carried by each sideband, the bandwidth may be halved by transmitting only one of them. Single-sideband suppressed-carrier AM is thus very efficient in the use of power and bandwidth. A compromise between DSB-AM and SSB-AM is VSB-AM.

Modulation of the phase or frequency of a CW wave with a voltage $v(t)$ gives

$$v_c = V_c \cos[\omega_c t + k_p v(t)]$$

for PM and

$$v_c = V_c \cos[\omega_c t + k_f \int v(t)\, dt]$$

for FM, where k_p and k_f are the carrier phase and frequency deviations per volt of modulating signal.

An FM signal carrying a tone ω_t is given by

$$v_c = v_c \cos[\omega_c t + \beta \sin(\omega_t t)]$$

where

$$\beta = d\omega_m(\max)/\omega_t(\max) = k_f v_t(\max)/\omega_t(\max)$$

is the frequency modulation index. If $\beta \ll 1$ then FM is the same as DSB-AM with the lower sideband shifted in phase by 180° and the modulated signal thus having twice the bandwidth of the modulating tone. As β increases extra frequency components proliferate non-linearly, the carrier signal being expressed by the series

$$v_c = \sum_{n=0}^{n} J_n\{\cos[(\omega_c - n\omega_t)t] + \cos[(\omega_c + n\omega_t)t]\}$$

The coefficients (J_n) are all functions of β and are Bessel functions.

The bandwidth of an FM signal may be estimated using Carson's rule, which states that $B=2\omega_t(\beta+1)$. If $\beta \ll 1$ or $\beta \gg 1$, then B becomes $2\omega_t$ or $2d\omega_m$ respectively.

A PM signal carrying a tone ω_t is given by

$$v_c = v_c \cos[\omega_c t + d\omega_m \cos(\omega_t t)]$$

The PM index ($d\omega_m$), unlike the FM index (β), does not decrease as ω_t increases, which makes FM more bandwith-efficient. Since the amplitude of FM signals is constant, the total power must be constant for all values of β, so power is transferred from the carrier to the sidebands as β increases. The extra information sent means that the receiver signal-to-noise ratio can be improved by increasing β rather than the transmitter output power.

Modulation of sinusoidal carriers by digital signals is used in data communications and finds application over the PSTN (using modems) and over multiplexed radio links (using satellites). The technique is known generally as shift keying (SK). Differential PSK is used at low bit rates but this has to give way to QPSK and QASK at the higher data rates.

Checks on exercises

4.1 $V_d = k_a V_p$. In Fig. 4.2, $V_d = 0.5V_c$ and $V_p = 0.6V_c$. So $0.5V_c = k_a \times 0.6V_c$, giving $k_a = 0.8$. The AM waveform for a pulse of $\pm V_p$ is shown in Fig. 4.29.

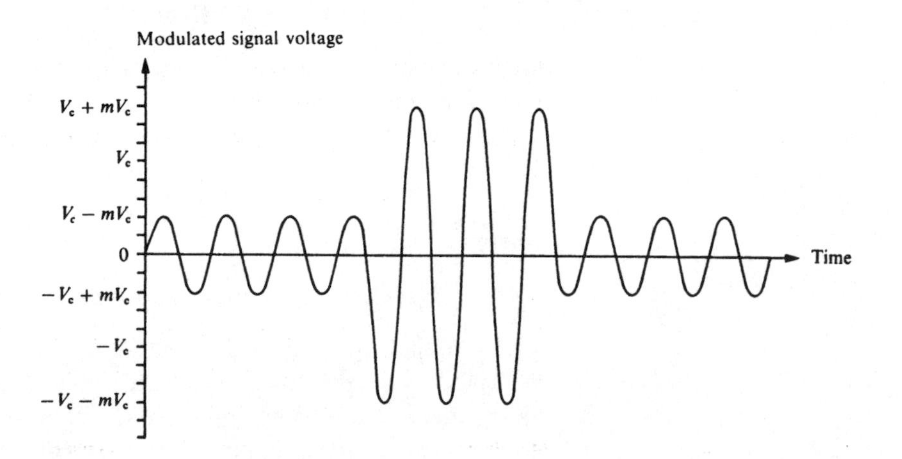

Figure 4.29. An example of AM by a bipolar pulse.

As V_p increases, the deviation in the carrier amplitude increases. It should be noted that, for this case where the carrier amplitude is reduced by the modulating signal, the deviation ($k_a V_p$) must not be greater than V_c otherwise *over-modulation* will occur—an undesirable effect because it will upset the demodulation process. In pulsed radar $k_a V_p = V_c$.

4.2 The waveform for single tone modulation at $m=0.5$ is shown in Fig. 4.30.

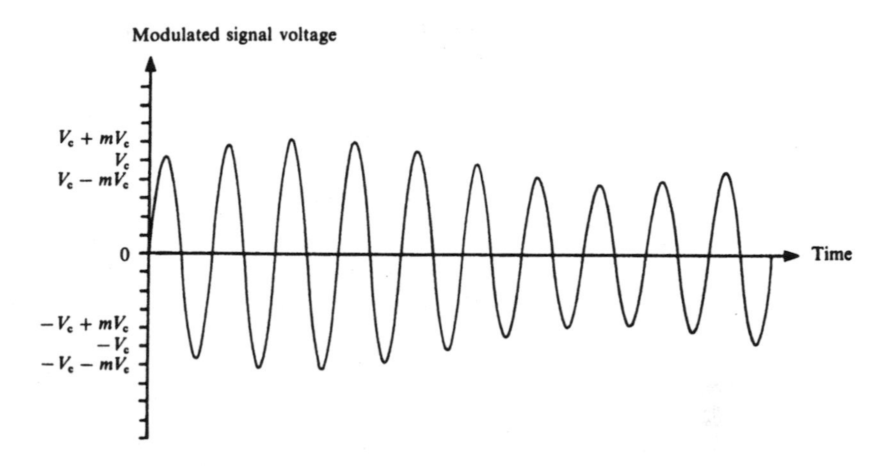

Figure 4.30. An example of AM by a single tone.

4.3 The speech baseband and AM signal spectrum diagrams for a DSB-AM signal which carries commercial-standard speech on a 3 MHz carrier are shown in Fig. 4.31.

Each speech channel with a 600 Hz guard band requires a bandwidth of 8 kHz, centred on its own carrier frequency. So the number of speech channels that could be accommodated in the 300 to 400 MHz band using FDM = 100/0.008 = 12,500.

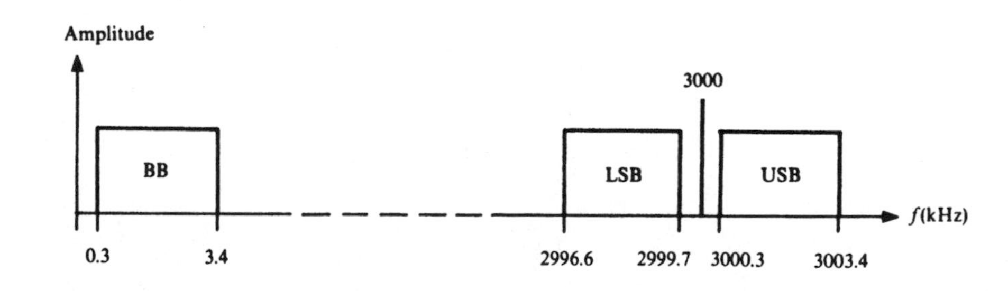

Figure 4.31. Speech baseband and AM signal spectrum diagrams.

4.4 The frequency spectra for DSB-SC-AM and SSB-AM transmissions of commercial-standard speech are shown in Fig. 4.32.

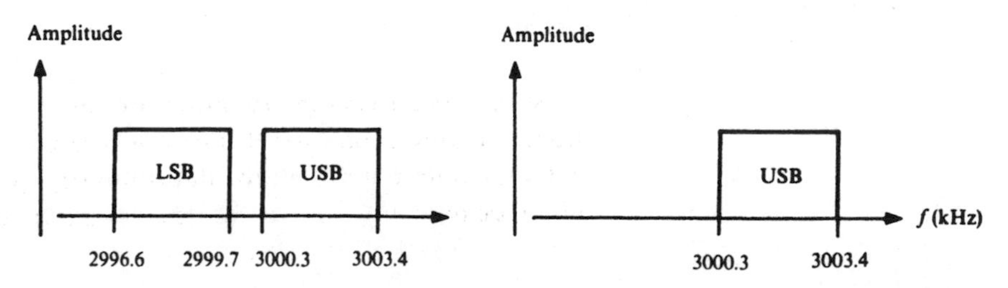

Figure 4.32. The spectra of DSB-SC-AM and SSB-AM signals modulated by speech.

4.5 The output of balanced modulator 1 is of the form $A\cos(\omega_t)\cos(\omega_c t)$. That of balanced modulator 2 is $A\sin(\omega_t)\sin(\omega_c t)$. The difference of these two will be $2A\cos(\omega_c+\omega_t)t$, which is SSB-AM.

The Hilbert transform of a signal $v(t)$ is the sum of the individual frequency components that have each been phase-shifted by 90°. This technique is, therefore, sometimes referred to as the Hilbert transform production of SSB-AM.

4.6 The waveforms are shown in Fig. 4.33.

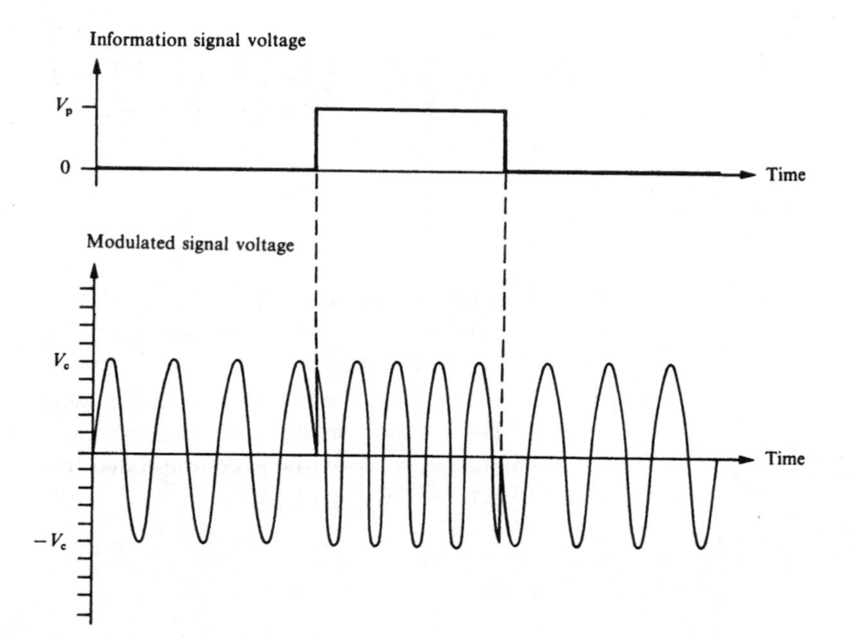

Figure 4.33. An example of FM.

4.7 The spectra are drawn in Fig. 4.34.

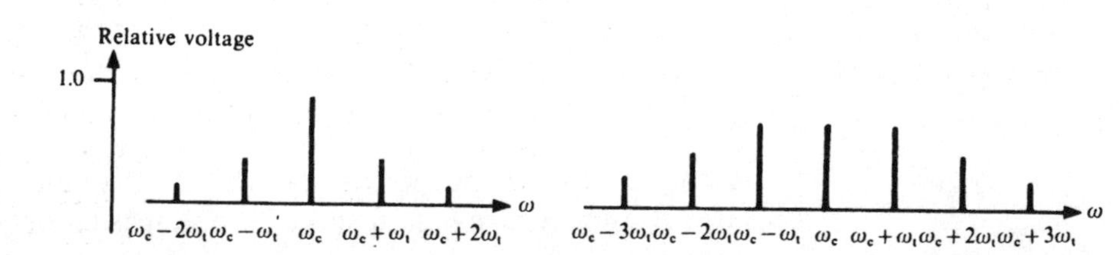

Figure 4.34. The spectra of FM signals modulated by a tone at $\beta=1$ and 2.

4.8 Since $\beta=d\omega_m/\omega_t=k_f V_t/\omega_t$ then, for tone frequencies of 25 Hz and 15 kHz at a carrier frequency deviation to 75 kHz, $\beta=3000$ and 5 respectively. To estimate the bandwidth required for the transmission of Hi-Fi sound use Carson's rule: $B=2\omega_t(1+\beta)=2(d\omega_m+\omega_t)$. This gives $B=2(75+15)=$ 180 kHz. These calculations imply that restricting the bandwidth has the effect of reducing the treble frequency response. Increasing the allowed carrier deviation above 75 kHz would allow for greater values of V_t (that is, a greater dynamic range) but this would increase the bandwidth.

4.9 Multiplication increases the carrier and deviation frequencies by the same factor and mixing shifts the FM spectrum only, without affecting the deviation or the modulation index β. Hence, the changes in β and carrier and deviation frequencies will occur as shown in Fig. 4.35.

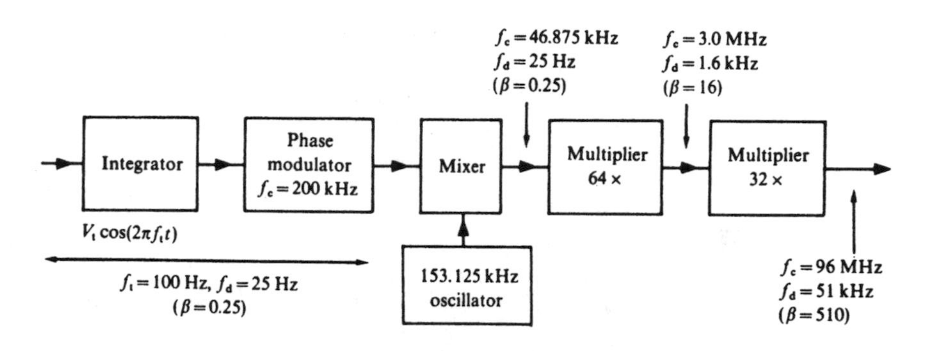

Figure 4.35. Frequency multiplication to increase β.

Self-assessment questions

4.1 What is meant by the terms 'frequency translation' and 'narrow banding'? State why these effects are important in radio transmission.

4.2 In an AM system a modulating voltage of 10 V changes a 20 V carrier by 5 V. Calculate the voltage deviation and the modulation index.

4.3 Calculate the carrier and total sideband power in a DSB-AM signal with a modulation index of 0.6 if the transmitted power is 100 kW. How could the transmitted power be reduced without affecting the sideband power?

4.4 Draw a block diagram of a high-power DSB-AM transmitter and explain how the modulated carrier is produced.

4.5 An impure tone of 440 Hz contains 40% third and 20% fifth harmonics. The tone amplitude-modulates a 30 kHz carrier. Sketch the frequency spectrum of the transmitted signal assuming that the carrier amplitude is twice that of the nearest side frequency.

4.6 What type of AM signal is produced by heterodyning and why is the balanced modulator preferred for generating this kind of signal?

4.7 Distinguish between the two forms of non-linear distortion and suggest how they may be reduced in non-linear channels.

4.8 What bandwidth would be required to transmit high-quality sound with a baseband of 30 Hz to 15 kHz using DSB-AM? Suggest how this bandwidth could be reduced without changing the modulating signal.

4.9 Explain the term 'selective fading' and state how it could be avoided.

4.10 What features of a carrier are affected in AM and FM by changes in the amplitude and frequency respectively of the modulating signal?
4.11 Show that FM may be regarded as PM with treble cut.
4.12 Define both amplitude and frequency modulation index and state how the channel bandwidth is affected by each of them.
4.13 Describe the principle of the varactor frequency modulator. Why is amplitude fluctuation in this modulator tolerable and how is frequency fluctuation obviated?
4.14 Refer to Fig. 4.15 and sketch the frequency spectra of an FM signal modulated by a 15 kHz tone when the tone amplitude is 0.6 of maximum.
4.15 Why is the VHF band used for FM broadcasting while AM is favoured in the LF and MF bands?
4.16 Why is the PSTN unsuitable for data communications at high bit rates?
4.17 Draw a waveform diagram to show how 1001 would appear in ASK, FSK, PSK and DPSK. State the relative disadvantages of these keying techniques.

5

Radio receivers

Recovering signals from carriers

Objectives

When you have completed this chapter you will be able to:

- Describe the signal recovery process
- State the kinds of amplification required in a radio receiver
- Describe the principle of signal selection using multiplicative mixing
- State the advantages of using amplifiers of fixed intermediate frequency
- Outline the action of a phase-locked loop
- State the principles of analogue and digital tuners
- Describe envelope detection of DSB-AM signals
- Outline the coherent detection of various AM signals
- Show how frequency demodulation may be accomplished
- Describe radio broadcasting in the UK
- Describe the principle of stereophonic radio
- Outline the techniques of automatic gain and frequency control

The aim of this chapter is to explain the ways in which modulating signals may be recovered from their modulated carriers and to describe how these techniques are applied in radio and TV broadcasting in the UK.

5.1 The signal recovery process

The operations carried out in a sound radio receiver are summarized in Fig. 5.1. The amplitude of the radio signal in the receiving aerial will usually be very small—a few microvolts—so it must be *amplified*. Having obtained a band of signals of workable amplitude, the required carrier signal must be *selected* and the modulating signal recovered or *detected*. That is, the modulated carrier must be *demodulated*—the analogue

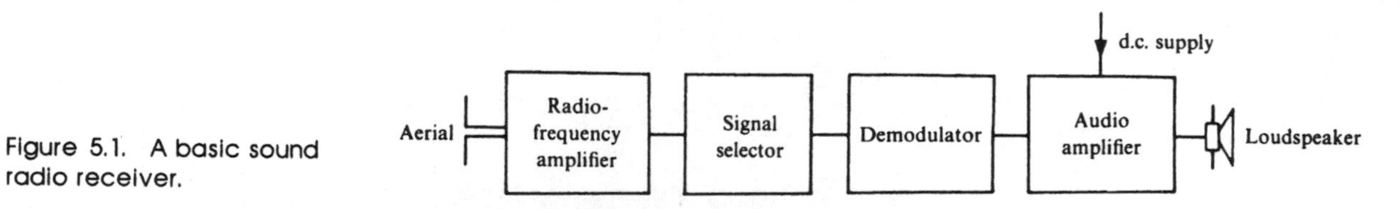

Figure 5.1. A basic sound radio receiver.

equivalent of decoding. The recovered audio signal would need amplification to develop enough power to drive a loudspeaker.

Amplification

The tasks of the two amplifiers in Fig. 5.1 can be dealt with quite easily, so we will look at this receiver aspect first before moving on to the more complicated tasks of selection and demodulation. The aerial will convert the electromagnetic wave energy of a band of carrier frequencies (and the inevitable noise) into a very small voltage. The function of the radio-frequency (RF) amplifier is to increase the level of this voltage, before more noise is added in the signal recovery process. This ensures a sufficiently high SNR at the demodulator output. The audiofrequency (AF) amplifier functions as a power converter, increasing the sound signal power to the level required by the loudspeaker and room acoustics. This power is obtained from a d.c. supply. The d.c. power required by the other receiver stages is negligible by comparison.

Signal selection

A radio receiver and its aerial will be designed for operation over one of the bands listed in Table 2.1. These bands are divided into frequency channels, each carrying information—a form of FDM. The MF band, for example, has part of its range from 300 kHz to 3 MHz divided into low-quality sound broadcasting channels, each 9 kHz wide, from about 600 to 1500 kHz, thus making available about $900/10 = 90$ channels (allowing for a 1 kHz guard band), which can be allocated to various broadcasting stations. By contrast, 625-line TV requires a channel width of 5.5 MHz, so the UHF band must be used in order to obtain a sufficient number of channels. Each geographical area is served by its own radio and TV stations, using several channels to which the receiver must be *tuned* in order to *select* the desired programme.

Channel selection could be achieved with very selective *band-pass* (BP) filters, but filter cost increases with selectivity, especially when the centre frequency has to be variable. In order to lower costs, BP filters of a fixed standardized centre frequency, intermediate between that of the carrier and the modulating signal, are used. The required carrier frequency (CF) is selected by shifting it down to this *intermediate frequency* (IF), using a multiplicative process similar to that used to shift from baseband up to

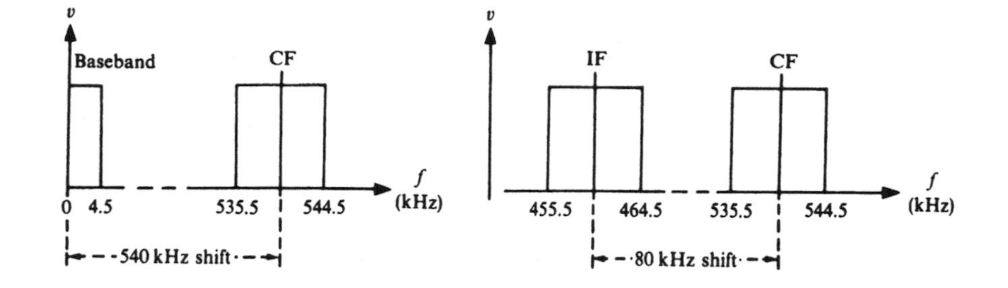

Figure 5.2. Comparison of frequency-shifting techniques.

carrier band in modulation. The up- and down-shifting processes are compared in Fig. 5.2.

The frequency down-shifting process is achieved by *heterodyning* or *mixing*. In this process, a *variable-frequency oscillator* (VFO) or *local oscillator* has its output (v_o) multiplied by the signal (v_c) from the RF amplifier as shown, for a tone-modulated AM signal, in Fig. 5.3.

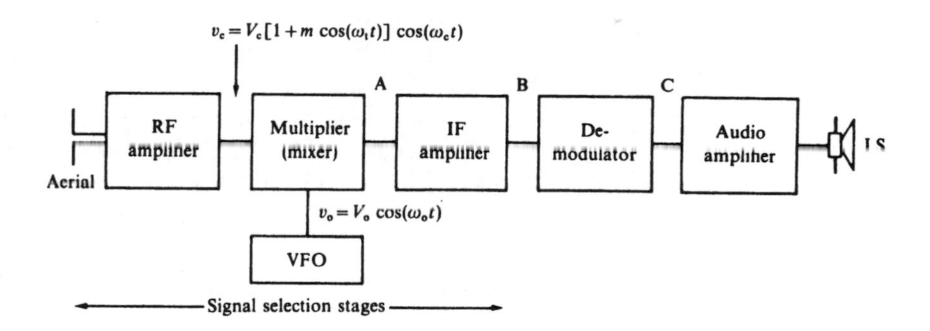

Figure 5.3. A radio receiver system.

In order to appreciate the mixer action, we assume initially that an unmodulated carrier, $V_c\cos(\omega_c t)$, is received. The output of the mixer will be

$$v_c v_o = \tfrac{1}{2}V_o V_c\{\cos[(\omega_o - \omega_c)t] + \cos[(\omega_o + \omega_c)t]\}$$

Thus the mixer outputs the sum and difference frequencies of the VFO and the carrier. The difference frequency, $\omega_o - \omega_c$, is made equal to the IF (ω_{if}) by adjustment of ω_o. The IF amplifier outputs $\tfrac{1}{2}V_o V_c\cos[(\omega_o - \omega_c)t] = \tfrac{1}{2}V_o V_c\cos(\omega_{if}t)$ at B in Fig. 5.3. If a modulated signal was received, the carrier signal and the sidebands would all be shifted by the same frequency $\omega_o - \omega_c$.

Exercise 5.1
Check on page 122

Show that the voltages at points A, B and C in Fig. 5.3 are given by

$$\begin{aligned} v_A = \tfrac{1}{2}V_c V_o\{&\cos[(\omega_o - \omega_c)t] + \cos[(\omega_o + \omega_c)t] \\ &+ \tfrac{1}{2}\cos[(\omega_o - \omega_c - \omega_t)t] + \tfrac{1}{2}\cos[(\omega_o + \omega_c + \omega_t)t] \\ &+ \tfrac{1}{2}\cos[(\omega_o - \omega_c + \omega_t)t] + \tfrac{1}{2}\cos[(\omega_o + \omega_c - \omega_t)t]\} \end{aligned}$$

$$v_B = \tfrac{1}{2} V_c V_o A_{if} \{\cos(\omega_{if} t) + \tfrac{1}{2}\cos[(\omega_{if} - \omega_t)t] + \tfrac{1}{2}\cos[(\omega_{if} + \omega_t)t]\}$$

$$v_c = \tfrac{1}{4} V_c V_o A_{if} \cos(\omega_t t) = V_t \cos(\omega_t t)$$

where A_{if} is the gain of the IF amplifier and $\omega_{if} = \omega_o - \omega_c$.

The standard IF for AM receivers is about 460 kHz and, in order to produce this, the VFO must have a frequency 460 kHz above or below the frequency of the carrier being selected. A variable frequency greater than the carrier frequency is chosen for the following reason. Assuming an MF range of 600 to 1500 kHz, the variable frequency range will have to be either 1060 to 1960 kHz or 140 to 1040 kHz—frequency ratios of about 1:2 or 1:7. By making the carrier frequency heterodyne with a higher VFO frequency, the VFO has to work over a range of only one octave, which makes it easier to design. The name *superheterodyne* or 'superhet' is given to receivers that work on this principle of frequency mixing or heterodyning. The term 'super' refers to the fact that the IF is superior to (greater than) the AF.

The action described in the last paragraph suggests a disadvantage of the superhet receiver, which is that, if there is another carrier frequency $2 \times 460 = 920$ kHz above the required carrier (the carrier image, as it were), this will also produce an IF signal, which will be heard interfering with the desired audio signal. To prevent *image interference*, the RF amplifier is designed to have a band-pass characteristic that rejects image signals. The higher the IF, the better will be the image rejection.

Exercise 5.2
Check on page 122

Stereophonic FM radio is broadcast in the UK on the VHF band over the range 88 to 108 MHz, using a standard IF of 10.7 MHz. What carrier frequency would produce image interference when the receiver was tuned to 88 MHz?

What are the maximum RF bandwidths that will give image rejection in MF and VHF radio? Assume that the IF used in MF radio is 460 kHz.

The great advantage of the superhet technique is that it enables fixed-frequency IF amplifiers to be used. This simplifies the tuning problem and makes for high selectivity and gain at low cost. The technique is used in all forms of radio, TV and radar receivers, the IF being appropriately chosen as in Table 5.1.

In AM receivers the IF is filtered from the intermodulation frequencies produced by the mixer by BP ceramic filters having a flat frequency response over the channel width of 9 kHz and a sharp roll-off on either side as shown in Fig. 5.4.

Table 5.1. Standard UK IFs

Application	IF
AM radio	460 kHz
VHF radio	10.7 MHz
UHF TV	39.5 MHz
Satellite TV	70 MHz

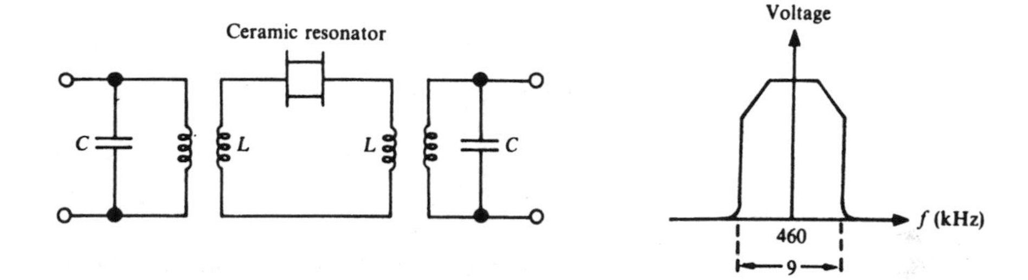

Figure 5.4. A ceramic filter unit and its frequency response.

The ceramic filter is fine-tuned by *LC* circuits to give the response shown in Fig. 5.4. The IF amplifier passes a band of frequencies centred on the IF, rejecting the many sum and product frequencies passed to it from the mixer. It is selective enough to reject the IFs of the adjacent carriers at 450 and 470 kHz. Poor selectivity at the IF stage would result in *adjacent channel interference.* Other forms of filter are used for the IFs given in Table 5.1, such as ferrite-core coils, crystals and solid-state devices.

Exercise 5.3
Check on page 122

The expression for a SSB-AM signal voltage is $V_t V_c \cos[(\omega_t + \omega_c)t]$. Derive an equivalent expression that is obtained when this signal is mixed with a local oscillator signal $V_o \cos(\omega_o t)$ and passed through an IF amplifier.

Channel selection

It should be appreciated that, in the superhet receiver, channel selection or *tuning* can be done by changing the local oscillator frequency. The RF stage is broadly tuned so that it passes the whole of the required range of carrier frequencies. If this range is greater than $2\omega_{if}$, the RF amplifier would have to be more selective in order to avoid image interference. This would imply a variably tuned RF amplifier. Analogue tuning is done by manually adjusting a continuously variable capacitor that determines the local oscillator frequency. In digital tuners the local oscillator consists of a *synthesizer*, which generates stepped frequencies by digital means. A popular form of synthesizer is based on the *phase-locked loop* (PLL).

5.2 The phase-locked loop

The PLL is a widely used circuit in telecommunications systems because of its versatility and cheap availability in IC form. The basic circuit consists of a phase comparator and a voltage-controlled oscillator (VCO) forming, in effect, a closed-loop control system as detailed in Fig. 5.5.

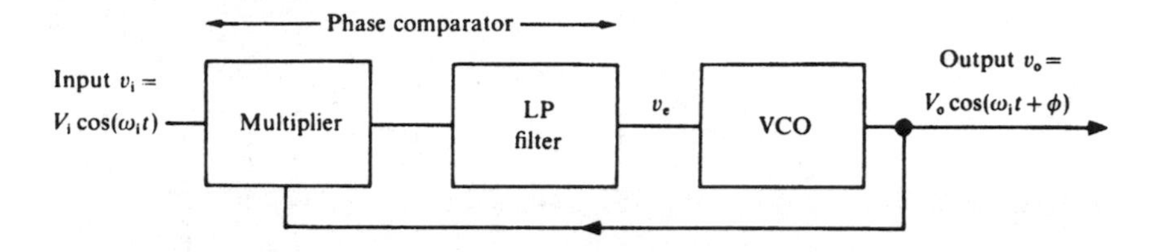

Figure 5.5. The basic PLL.

The action of the PLL is as follows. Initially there is likely to be a phase difference ϕ between the input and output signals, so the output of the multiplier is $V_i \cos(\omega_i t) V_o \cos(\omega_i t + \phi)$. This can be written in the form $\frac{1}{2} V_i V_o [\cos(2\omega_i t + \phi) + \cos \phi]$. The filter removes the component $\cos(2\omega_i t + \phi)$ to give an error voltage (v_e) equal to $\frac{1}{2} V_i V_o \cos \phi$. The error voltage changes the frequency of the VCO, causing the phase of v_o to adjust towards the phase of v_i until v_e reaches zero. Then the phase of v_o remains locked to the quadrature phase of v_i, following any subsequent changes in the phase of v_i that may occur. Since frequency is the rate of change of the angle of a signal, the output signal frequency follows changes in the input signal frequency.

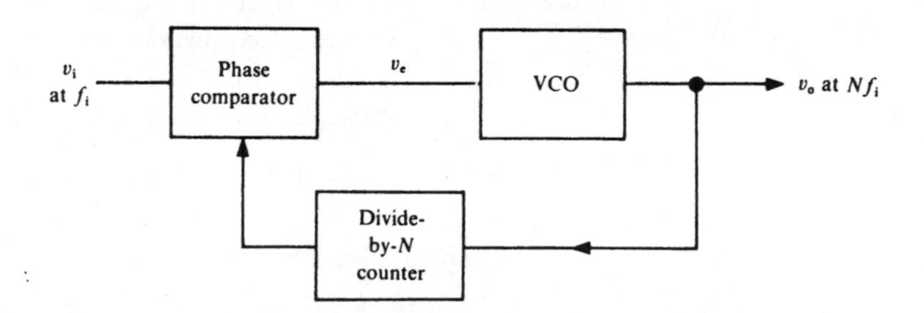

Figure 5.6. A basic synthesizer.

This PLL principle is applied in the basic synthesizer shown in Fig. 5.6. In this arrangement, the VCO has to output a frequency Nf_i in order to reduce v_e to zero. This synthesizer could be used as a digitally controlled local oscillator whose frequency would be varied by changing N. This means that tuning to the required programme using a *digital tuner* simply implies pressing a button that adds to or subtracts from the counter a number of pulses until the tuning is effected. This programme search can be done semi-automatically by monitoring the signal strength and stopping the counter when a strong signal is received. A programmable counter would give a preset facility by allowing specific carrier frequencies to be programmed into a non-volatile RAM (random access memory).

Note that v_e changes in proportion to the phase of the input signal frequency variations. This enables the PLL to function as a frequency-to-voltage converter. In this respect it finds wide use as a demodulator in FM receivers.

5.3 Demodulation

The method used to recover the information in a modulated signal depends on the type of modulation used. We will now consider, in turn, the demodulation of DSB-AM, DSB-SC-AM, SSB-AM, VSB-AM and FM signals.

Double-sideband demodulation

The traditional method of detecting DSB-AM, known as *envelope detection*, is very simple and cheap. It uses a diode as a half-wave rectifier and a LP filter to recover the envelope of the rectified waveform (Fig. 5.7).

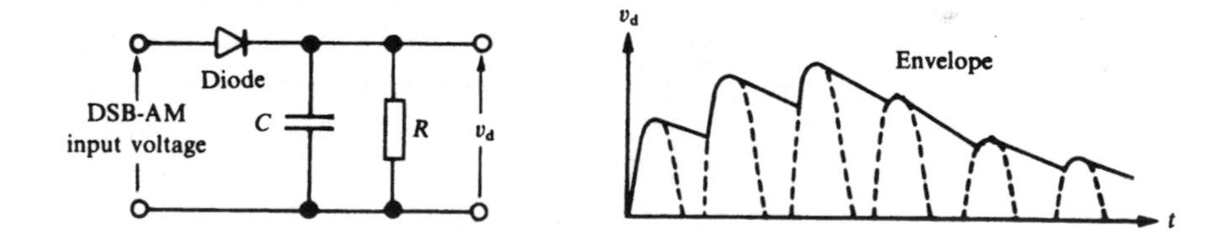

Figure 5.7. Envelope detection.

The CR circuit forms a LP filter with a cut-off frequency (ω_c) equal to $1/CR$. This frequency must be chosen so that the baseband is passed and the carrier frequency rejected. In effect this means that the capacitor must lose negligible charge between the half-wave peaks.

Exercise 5.4
Check on page 122

Draw the half-wave rectified waveform of a DSB-AM signal with $m > 1$.

Another form of demodulation works as the inverse of modulation, the signal frequencies being translated back from IF to baseband by frequency multiplication or mixing as exemplified for DSB-AM in Fig. 5.8.

A DSB-AM IF signal is given by

$$v_{if} = V_{if}[1 + m\cos(\omega_t t)]\cos(\omega_{if} t)$$

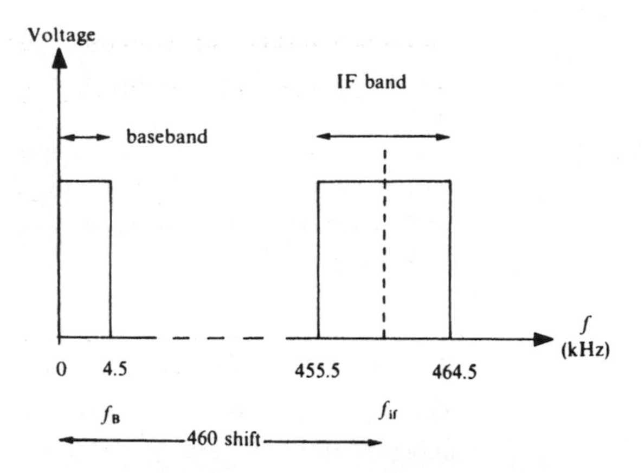

Figure 5.8. Demodulation by frequency shifting.

for each tone component ω_t present within the baseband 0 to ω_B. Multiplying this by the IF amplifier output signal, $V_{if}\cos(\omega_{if}t)$, gives a demodulated signal

$$
\begin{aligned}
v_d &= V_{if}^2[1+m\cos(\omega_t t)]\cos^2(\omega_{if}t) \\
&= V_{if}^2[1+m\cos(\omega_t t)]\tfrac{1}{2}[1+\cos(2\omega_{if}t)] \\
&= \tfrac{1}{2}V_{if}^2[1+m\cos(\omega_t t)] + \text{higher-frequency terms}
\end{aligned}
$$

A baseband filter will remove the zero-frequency and higher-frequency terms and pass the required demodulated signal $\frac{1}{2}mV_{if}^2\cos(\omega_t t)$. This system requires a replica of the carrier signal, shifted down to the IF. This replica can be obtained from the incoming signal by using a PLL as shown in Fig. 5.9.

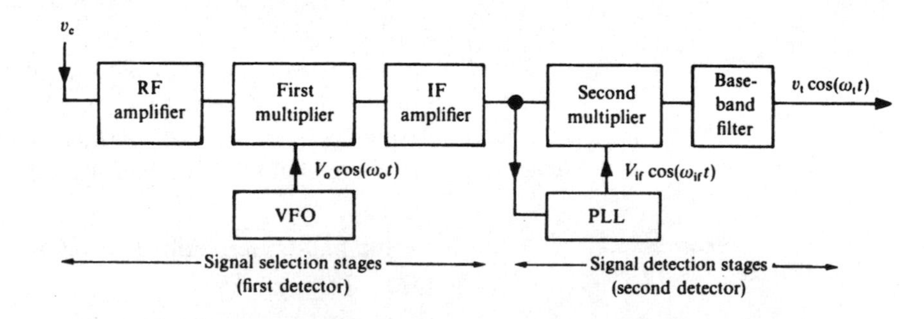

Figure 5.9. Selection and detection.

The similarity of the signal selection and detection stages in this scheme has led to them being termed first and second detectors. Note that, if the carrier frequency drifts, the IF will drift similarly and the second detector will use a shifted carrier replica that is *synchronized* or *coherent* with the carrier frequency. A diode demodulator, by contrast, uses *non-coherent* detection: it does not need to know the carrier phase for its action.

Double-sideband suppressed-carrier demodulation

Coherent detection of DSB-SC-AM requires the carrier to be regenerated in the receiver. This can be done as shown in Fig. 5.10.

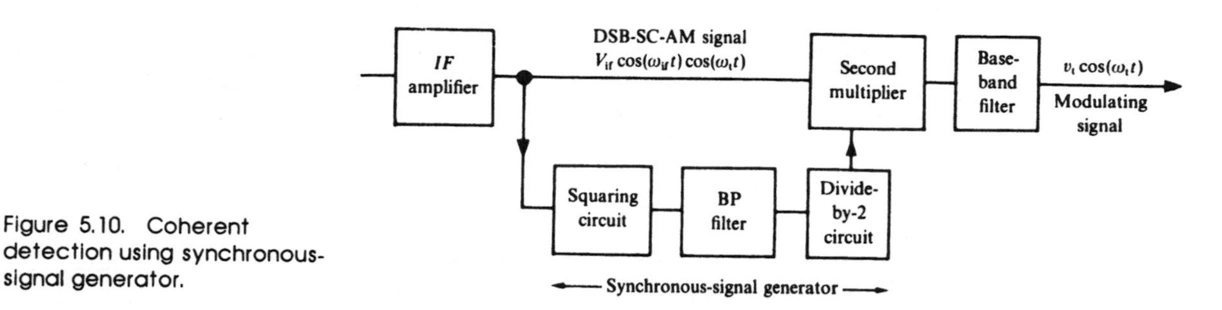

Figure 5.10. Coherent detection using synchronous-signal generator.

The output of the squaring circuit will be $[V_{if}\cos(\omega_{if}t)\cos(\omega_t t)]^2$. This can be written $\frac{1}{4}V_{if}^2[1+\cos(2\omega_{if}t)][1+\cos(2\omega_t t)]$, which expands further to

$$\tfrac{1}{4}V_{if}^2\{1+\tfrac{1}{2}\cos[2(\omega_{if}+\omega_t)t]+\tfrac{1}{2}\cos[2(\omega_{if}-\omega_t)t]$$
$$+\cos(2\omega_{if}t)+\cos(2\omega_t t)\}$$

The band-pass (BP) filter selects the component $\frac{1}{4}V_{if}^2\cos(2\omega_{if}t)$, which is then frequency-divided by 2 to obtain a synchronous shifted carrier signal.

An ingenious PLL arrangement for synchronous-signal generation, uses the *Costas loop*, shown in Fig. 5.11. The signal from the IF amplifier will be

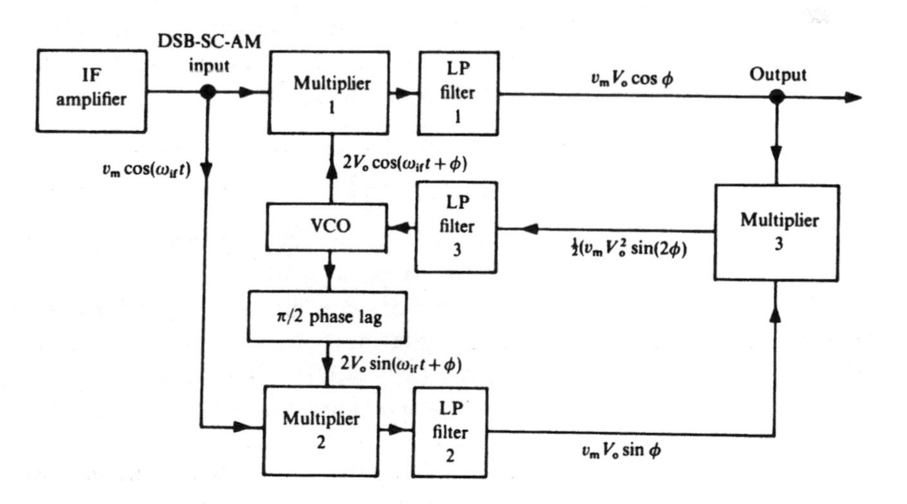

Figure 5.11. DSB-SC-AM detection using the Costas loop.

of the form $V_{if}\cos(\omega_t t)\cos(\omega_{if}t)$, which may be written $v_m\cos(\omega_{if}t)$. The VCO is set to oscillate at ω_{if} but it will probably be out of phase with the incoming IF carrier signal. The Costas loop brings the VCO into phase with the IF as follows. Suppose the VCO output is $2V_o\cos(\omega_{if}t+\phi)$, then the inputs to multipliers 1 and 2 will be, respectively:

$$v_m \cos(\omega_{if})2V_o \cos(\omega_{if}t+\phi)=v_m V_o[\cos(2\omega_{if}+\phi)+\cos\phi]$$

$$v_m \cos(\omega_{if})2V_o \sin(\omega_{if}t+\phi)=v_m V_o[\sin(2\omega_{if}+\phi)+\sin\phi]$$

The LP filters 1 and 2 will remove the $2\omega_{if}$ components, leaving the $\cos\phi$ and $\sin\phi$ components to be applied to multiplier 3, where they yield the product

$$(v_m V_o)^2 \sin\phi\cos\phi=(v_m V_o)^2\tfrac{1}{2}\sin(2\phi)=[V_{if}\cos(\omega_t t)V_o]^2\tfrac{1}{2}\sin(2\phi)$$
$$=(V_{if}V_o)^2\tfrac{1}{2}[1+\cos(2\omega_t t)]\tfrac{1}{2}\sin(2\phi)$$

A narrow-band LP filter 3 will remove the component of frequency $2\omega_t$, leaving a voltage $\frac{1}{4}(V_{if}V_o)^2\sin(2\phi)$ to be used to alter the VCO frequency until ϕ becomes zero. Once the VCO frequency is thus locked to $v_m\cos(\omega_{if}t)$, the output voltage will be maintained at $v_m V_o$, which is the demodulated signal. The action of the Costas loop is similar to the PLL.

Single-sideband demodulation

A DSB-SC-AM IF signal, $V_{if}\cos(\omega_t t)\cos(\omega_{if}t)$, may be written

$$\tfrac{1}{2}V_{if}\{\cos[(\omega_{if}-\omega_t)t]+\cos[(\omega_{if}+\omega_t)t]\}$$

If the lower sideband is suppressed, the tone-modulated IF voltage will be

$$\tfrac{1}{2}V_{if}\cos[(\omega_{if}+\omega_t)t]$$

Demodulation can be achieved, as in other forms of AM, by multiplying with the IF amplifier output signal, $V_{if}\cos(\omega_{if}t)$, to obtain the demodulated signal. Thus

$$v_d=V_{if}\cos[(\omega_{if}+\omega_t)t]V_{if}\cos(\omega_{if}t)$$
$$=\tfrac{1}{2}V_{if}^2\{\cos(\omega_t t)+\cos[(2\omega_{if}+\omega_t)t]\}$$

A LP filter passes the required signal $\frac{1}{2}mV_{if}^2\cos(\omega_t t)$. This system also requires a replica of the carrier signal, at the IF. The replica cannot be obtained from the incoming signal by a synchronous-signal generator because the output of the squaring circuit of Fig. 5.10 will be $\{\frac{1}{2}V_{if}\cos[(\omega_{if}+\omega_t)t]\}^2$, which, when written $\frac{1}{4}V_{if}^2\{1+\cos[2(\omega_{if}+\omega_t)t]\}$, clearly has no IF carrier component to recover for synchronous detection.

Exercise 5.5
Check on page 123

Verify that, for SSB-AM, the narrow-band LP filter 3 will remove both the component frequencies $2\omega_t$, so the Costas loop cannot be used in SSB demodulation.

In SSB receivers, a crystal-controlled local oscillator is used often for synchronous detection, its frequency stability being about 1 Hz drift at 1 MHz. Once the oscillator frequency is set, only occasional manual adjustment is normally required. An alternative technique transmits a low-power reference signal or *pilot tone* at the carrier frequency, which can be used in detection.

Vestigal-sideband demodulation

Detection of VSB-AM signals is accomplished by the same methods as those used for DSB detection. However, the receiver frequency response must compensate for the asymmetry of the VSB signal as shown in Fig. 5.12. Both sidebands are transmitted over the video range 0 to 1.25 MHz, but only one over the range 1.25 to 5.5 MHz. This means that the received signal power in the band 0 to 1.25 MHz will be twice that of the band 1.25 to 5.5 MHz. If the receiver response is shaped as shown in Fig. 5.12 so that $|a|+|b|=|c|$, the receiver detector will output a constant amplitude response from 0 to 5.5 MHz.

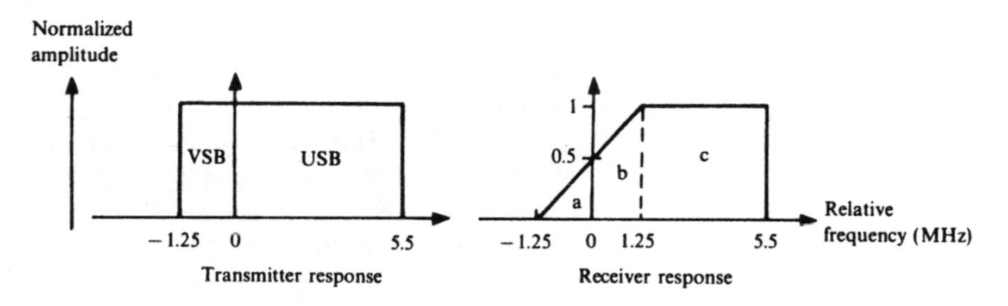

Figure 5.12. The transmitter and receiver responses for VSB TV video signals.

Frequency demodulation

The FM detector is a frequency-to-voltage converter, so the PLL is often used as an FM detector. The same circuit, as we have seen, can also be used as a synchronous detector for DSB-AM, which makes it doubly useful since most broadcast receivers are required to work with both AM and FM. The demodulation stages of an FM monophonic receiver are shown in Fig. 5.13. An amplitude limiter is included to remove any AM due to noise.

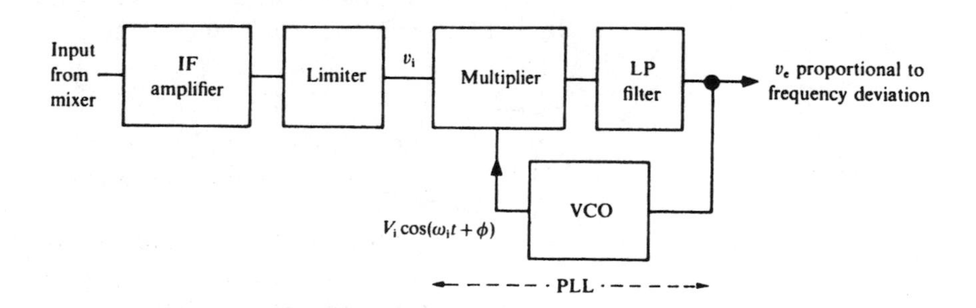

Figure 5.13. FM detection using a PLL.

5.4 Sound broadcasting

World-wide AM radio broadcasting uses VLF and LF surface waves and HF sky waves (see Sec. 2.3). In the UK, radio programmes are broadcast by the BBC and by commercial radio stations using DSB-AM or FM. The BBC transmit their AM Radio 4 programmes using an LF carrier at 198 kHz. Two transmitters, one of 400 kW at Droitwich and another of 50 kW at Westerglen, cover most of the country, although a few small 'shadow' areas have to be served at MF. Other AM programmes are transmitted using carrier frequencies in the MF band. The shorter effective range at MF necessitates a greater number of transmitters to achieve national coverage. These must be allocated channels within a relatively small MF band (600 to 1500 kHz) and this restricts the baseband to only 4.5 kHz (an internationally agreed value). The transmitted power at MF is typically between 2 and 150 kW according to the size of the area to be served. The sound quality of AM radio can be very poor, due mainly to interference and fading and partly to the restricted bandwidth. Great improvement in these respects is achieved by increasing the baseband to 15 kHz and using FM.

Chapter 4 shows that FM requires a much larger bandwidth than AM for the same baseband. In order to accommodate enough channels, all FM broadcasting uses the VHF band over the internationally agreed range 88 to 108 MHz. In the UK the range 88 to 98 MHz is used by the BBC and commercial stations, each station being assigned a carrier frequency in this band at 200 kHz intervals. At VHF, received signals, which are not line-of-sight, are weak and well rejected by the receiver. This means that broadcast service areas are small, so more transmitters are required for national coverage than for MF. Extra low-power transmitters are provided in hilly areas where the line of sight is short. VHF transmitters operate at about the same power levels as those using MF.

A further improvement in sound quality is made by FM *stereophonic* broadcasting. Stereophonic reproduction uses two separated loudspeakers fed by left (L) and right (R) audio channels. The FM baseband signal must contain both L and R signals, which can be decoded and passed to the L and R audio amplifiers. The stereophonic system adopted had to be compatible with monophonic receivers. The baseband spectrum arrangement that met this compatibility requirement is shown in Fig. 5.14.

The monophonic band (L + R) covers the audio range from about 30 Hz to 15 kHz, with the maximum signal amplitude producing a frequency

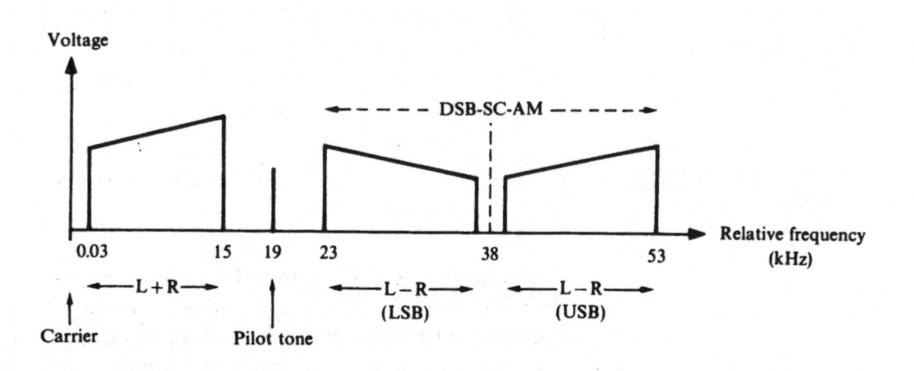

Figure 5.14. FM stereo baseband spectrum.

deviation of ± 75 kHz. Note that the amplitude of the signal increases towards the treble. This so-called *pre-emphasis* is employed to improve the SNR: de-emphasis or treble cut in the receiver will reduce 'hiss' while restoring the signal to its original spectral amplitude distribution (see Chapter 8 for further explanation).

The stereophonic information is transmitted as a L – R signal carried by DSB-SC-AM. The carrier frequency information, which is needed for synchronous detection, is supplied by a low-power pilot tone of 19 kHz. This tone frequency is doubled at the receiver to provide a 38 kHz *sub-carrier*, which is combined with the L – R lower and upper sidebands. The whole of this composite baseband signal is frequency-shifted, by FM, to the VHF range, prior to transmission. The whole process is illustrated in Fig. 5.15.

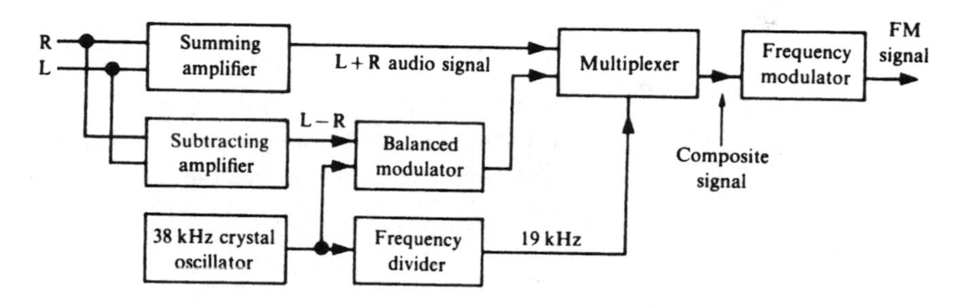

Figure 5.15. FM stereo encoder and frequency modulator.

The two amplifiers in Fig. 5.15 pre-emphasize the audio signal and output the sum and difference signals of the two sound channels. A balanced modulator produces a DSB-SC-AM version of the L – R signal. The 19 kHz pilot tone is derived from a very stable crystal oscillator and the composite baseband spectrum of Fig. 5.14 is put together by a multiplexer. The resulting composite voltage signal forms the input to a frequency modulator.

The bandwidth (B_s) required for stereophonic FM transmission is given by Carson's rule as $B_s = 2(f_{dm} + f_t)$ where f_t is the baseband width of 53 kHz and f_{dm} is the maximum frequency deviation of 75 kHz. Therefore, $B_s = 2(75 + 53) = 256$ kHz. This is greater than the 200 kHz spacing allowed so, to avoid adjacent channel interference, advantage must be taken of the short range of VHF signals by allocating adjacent channels to widely separated geographical areas.

Exercise 5.6
Check on page 123

Estimate the bandwidth required for monophonic FM broadcasting.

A stereo signal is decoded by the arrangement shown in Fig. 5.16. The demodulated signal is decomposed by filters into (L + R) baseband, (L – R) DSB-SC-AM and pilot tone. A synchronous demodulator shifts the L – R

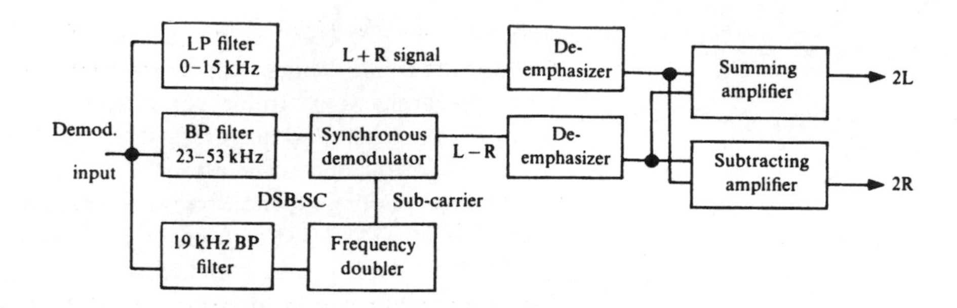

Figure 5.16. FM stereo decoder.

signal to baseband. After removing the HF emphasis with de-emphasis circuits (see Sec. 8.3), the R signal is obtained by subtracting (L − R) from (L + R) and the L signal is formed by adding (L − R) to (L + R).

5.5 Automatic gain and frequency control

The input signal to a radio receiver tends to vary in power according to atmospheric conditions. In order to maintain a constant signal level at the demodulator, *automatic gain control* (AGC) is used as shown in Fig. 5.17.

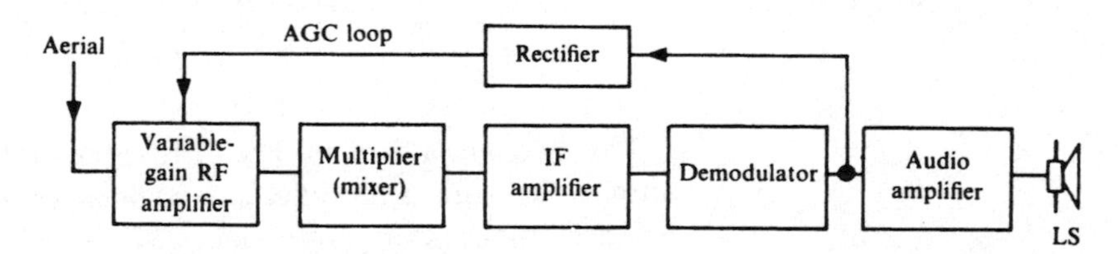

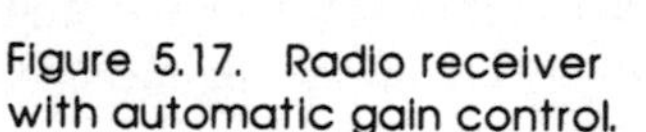
Figure 5.17. Radio receiver with automatic gain control.

If the signal from the aerial is weak, the audio signal from the demodulator will also be weak. A direct voltage, proportional to the amplitude of the audio signal, is fed back to the RF amplifier to increase its gain. The reverse action will occur when the signal level increases, thus giving a fairly constant signal level under a wide range of reception conditions. If the signal becomes too weak, the limited variable gain range of the RF amplifier prevents AGC operating. The output of the rectifier can also be used to drive a voltmeter to provide a visual indication of tuning.

In VHF receivers, signal strength variations are small because short-distance line-of-sight reception is used. However, frequency stability in the local oscillator is important, since fluctuation in the IF can have a considerable effect on the output of the demodulator. AM systems are more tolerant in this respect. A similar technique to AGC, called *automatic frequency control* (AFC), is used to vary the local oscillator frequency so that a constant IF is obtained (Fig. 5.18).

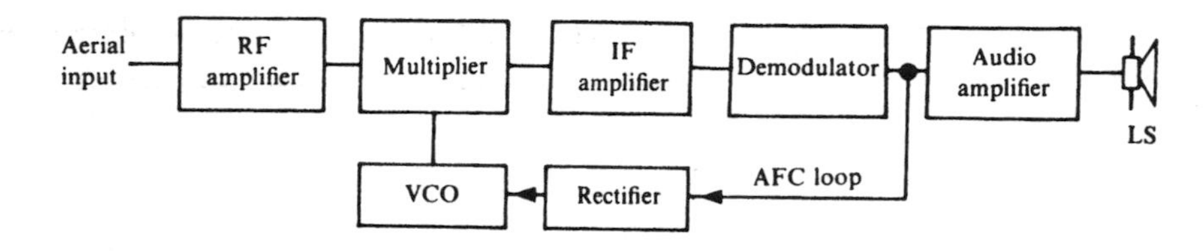

Figure 5.18. Radio receiver with automatic frequency control.

Summary

The recovery of information from a modulated radio signal involves three basic processes: selection, amplification and demodulation. Amplification is required at various stages in a receiver in order to keep the SNR at a tolerable level and to develop sufficient power to drive the output transducer. Selection is needed because many different radio signals are present in the aerial. Demodulation recovers the modulating signal from the modulated carrier.

Selection is achieved by varying the frequency of a local oscillator, which 'superheterodynes' the relatively lower carrier frequency to produce an intermediate frequency. This is presented to an IF amplifier, which is, in effect, a BP filter with a fixed, standardized centre frequency, intermediate between that of the carrier and the modulating signal. This stage removes all the unwanted difference signals and all the intermodulation components, passing only the difference between the local oscillator and wanted band of frequencies. Highly selective IF stages reduce adjacent channel interference. Image interference may be reduced by selective RF amplifiers or by restriction of the allocated band to less than twice the IF. Selection or tuning is done by varying the local oscillator frequency using either analogue or digital means.

DSB-AM signals may be coherently or incoherently (non-synchronously) demodulated. Other forms of AM, where the carrier is suppressed, require coherent detection. FM demodulation involves frequency-to-voltage conversion. The phase-locked loop is a versatile circuit that can be used in both coherent amplitude demodulation and frequency demodulation, as well as in tuning.

Radio signals tend to vary in power at the receiver according to atmospheric conditions, so AGC is needed to maintain a constant signal level. In FM receivers, stability of the IF is important, so AFC is used to vary the local oscillator frequency and so maintain a constant IF.

Domestic radio programmes are broadcast using DSB-AM or FM with base bandwidths of 4.5 kHz and 15 kHz respectively. AM carrier frequencies are in the LF and MF band. Improved noise immunity is possible with FM, but FM requires a much larger bandwidth than AM for the same baseband. In order to accommodate sufficient channels, all FM broadcasting uses the VHF band. VHF transmitters operate at about the same power levels as MF transmitters but have a more restricted range.

Improved sound quality may be achieved by FM stereophonic broadcasting.

Checks on exercises

5.1 The input signal expression $v_c = V_c[1 + m\cos(\omega_t t)]\cos(\omega_c t)$ represents DSB-AM and expands to $V_c\cos(\omega_c t) + \frac{1}{2}mV_c\cos[(\omega_c - \omega_t)t] + \frac{1}{2}mV_c \cos[(\omega_c + \omega_t)t]$. When this is multiplied by $V_o\cos(\omega_o t)$, we get a further similar expansion to give the expression quoted for v_A. The IF amplifier then selects the band of frequencies from $(\omega_{if} - \omega_t)$ to $(\omega_{if} + \omega_t)$ from which the demodulator recovers $V_t\cos(\omega_t t)$.

5.2 The IF is 10.7 MHz, so a carrier at $88 + (2 \times 10.7) = 109.4$ MHz would produce image interference when the receiver was tuned to 88 MHz. Thus, even when the receiver is tuned to the extreme end of its range, the image signal lies beyond the range allocated to VHF broadcasts. This implies that a VHF RF stage does not have to be tuned to reject the image signal.

The maximum RF bandwidth for image rejection is twice the IF, that is 920 kHz for AM and 21.4 MHz for FM.

Since the VHF range used for FM radio broadcasting in the UK is restricted to a 20 MHz band, the RF stage need not be variably tuned. This is also the case with the MF range, 600 to 1500 kHz, which is less than 920 kHz.

5.3 When a SSB-AM signal is mixed with the local oscillator frequency we get

$$v = V_t V_c \cos[(\omega_c + \omega_t)t]\, V_o\cos(\omega_o t)$$
$$= \tfrac{1}{2}V_t V_c V_o\{\cos[(\omega_c + \omega_t - \omega_o)t] - \cos[(\omega_c + \omega_t + \omega_o)t]\}$$

After IF amplification we get

$$v_{if} = V_{if}\cos[(\omega_c + \omega_t - \omega_o)t]$$

Thus, the SSB-AM signal has been frequency-shifted downwards from $(\omega_c + \omega_t)$ to $(\omega_c + \omega_t - \omega_o)$.

5.4 A half-wave rectified waveform of a DSB-AM signal with $m > 1$ would be as shown in Fig. 5.19.

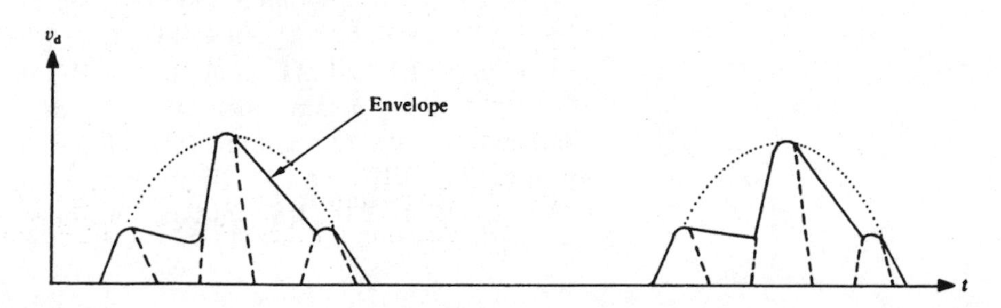

Figure 5.19. Envelope detection when $m > 1$.

5.5 The inputs to multipliers 1 and 2 will be, respectively:

$$V_{if}\cos[(\omega_{if}+\omega_t)t]\; V_o\cos(\omega_{if}t+\phi)=V_{if}V_o[\cos(2\omega_{if}+\omega_t+\phi) + \cos(\omega_t+\phi)]$$

$$V_{if}\cos[(\omega_{if}-\omega_t)t]\; V_o\sin(\omega_{if}t+\phi)=V_{if}V_o[\sin(2\omega_{if}+\omega_t+\phi) + \sin(\omega_t+\phi)]$$

The LP filters 1 and 2 will remove the $2\omega_{if}$ components, leaving the $\cos(\omega_t+\phi)$ and $\sin(\omega_t+\phi)$ components to be applied to multiplier 3 to yield the product

$$(V_{if}V_o)^2\sin(\omega_t+\phi)\cos(\omega_t+\phi)=(V_{if}V_o)^2\tfrac{1}{2}\sin[2(\omega_t+\phi)]$$
$$=\tfrac{1}{2}(V_{if}V_o)^2[\sin(2\omega_t)\cos(2\phi) + \cos(2\omega_t)\sin(2\phi)]$$

Thus the narrow-band LP filter 3 will remove both the component frequencies $2\omega_t$, so the Costas loop cannot be used in SSB demodulation.

5.6 The bandwidth required for monophonic FM broadcasting $=2(75+15)=$ 180 kHz.

Self-assessment questions

5.1 Draw a block diagram of a typical FM radio receiver.
5.2 What advantages are gained by using an RF amplifier in a radio receiver?
5.3 List the frequency components in the output of a multiplicative mixer.
5.4 How is a radio programme selected by a receiver?
5.5 What are the advantages of using a standardized IF?
5.6 Show the effect of heterodyning a narrow-band FM signal, which has a 30 Hz to 15 kHz baseband, by sketching a voltage/frequency graph.
5.7 Why is the VCO frequency in a superhet receiver normally greater than the carrier frequency?
5.8 Distinguish between adjacent channel and image interference and state how they are reduced.
5.9 Distinguish between coherent and non-coherent detection.
5.10 Show that a PLL produces an output from its VCO proportional to the phase change in the input to its multiplier.
5.11 State the advantages of using a PLL in an AM/FM digital tuner.
5.12 Compare the functions of the first and second detectors in DSB-AM receivers.
5.13 Describe two methods by which SSB-AM signals may be demodulated.
5.14 List the advantages of FM broadcasting on the VHF band and AM broadcasting on the LF, MF and SF bands.

6

Television principles

Encoding and decoding video signals

Objectives

When you have completed this chapter you will be able to:

- Outline the operation of a TV broadcasting system
- Describe how an image is scanned to give an electrical signal
- Draw a simplified cross-sectional view of a TV camera tube system
- Describe how electrical signals are converted into light to build up a picture
- Draw a monochrome TV receiver tube
- Describe the principle of interlacing
- Describe how vision and sound signals are combined for transmission
- Describe a TV receiver system
- Describe the principle of the shadow-mask colour TV tube
- Outline how colour information is encoded
- Describe TV broadcasting by satellite

This chapter is concerned mainly with the generation, transmission, reception and reproduction of video and data television signals. Broadcast TV uses radio propagation techniques and modulated waves to send moving pictures, sound and data to TV receivers in a service area.

A TV service area is quite small when terrestrial transmitting systems are used. Satellite-based TV relay stations increase the service area considerably and make possible world-wide TV broadcasting. Cable TV can offer a high-quality, low-cost distribution service in metropolitan areas.

An audio signal normally accompanies a TV video signal and occupies a small part of the video baseband. FM or companded digital modulation is used to carry the sound information, but bandwidth restriction necessitates the use of vestigial-sideband AM to carry the picture information.

A TV may be used to display data as well as pictures. Data telecommunication is considered in more detail in Chapter 11.

6.1 Television broadcasting systems

To appreciate how moving coloured pictures can be telecommunicated, it is necessary to understand the principles of signal/picture transducers, that is the TV camera and picture tube. The TV image is made up of a large number of dots of varying intensity and colour combination, and relies on the principle that the eye is unable to resolve these separate dots if they are close enough together and the viewing distance is not too short. Moving TV pictures rely for their effect on persistence of vision, as does cinematography, presenting the viewer with a rapid series of still pictures—each changing slightly from the previous one—to give the illusion of smooth movement. A TV baseband signal must contain light intensity and colour information encoded so that the original moving image can be re-formed in the receiver. This information requires a bandwidth of several megahertz so the baseband signal must be transmitted on UHF or SHF carriers in order to squeeze in the required number of channels.

TV may be used for the display of data. If pages of text are broadcast, all that is required is a Teletext decoder, added to the receiver, so that database services such as Ceefax and Oracle can be accessed. For two-way data communication, private companies offer data exchange facilities, which use telephone lines instead of radio waves. A TV or special monitor may be used as the visual display unit (VDU). BT's Prestel is an example of this type of non-broadcast data service known generally as Viewdata.

A TV broadcasting transmitter, working within a range of frequencies allocated for its service area, disperses picture and voice signal power over that area. Within this area, viewers can select the entertainment or data service they require by tuning their TVs to the frequency band carrying the required information. Each programme is allocated a carrier frequency in the UHF band or SHF band in the case of satellite links. TV may be used for data transmission as well as entertainment, with broadcast pages of data called Teletext being available from the BBC (Ceefax) and ITV (Oracle). TV broadcasting services are often combined with radio broadcasting to serve a similar area, using the same mast to support the transmitter aerials.

A TV channel has to accommodate both vision and sound within the same frequency channel, so a TV set consists of two receiving systems: one for vision and the other for sound. In the UK, the picture information is sent using vestigial-sideband AM, while monophonic sound is broadcast using FM and stereophonic sound is transmitted using the NICAM system.

Exercise 6.1
Check on page 143

Suggest what advantage might be obtained by transmitting vision and sound using different modulation systems.

Before describing how a TV signal is modulated and demodulated, it is necessary to consider how the baseband is made up.

6.2 The TV camera

To create a TV picture, the image must be *scanned*. This is rather like reading the page of a book: the eye goes from left to right along a line and then flies back to the beginning of the next line and so on until a complete page has been read. This scanning process is done with an electronic camera whose principle of operation depends on forming an image on a screen made up of a matrix or *raster* of small *photocells* as shown in Fig. 6.1.

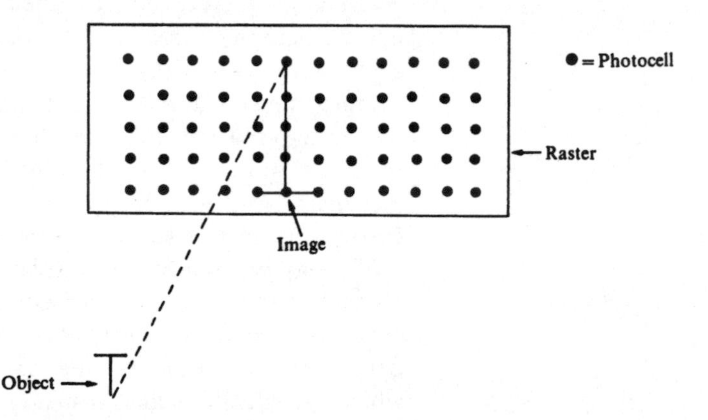

Figure 6.1. Image formation.

The photocells, which are the transducers that convert light into electrical energy, are made up of very small alkali-metal blobs distributed on a sheet of insulating material which is backed with a sheet of copper (see Fig. 6.3). Each cell emits electrons in proportion to the intensity of illumination or *luminance* of the light absorbed by the cell. This leaves the cell with positive charge proportional to the luminance of that area of the image corresponding to the cell's position in the raster.

In order to convert the image into an electrical signal, a beam of electrons (cathode ray) is directed towards the raster so that it scans the cells line by line as shown in Fig. 6.2. Each photocell absorbs electrons according to its positive charge, so a sequence of current pulses flow, from

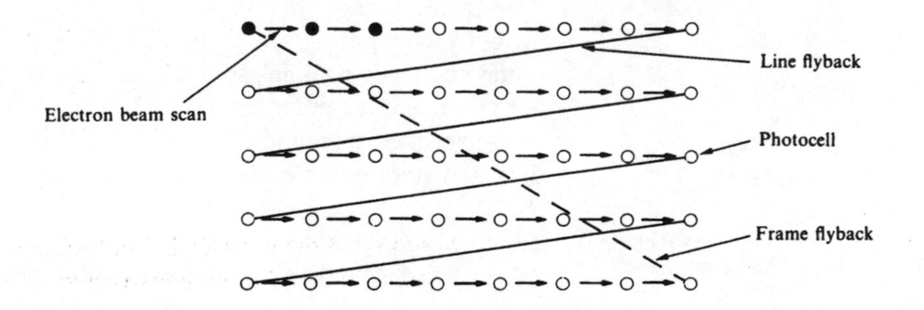

Figure 6.2. Scanning.

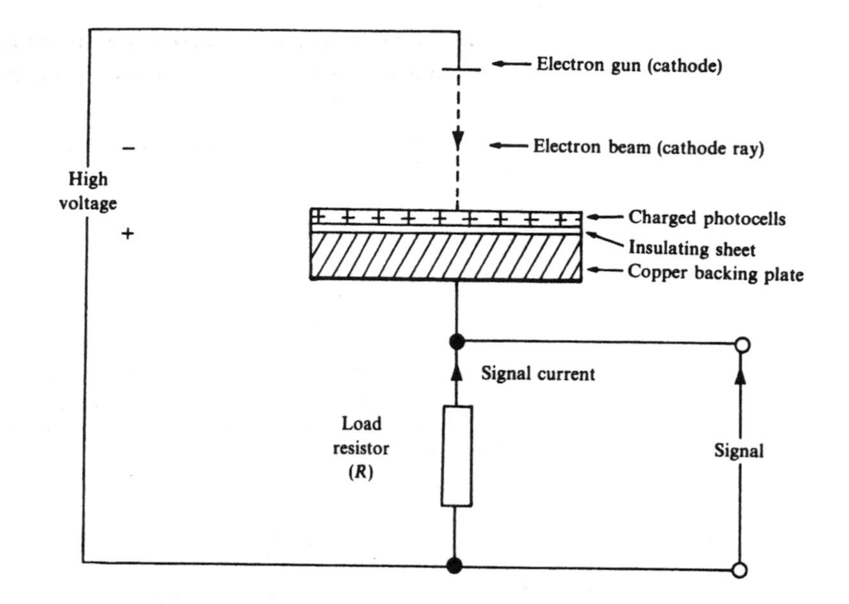

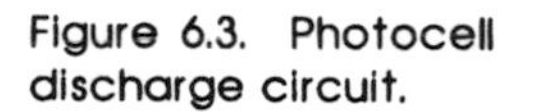
Figure 6.3. Photocell discharge circuit.

each cell in turn, in the circuit connecting the copper backing plate as the raster is scanned. The circuit is shown in Fig. 6.3.

On each successive raster scan, a signal voltage is obtained across the load resistor, so an image could be converted into a video signal voltage as shown in Fig. 6.4. The scanning process converts spatial information (the image) into temporal information (the signal). A complete set of raster lines constitutes a *frame*, and in the UK TV system there are 625 lines per frame.

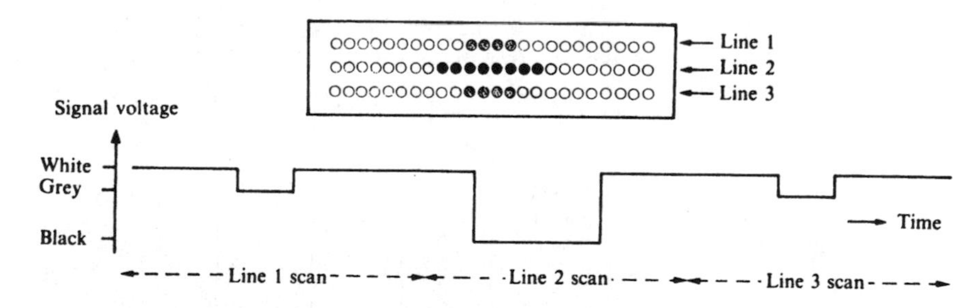

Figure 6.4. Image-to-signal conversion.

Exercise 6.2
Check on page 143

Draw the video signal that would produce a white diagonal line running from the top left-hand corner of the screen to the bottom right-hand corner and gradually increasing in whiteness. Suggest how the illusion of picture movement might be achieved.

Raster formation

We now consider how a raster is built up, and to do this we must look at the way a camera tube is constructed. The electron beam must be produced in a vacuum and made to scan the raster of photocells by being deflected by two sets of magnetic deflection coils at right angles to one another. The TV camera tube is a form of *cathode ray tube* (CRT) with a photocell screen and a deflection system, shown diagrammatically in Fig. 6.5.

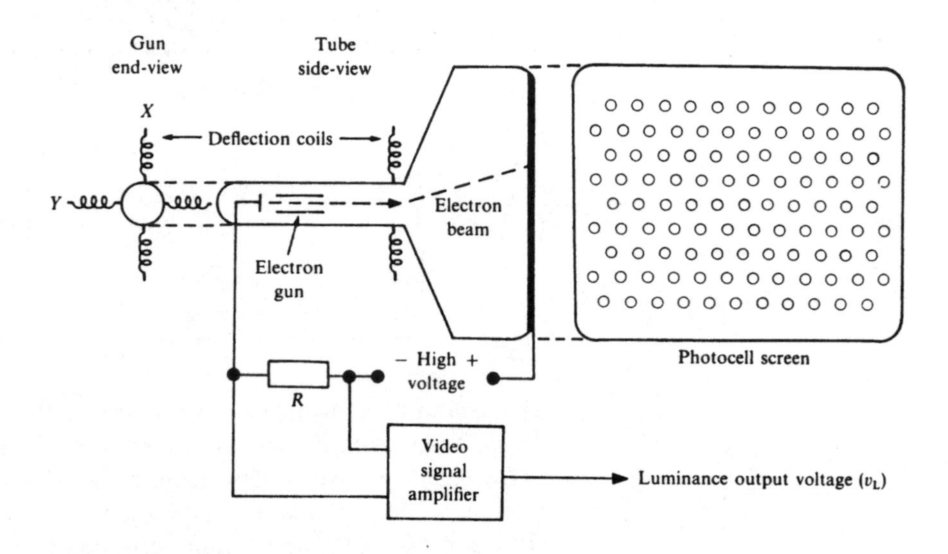

Figure 6.5. A basic TV camera tube system.

The current in the deflection coil used for the line scan will have to increase steadily as the electron beam is deflected from one end of the line to the other. When the beam reaches the end of the line it must *fly back* as quickly as possible to the beginning of the next line to be scanned. When the frame has been completed the beam must fly back to the first line starting position for the next frame. Line and frame scan current waveforms that will do this are shown in Fig. 6.6.

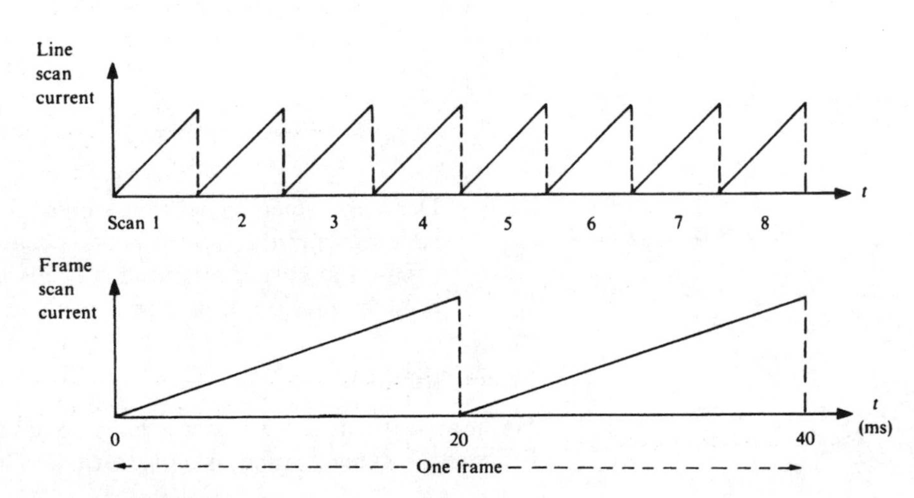

Figure 6.6. Line and frame scan current waveforms for a four-line raster.

A tolerable picture flicker is obtained by refreshing the whole screen about 50 times per second (once every 20 ms), giving a line scan period of 20 ms/625 = 32 μs. If this period can be lengthened, a smaller transmission bandwidth can be used. To this end, alternate lines are displayed first, after which the beam is deflected back to give a second sequence of line scans *interlaced* with the first. Interlacing gives two *fields* per picture frame, each one repeated 25 times per second, thus giving the appearance of the whole screen being refreshed 50 times per second. The line scan period would now be 64 μs and bandwidth will be saved.

Exercise 6.3
Check on page 144

Draw the raster that would be formed by the timebase voltages shown in Fig. 6.7.

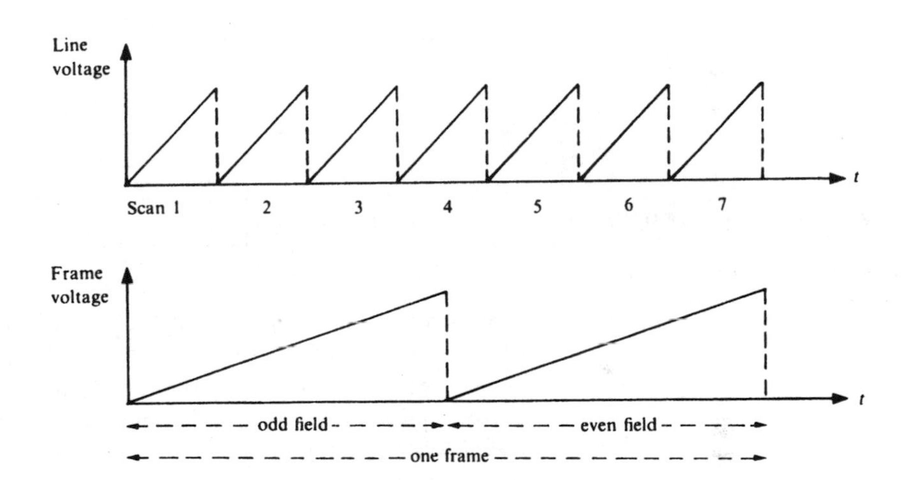

Figure 6.7. Line and frame timebases for a seven-line raster.

Line and field synchronization

Synchronization of the video signal with the line timebases is achieved with *sync pulses* as shown (simplified) in Fig. 6.8. Negative line-sync pulses trigger the line timebase generator at the beginning of each line scan. Similar field-sync pulses trigger the field timebase generator at the beginning of each field. More details of sync pulses appear in Fig. 6.14.

In monochrome TV, the image is divided into picture elements (pels) by individual photocells. The luminance at each pel is proportional to the

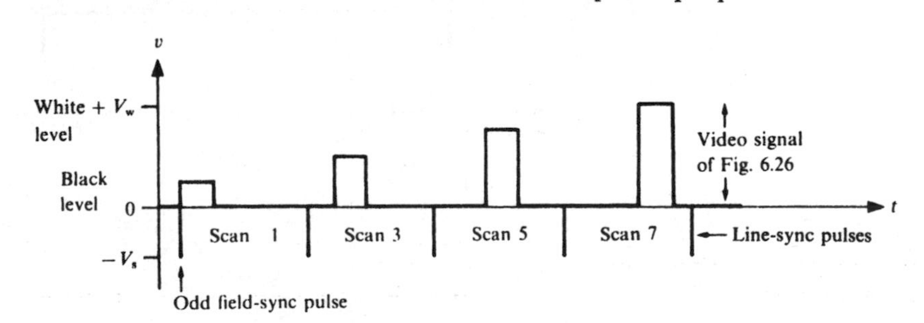

Figure 6.8. Simplified sync pulses for video signal of Fig. 6.26.

amplitude of the video signal produced in the camera tube, black being represented by zero voltage and the lighter shades of grey by higher voltages. The baseband video signal consists ofanalogue luminance information together with sync pulses. The video and sound signals are transmitted by two separate systems sharing the same aerial as shown in Fig. 6.9.

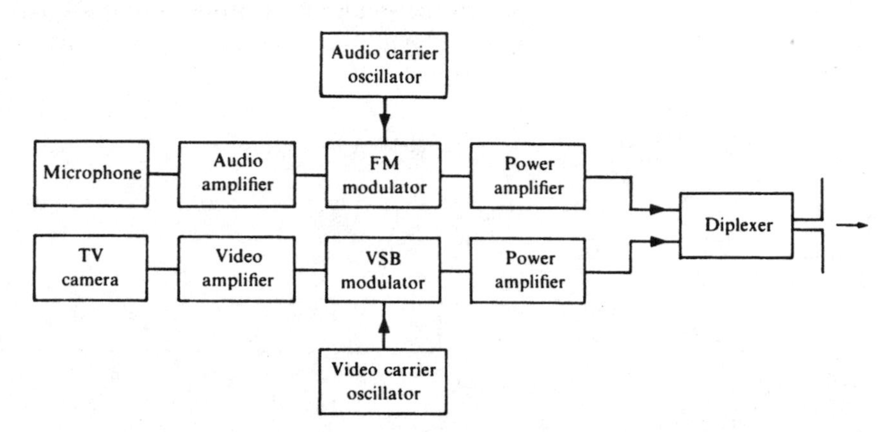

Figure 6.9. A TV transmitting system.

We shall see, later in the chapter, that 625-line TV requires a baseband from 0 to 5.5 MHz for acceptable picture resolution. Since a zero-frequency component must be transmitted, SSB-AM cannot be used, DSB-AM would require a transmission bandwidth of 11 MHz, so the LSB is reduced from 5.5 to 1.25 MHz to give a vestigial LSB and a reasonable compromise between picture quality and bandwidth economy. The audio signal is frequency-modulated onto a carrier (f_{ac}), which is 6 MHz above the video carrier frequency (f_{vc}). The audiofrequency baseband is 30 Hz to 15 kHz, with a maximum frequency deviation of 50 kHz (for FM radio this is 75 kHz), giving an acceptable SNR and the spectrum of Fig. 6.10.

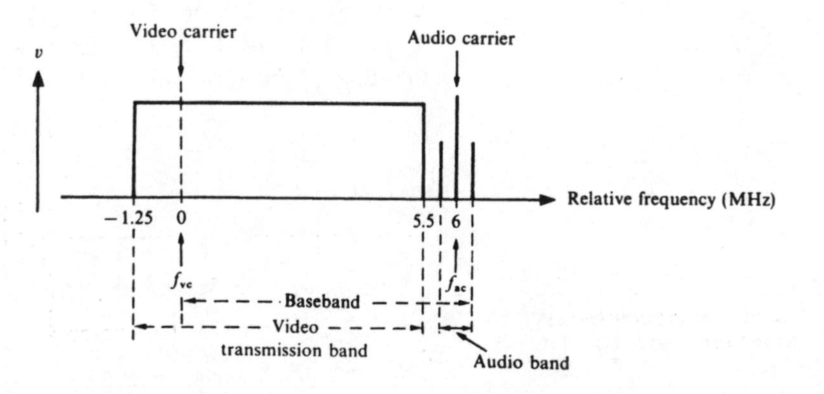

Figure 6.10. The 625-line monochrome TV spectrum.

Exercise 6.4
Check on page 144

Calculate the overall bandwidth of 625-line TV transmission.

Vestigial-sideband demodulation

Detection of VSB-AM signals is accomplished by the same methods as those used for DSB detection. However, the receiver frequency response must compensate for the asymmetry of the VSB signal as shown in Fig. 6.11.

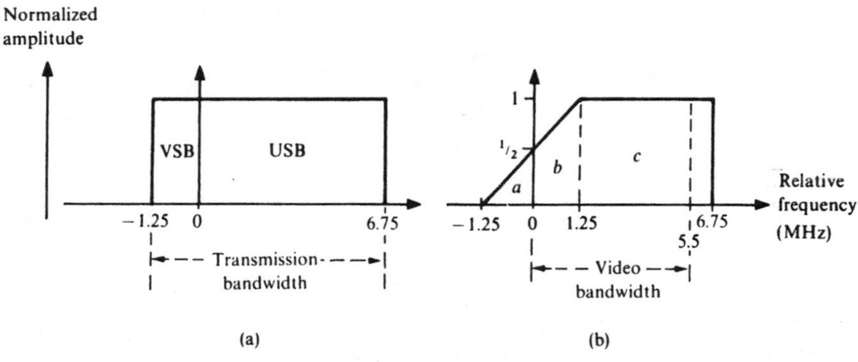

Figure 6.11. Transmitter and receiver responses for VSB TV video signals: (a) transmitter response; (b) receiver response.

If the IF amplifier response is shaped as shown in Fig. 6.11, the video detector will output a constant-amplitude response from 0 to 6.75 MHz.

TV with digital sound

In order to transmit Hi-Fi stereo sound, the BBC has developed a digital system using *near-instantaneously companded audio multiplex* (NICAM), which is a bandwidth-efficient form of *pulse code modulation* (PCM). PCM is described in Chapter 7. The stereo sound signal is transmitted on a channel in the frequency gap between the FM carrier and the adjacent TV channel as shown in Fig. 6.12.

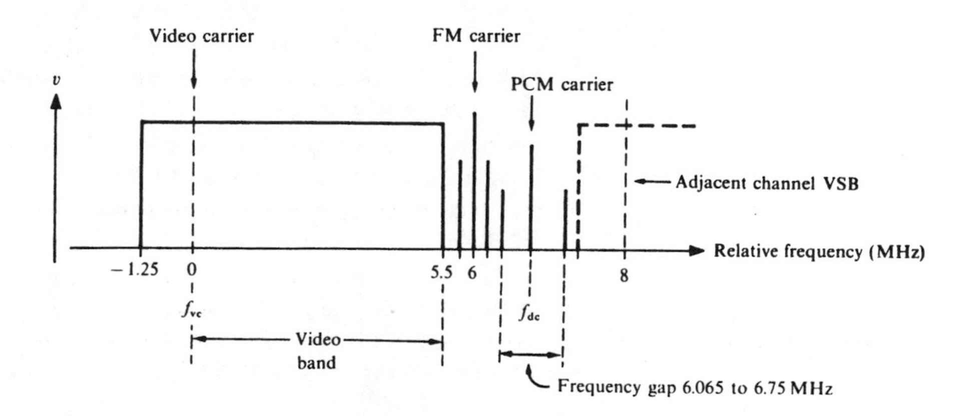

Figure 6.12. The 625-line NICAM monochrome TV spectrum.

The frequency gap from 6.065 MHz to 6.75 MHz accommodates a digital audio signal on a carrier frequency (f_{dc}) at 6.552 MHz above the video carrier and thus gives a system that is compatible with 625-line TV using FM analogue audio signals.

6.3 The action of a monochrome receiver tube

The receiver tube converts a luminance signal into an image using a CRT, similar to the camera tube, but with a *phosphor-coated screen* instead of a photocell screen (Fig. 6.13). The phosphor dots are the transducers that convert electrical energy into light. The spatial information in the signal is contained in the relation between the luminance signal and the sync pulses. These pulses are used to build up the raster by triggering line and field timebase voltages generated in the receiver.

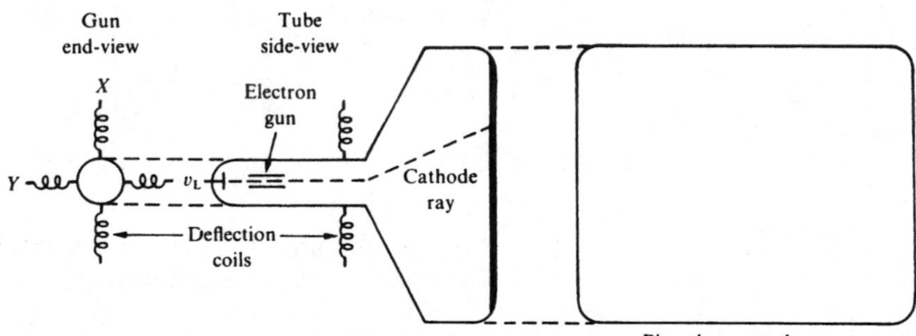

Figure 6.13. The TV receiver tube.

Electrons are formed into a ray by an *electron gun* and focused on to the phosphor dots, which glow when hit by the electrons. The brightness of the glow depends on the energy of impact, which, in turn, depends on the voltage between the gun and the phosphor screen. This voltage is varied by the video signal voltage applied between the *grid electrodes* and the cathode in the electron gun.

In order to create sufficient energy to excite the phosphor, the electrons are accelerated towards the screen by a voltage difference of about 15 kV between the screen and the cathode. The point on the screen where the electrons give up their energy is controlled by a magnetic field produced by coils, which deflect the cathode ray, as it passes through them, as in the camera tube (Fig. 6.5).

The luminance signal is applied only while the electron beam is scanning from left to right: during line and field flyback the beam intensity is held at the black level. This is done with *blanking pulses* as shown in Fig. 6.14. A relatively long field blanking period allows plenty of time for the beam to return to the top of the screen to start another field. This period, which is equal to 25 line scans ($25 \times 64 = 1600\ \mu s$), is devoid of picture information and is used to transmit *Teletext* data (see Chapter 11). This means that the actual picture in view is made up of $625 - 50 = 575$ lines.

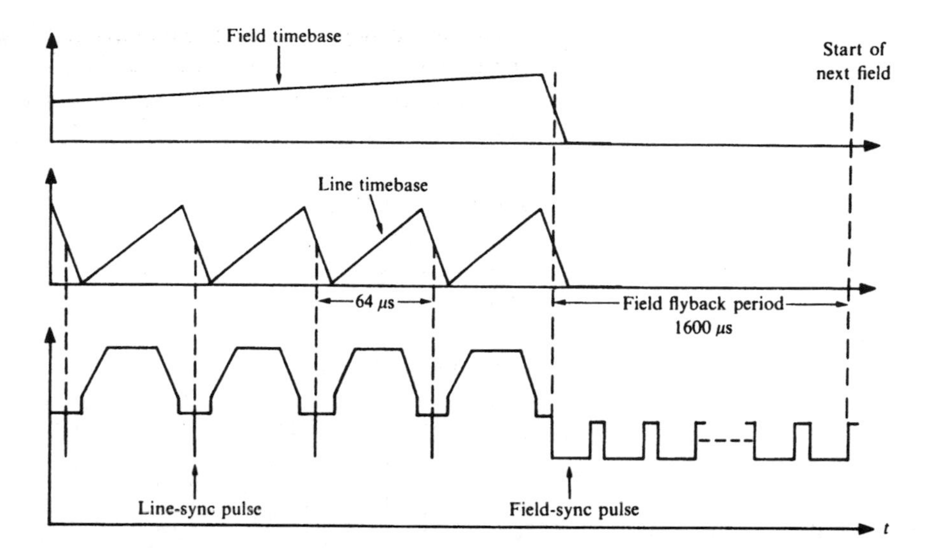

Figure 6.14. Timebase and video waveforms at end of field.

The sync pulses, being of opposite polarity to the luminance signal, are easy to separate from this signal. The field pulses are then separated from the line pulses by applying them to an integrator, which builds up a field-flyback blanking pulse from the relatively wide field-sync pulses. The trailing edge of this blanking pulse is used to trigger the field timebase circuit.

Synchronization pulse details

The line flyback is triggered by the leading edge of a sync pulse, which, in order to achieve precise timing, should have a short risetime (t_r). A t_r of 0.2 μs is specified for 625-line TV, giving the timings shown in Fig. 6.15. The visible line period is about $64-12=52$ μs. A sync-pulse

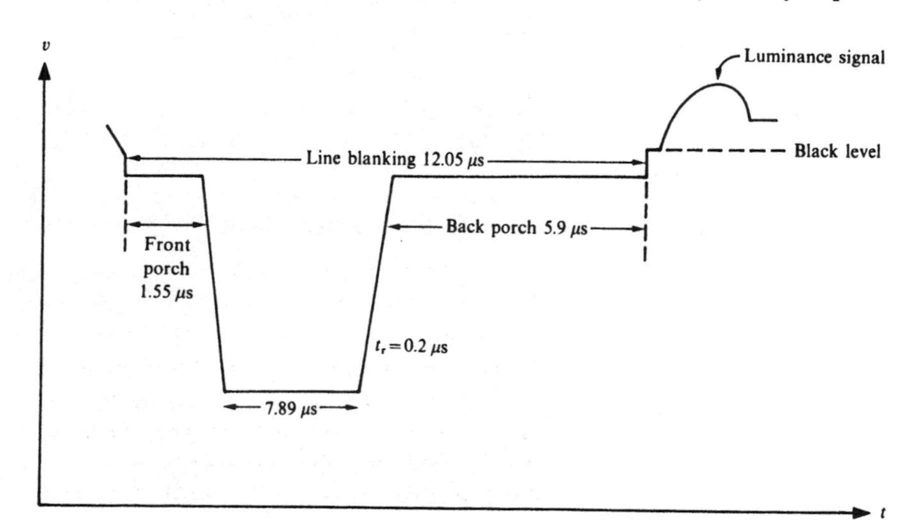

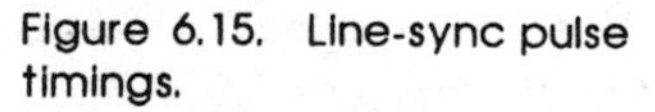

Figure 6.15. Line-sync pulse timings.

risetime of 0.2 μs implies a period of 0.4 μs (a frequency of about 2.5 MHz).

The 625-line TV transmissions in the UK amplitude-modulate a video carrier so that an increase in brightness decreases the signal power. This means that the envelope of the AM video signal of Fig. 6.14 will be as shown in Fig. 6.16. This arrangement is called *negative AM* and it is used because it produces black spots from interference, which are less objectionable than the white spots that would be produced using positive AM.

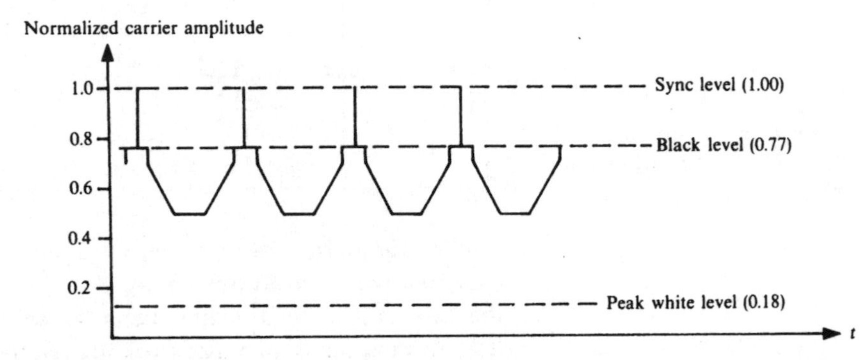

Figure 6.16. AM carrier envelope.

The bandwidth needed for a 625-line TV picture

The bandwidth of a TV channel depends on the number of pels transmitted per second. The number of pels per line determines the horizontal resolution, and this should be similar to the vertical resolution, which is fixed by the 575 displayed lines. The aspect ratio (width to height) of 625-line TV is 4:3, so the number of pels per line is about $575 \times 4/3 = 767$, giving a pel transmission rate of 767 pels per $52\ \mu s = 14.75$ Mpel/s. The maximum rate of change of the luminance signal will occur when adjacent pels alternate between black and peak white, that is, when there are $14.75 \times 10^6/2$ alternations of the signal per second. Thus the bandwidth required is about 7.38 MHz. In practice, a reduction in horizontal resolution can be tolerated, allowing the video bandwidth to be restricted to 5.5 MHz. This corresponds to $767 \times 5.5/7.38 = 572$ pels per line.

6.4 The monochrome TV receiver

The VSB-AM UHF signal is selected using the superhet principle, as shown in Fig. 6.17, using a local oscillator frequency 39.5 MHz above the required carrier frequency. The AM output from the IF amplifier is demodulated with a simple diode envelope detector. The recovered composite baseband signal (shown in Fig. 6.10) is amplified and the video signal routed to the cathode of the CRT via a sync separator. The sound IF amplifier, tuned to 39.5 MHz, extracts the FM sound signal, which is then demodulated and amplified as shown in Fig. 6.17.

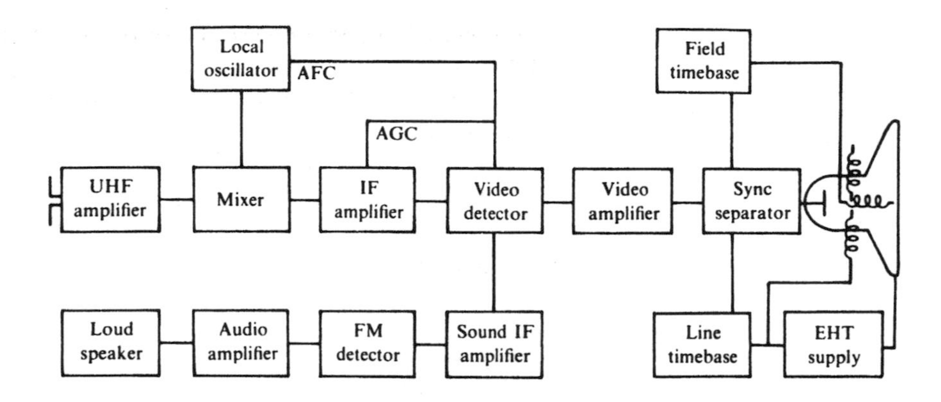

Figure 6.17. A superheterodyne monochrome TV receiver system.

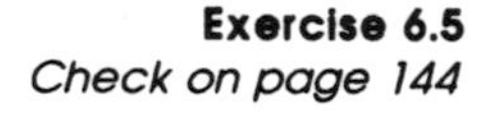

Exercise 6.5
Check on page 144

Refer to Fig. 6.10 and draw the IF spectrum of a 625-line TV receiver that this figure would suggest.

In practice the energy distribution in the video spectrum is not uniform but tends to be greater at the lower baseband frequencies, that is, near the IF. Also, the IF amplifier must have an asymmetric response to compensate the asymmetry of the transmitted VSB-AM signal spectrum.

The sync separator feeds the luminance signal to the cathode of the TV tube and the line-sync and field-sync pulses to the appropriate timebase generators. The TV tube requires an *extra-high-tension* (EHT) supply of about 15 kV to achieve high brightness. This is provided by half-wave rectifying the 15.625 kHz line timebase output whose current changes very rapidly during flyback and so is capable of producing the necessary EHT when applied to a step-up transformer.

6.5 Colour television

To obtain a coloured picture, different-coloured phosphors must be used. Reproduction of a coloured image on a CRT requires the use of *additive colour mixing*, which depends on the fact that red (R), green (G) and blue (B) when mixed together in various proportions produce all the many hues required. For example, a shade of yellow might be 50% R and 50% G; a magenta might be 40% R, 20% G and 40% B. An acceptable white can be obtained with 30% R, 59% G and 11% B. The effect is *not* the same as that used in painting, where subtractive colour mixing is used, giving different hues to those obtained in additive mixing.

A colour TV signal must contain *luminance* information, which can be used by a monochrome receiver so that broadcasts are compatible with both monochrome and colour receivers. Colour information or *chrominance* is also carried by the signal, and this information must be routed to the appropriate electron gun of the colour TV receiver tube. This is a

CRT with R, G and B patterns of phosphor dots forming the screen. All the R dots are illuminated by a 'red electron gun', the G dots by a 'green gun' and the B dots by a 'blue gun'. This is the basis of the *shadow-mask* tube shown in Fig. 6.18.

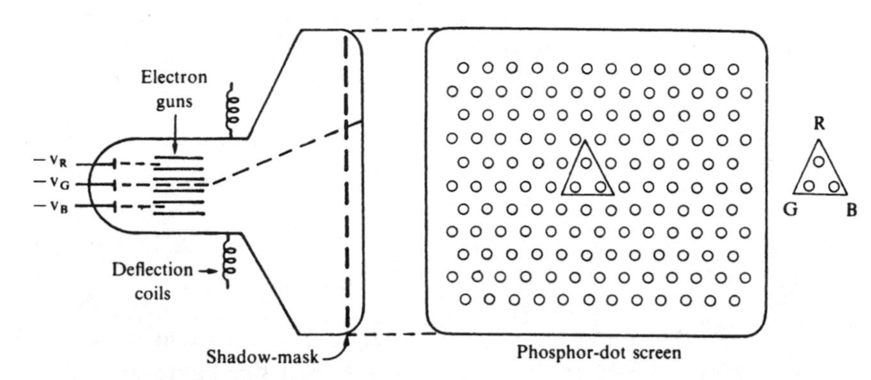

Figure 6.18. The principle of the shadow-mask tube.

The phosphor dots, arranged in RGB colour triangles, are so small and densely distributed on the screen that the appearance, at normal viewing distance, is that of a smooth mix of colours when the dots are energized. A shadow mask, having holes which line up with the phosphor dots, ensures that the electrons from the G gun impinge on the G dots only and similarly for the R and B dots. As the intensities of the beams from the three guns are varied, so the hue changes.

Colour TV systems

Colour TV systems use the same principles as monochrome systems, but there are variations in the way the colour information is processed. There are three principal colour systems operating in the world: the *National Television Standards Committee* (NTSC) system, invented in the USA and now used throughout the North American continent; the *Phase Alternation Line-by-line* (PAL) system, which was developed from the NTSC system in West Germany and adopted in the UK and most of Europe; and the *Séquential Couleur à Mémoire* (SECAM) system, which is used in France and the USSR. NTSC is a 30-frame-per-second, 525-line system: the others use 625 lines and 25 frames per second, the frame rates being based on the national mains frequency.

The chrominance signal

A TV camera consists of an optical system that resolves the original scene into red, green and blue images, which are separately focused on to three camera tubes, of the kind shown in Fig. 6.5. The photocell screens of these are scanned synchronously and signal voltages, v_R, v_G and v_B, are produced. These voltages could be transmitted as three separate video signals, but this would require a bandwidth of 3×5.5 MHz and would not give a monochrome-compatible signal. Instead, a compatible luminance signal and two chrominance signals are sent. In the PAL system, these three composite signal voltages are formulated according to the equation

$$v = xv_R + yv_G + zv_B \tag{6.1}$$

The luminance signal is given by

$$v_L = 0.30v_R + 0.59v_G + 0.11v_B$$

because this mix gives an acceptable white. If any two of the three colour voltages are transmitted with v_L, then the third can be derived by the receiver.

In practice, use is made of the feature that the chrominance information can be expressed by any two of the three colour difference signals: $v_R - v$, $v_G - v$ and $v_B - v$. This can be verified as follows. Rearranging Eq. (6.1) gives

$$v_G = v/y - xv_R/y - zv_B/y$$

Subtracting v from both sides of the equation gives

$$\begin{aligned} v_G - v &= (1-y)v/y - xv_R/y - zv_B/y \\ &= (x+z)v/y - xv_R/y - zv_B/y \\ &= (v - v_R)x/y + (v - v_B)z/y \end{aligned}$$

Thus, in this example, the green difference signal has been expressed in terms of the other two difference signals. There are three chrominance signals, but if the particular value of v transmitted is the luminance signal v_L, only two chrominance signals need be transmitted. It so happens that the green difference signal has a lower mean-squared value than the other two, so the greatest SNR is obtained by transmitting the red and blue difference signals in preference to the green difference signal.

The chrominance signal $v_{CR} = v_R - v_L$ is carried on a cosine carrier while the signal $v_{CB} = v_B - v_L$ is carried on a sine carrier, both using DSB-SC-AM. The two carriers are added to produce a *quadrature amplitude-modulated* (QAM) signal, given by

$$v_C = v_{CR}\cos(\omega_{Cc}t) + v_{CB}\sin(\omega_{Cc}t)$$

which can be synchronously detected to give v_{CR} and v_{CB}. QAM is also used in data transmission where it is called QASK (see Sec. 4.6). The colour information can then be obtained using the luminance signal v_L and a decoder as shown in Fig. 6.19.

The local oscillator, which supplies the synchronous chrominance carrier frequency (ω_{Cc}) needed for coherent detection, is synchronized to the trans- mitted chrominance carrier frequency by 10-cycle *colour bursts* of ω_{Cc}. These occur during the back porch of each line-sync pulse as shown in Fig. 6.20.

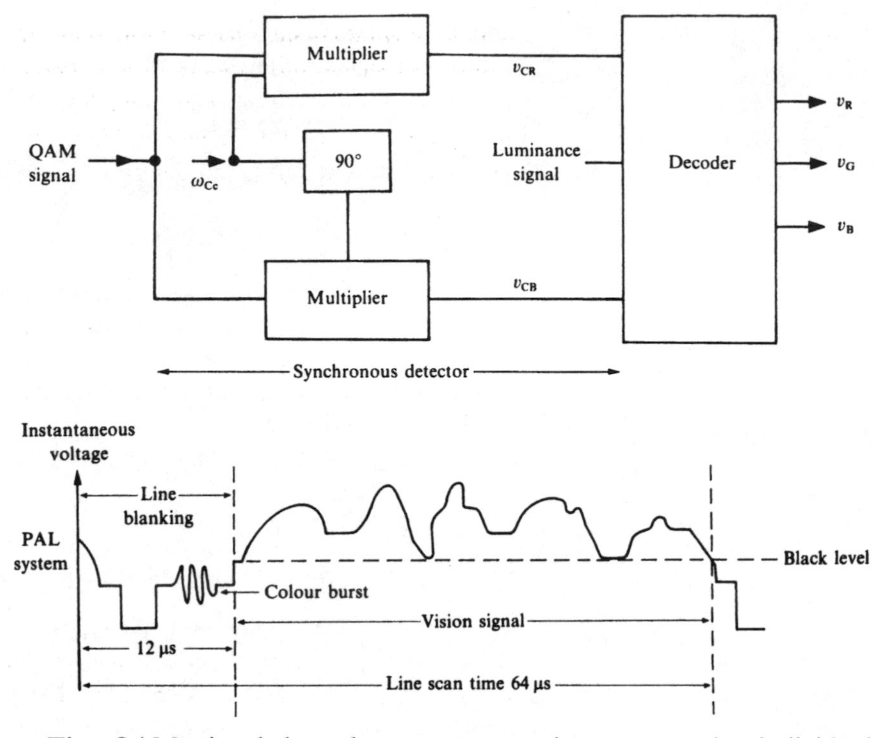

Figure 6.19. Recovery of chrominance information from QAM signal.

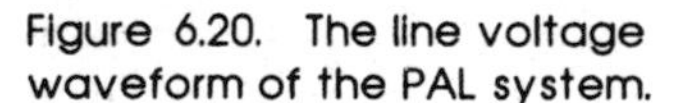

Figure 6.20. The line voltage waveform of the PAL system.

The QAM signal has the same spectral range as the individual DSB-SC-AM signals, but the problem is to accommodate the chrominance signals within the 5.5 MHz monochrome bandwidth. The solution to this problem lies in the fact that the luminance signal does not completely occupy the bandwidth allocated to it. This is because this signal is effectively transmitted in samples at the line scan frequency (f_s) of 15.625 kHz. This means that upper and lower sidebands of the luminance signal, centred on harmonics of the line (or sampling) frequency, will be transmitted. Sampling is described in Sec. 7.3. Since most of the energy of the luminance signal is concentrated towards the lower end of the baseband, most of the USB and LSB energy is concentrated near the harmonics of the line frequency, leaving gaps (shown in the spectrum of Fig. 6.21) that contain very little luminance information.

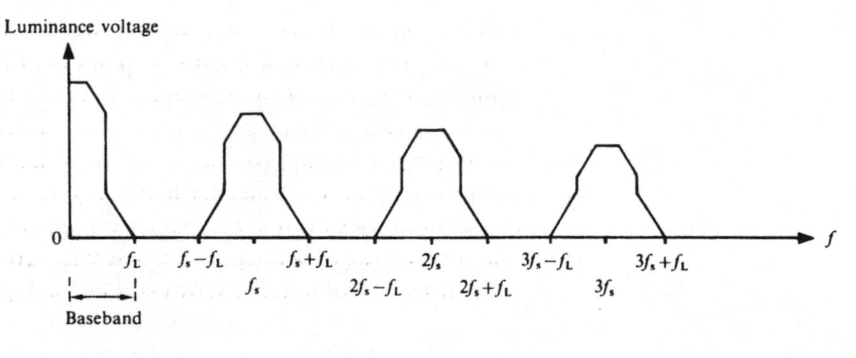

Figure 6.21. Luminance signal spectrum.

The chrominance carrier frequency has been chosen so that the modulated chrominance signal spectral components coincide with the gaps in the luminance signal spectrum. The baseband luminance and chrominance signals may thus be combined, as shown in Fig. 6.22, to form a composite, monochrome-compatible colour video signal using the monochrome bandwidth.

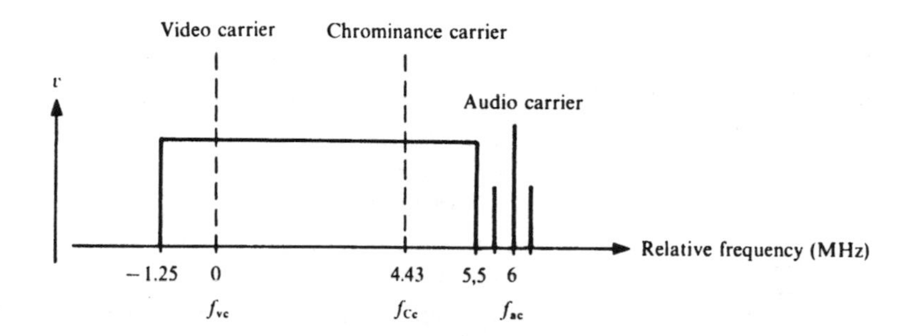

Figure 6.22. The 625-line colour TV spectrum.

The eye can accept inferior resolution of colour compared with luminance. This means that the chrominance bandwidth can be restricted to about 1 MHz, so f_{Cc} is set to the f_s harmonic nearest to 4.5 MHz (refer to Fig. 6.22). That is the 4.5 MHz/15.625 kHz = 29th harmonic. This would give $f_{Cc} = 29 \times 15.625$ kHz = 4.53 MHz. The PAL system actually uses a relative chrominance carrier frequency of 4.43 MHz because this has been found to minimize carrier phase error, which is caused mainly by phase distortion in the transmission channel.

Phase distortion of the carrier gives rise to colour variations. The PAL system corrects this by reversing the phase of one of the chrominance signals on alternate line scans—hence the name of the system, 'phase alternation line-by-line'. The SECAM system also transmits the colour-difference signals sequentially on alternate line scans, but uses frequency modulation of the chrominance carrier.

6.6 Satellite television

Satellites and TV were first combined in 1962 when the satellite 'Telstar' was used to relay programmes around the world. The BBC and the private television companies are restricted in their number of programmes because the UHF transmitters, with their inherently small service area, have used up most of the allocated bandwidth. More channels will become available using *direct broadcasting by satellite* (DBS), in which a satellite-based *transponder* relays programmes, sent to it from a ground-based transmitter, to TV receivers within a specified service area (see Fig. 11.37).

Unlike terrestrial TV broadcasting, DBS tends to provide coverage over almost a whole hemisphere. Such wide coverage is generally unnecessary and wasteful of energy. However, achieving coverage over a specified area is difficult since it requires the use of very directional antennae with particular ground-cover patterns or *footprints*. This involves the use of aerial arrays using complex feed techniques, which increase the payload of the satellite but have the advantage of allowing the footprint to be varied from the ground when service area requirements change. The general advantages of DBS are such, however, that the technical difficulties are being overcome, and this seems to be how broadcast TV will develop.

A ground-based antenna (dish) is permanently focused on the satellite and the signals received by the satellite are amplified, lowered in frequency by a transponder on board and transmitted back to the service area using a satellite-based dish, which continuously points to the required service area.

Small and directional dishes imply the use of microwave signals. The allocated wavelengths used correspond to carrier frequencies in the range 12 to 14 GHz. In order to allow the transmitter and receiver dishes to be stationary, the satellite is placed in a *geostationary* orbit (that is, one in which it has no movement relative to the Earth).

FM has a SNR advantage over AM, which is obtained at the expense of bandwidth (see Sec. 8.3). This has led to the adoption of FM for satellite TV. In satellite communications the prime consideration is to minimize the satellite transmitted power. At microwave carrier frequencies, bandwidth can be readily traded for power reduction to maintain an acceptable SNR.

The use of FM for the video as well as the audio signal has enabled improvements to be made to the present AM/FM TV systems. In the UK a PAL-compatible system has been used. It is a *multiplexed analogue component* (MAC) system, called *D-MAC*, in which digitally encoded sound and data are combined with the analogue video signal. PAL and D-MAC line signals are compared in Fig. 6.23.

The transmitted signal consists of a time division multiplex (TDM) of three components, two analogue and one digital burst of data, which carries the sound plus data plus sync information. The clamp period T_c is a break in transmission used by the receiver for clamping the digital component.

Each of the chrominance components is transmitted on alternate lines with each luminance component. The luminance component is carried on lines 24 to 310 and 336 to 623 together with alternate chrominance components (which are also carried on lines 23 and 335). The luminance bandwidth of 5.6 MHz is virtually the same as that of PAL, but the chrominance bandwidth is increased from 1 MHz to 2.8 MHz to increase the colour SNR and hence the colour quality. In the PAL system the luminance and chrominance components each occupies 52 μs. Squeezing them into 34 μs and 18 μs respectively necessitates time compressions of 52:34 and 52:18 (about 3:2 and 3:1) before transmission. A decoder in the TV receiver expands these components and also processes the digital information.

The digitized sound (baseband 40 Hz to 15 kHz) is sent with the line and

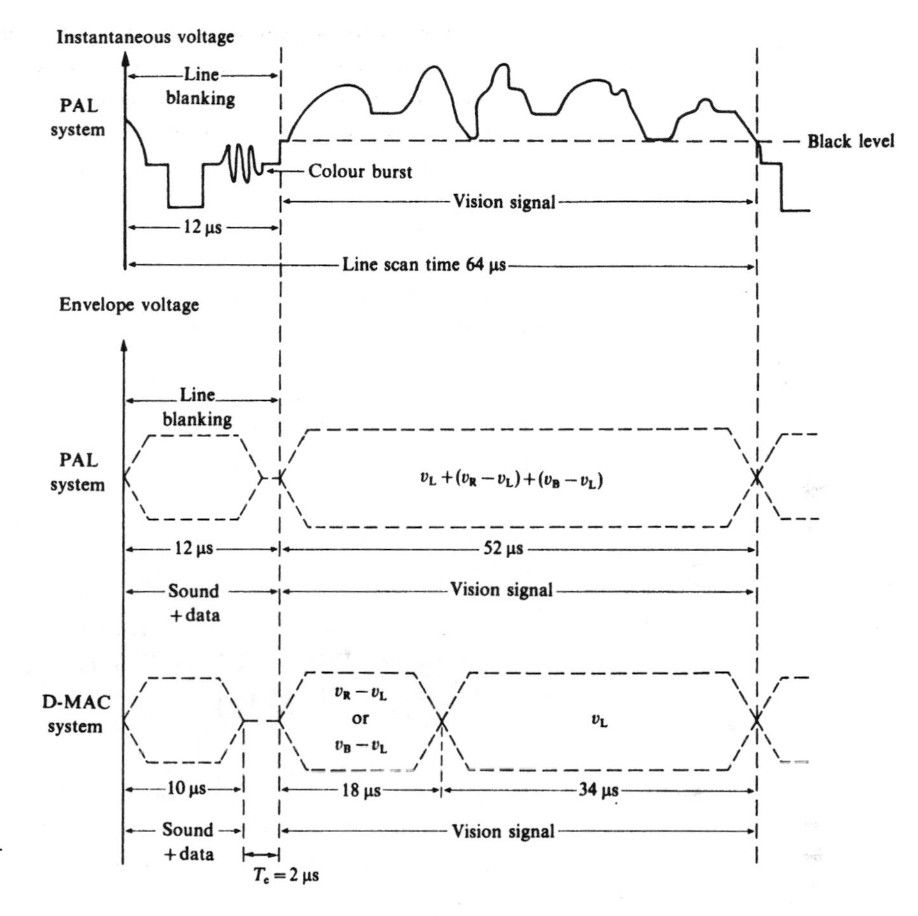

Figure 6.23. Line voltage waveforms of the PAL and D-MAC systems.

frame data and teletext (and much other information if required) in a 10 µs pulse burst using QPSK (see Sec. 4.6). The 52 µs of unblanked line consists of about 572 pels, giving a maximum bit rate of $572 \times 2/52\ \mu s = 22$ Mb/s. In practice a bit rate of 20.25 Mb/s is adopted for the digitized information.

Each line carries a 10 µs burst of data, which constitutes one data *packet* (see Sec. 11.4). The data in these packets are assembled at the end of each frame to make up the sound, data and sync information. The packet format is shown in Fig. 6.24. The amount of useful data carried per frame is about $728 \times 625 = 455$ kb, which can be used to provide a range of services such as several high-quality sound channels, multi-lingual voice options and thousands of pages of teletext.

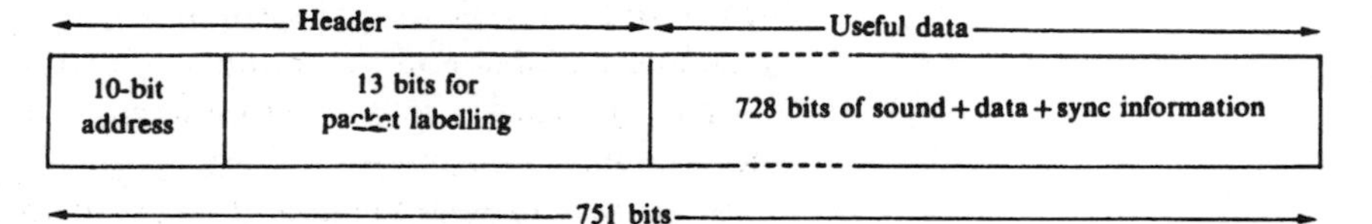

Figure 6.24. The MAC packet format.

Exercise 6.6
Check on page 144

1. The sampling rate in D-MAC is 32 kHz. Suggest why this rate has been chosen.
2. How many of the 728 data bits are needed for a sound dynamic range of just over 80 dB?

A satellite TV receiver system

Satellite broadcasts can be received on any PAL receiver by using a dish antenna and a D-MAC decoder as shown in Fig. 6.25. The signal power received by the dish is so small that low-noise SHF amplification must take place at the focus of the dish in order to establish an acceptable input SNR at the D-MAC interface. Signal gain is further increased by down-frequency conversion and first IF amplification before the signal is fed down from the dish to the interface.

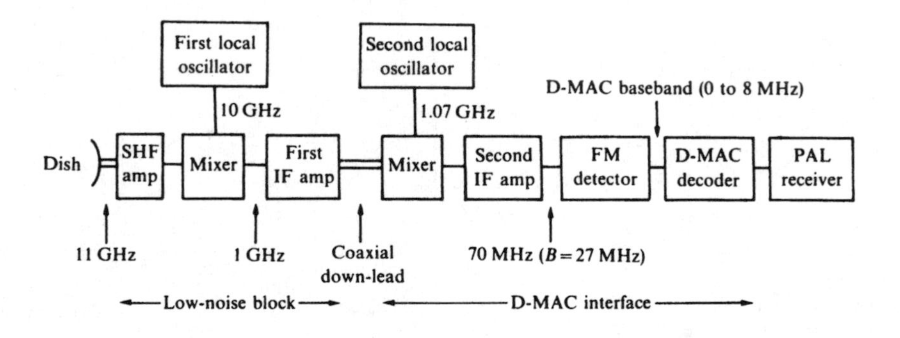

Figure 6.25. A satellite TV receiver system.

Summary

A TV broadcasting transmitter works within a range of frequencies allocated to its service area.

A TV picture is created by the CRT of a TV receiver, which obtains its signal from a TV camera. The original image is optically formed on the photocell-matrix screen of a TV camera tube, which is scanned by an electron beam. Scanning produces a signal, which ultimately creates the original image on the phosphor screen of a CRT. The scanning process is reproduced in the receiver CRT, using line and frame timebase waveforms that are synchronized to those of the TV camera. Sync pulses are combined with a luminance signal to form a video signal. Moving pictures are obtained by rapid, repeated scanning and interlacing.

Colour TV makes use of chrominance as well as luminance information carried by the video signal. Closely packed triangles of red, green and blue phosphor dots in a receiver CRT allow many hues to be produced by additive colour mixing. Three separate electron guns are used, each giving a beam of electrons that, by means of a shadow-mask, can only activate phosphor dots of one colour. The individual dots are so small and numerous that the appearance from a distance is that of a mix of colours creating the required hue. The hue changes as the intensity of the ray from each gun varies.

In the PAL system, the video signal carries the colour information on a DSB-SC-AM chrominance carrier in the form of two colour difference signals. This chrominance signal is incorporated with the 5.5 MHz baseband luminance signal to form a composite, monochrome-compatible signal, which is transmitted on a VSB-AM UHF carrier at an allocated bandwidth of 8 MHz per channel. The composite video signal is synchronously detected using the 10-cycle colour burst of chrominance carrier, which is transmitted during the back porch of each line-sync pulse. The recovered baseband signal is decoded to extract the luminance and chrominance information.

TV broadcast systems used for entertainment are accompanied by sound. In the PAL system, monophonic audio information is sent on a frequency-modulated audio carrier, which is added to the high-frequency end of the composite baseband video signal. A stereophonic sound system is available in which the audio signal is squeezed into the existing 8 MHz channel using digitally encoded sound and near-instantaneously companded audio multiplexing (NICAM).

More channels of potentially higher quality are made available if frequency-modulated microwave carriers are used. These require satellite-based relay equipment to provide large service areas.

TV may be used for data transmission as well as entertainment, with broadcast pages of data (Teletext) being available in the UK as Ceefax or Oracle. The TV receiver may also be used as a computer monitor or VDU.

► Checks on exercises

6.1 The advantage obtained by transmitting vision with AM and sound with FM is that the sound FM receiver in the TV set does not respond to the picture signal nor does the picture AM receiver in the same set respond to sound signals. There is no mutual interference between sound and vision unless the receiver is mistuned.

6.2 The video signal that would produce a white diagonal line running from the top left-hand corner of the screen to the bottom right-hand corner gradually increasing in whiteness would be as shown in Fig. 6.26.

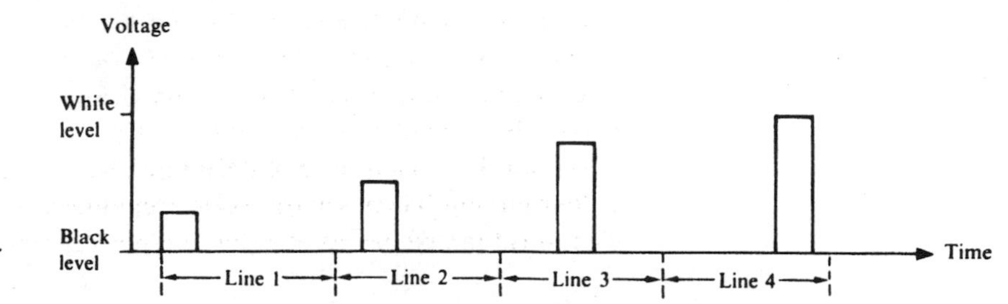

Figure 6.26. Video signal for diagonal line image.

In order to obtain the illusion of picture movement, the scanning process must be repeated in rapid succession with each of the video voltage pulses in Fig. 6.26 being advanced or delayed, according to the direction of movement, in each successive frame. In practice, to get the appearance of movement without undue flicker, the eye needs to see 40 to 50 frames per second at least. At this frame frequency rate, persistence of vision ensures a smooth-moving picture.

6.3 The raster formed by the timebase voltages of Fig. 6.7 will look something like that shown in Fig. 6.27. An odd field starts at A and the beam is deflected to B (scan 1 of Fig. 6.7). Flyback occurs from B to C. The next full lines of the odd field go from C to D (scan 2) and E to F (scan 3) with a

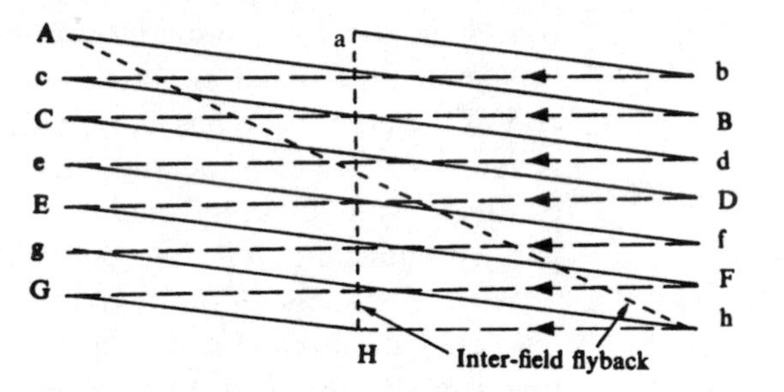

Figure 6.27. Interlaced scanning.

final half-line from G to H (half of scan 4). This completes the first odd field of $3\frac{1}{2}$ lines. The beam is deflected back to begin the next field at a. After the even field is scanned the electron beam is at h, from which point it is returned to A to begin the next odd field scan.

This half-line finish of odd fields and half-line start of even fields requires an odd number of lines for each frame. This ensures that interlacing is achieved with the same field timebase voltage for both odd and even fields. In practice, the deflection coils around the neck of the CRT are twisted until the scan lines appear horizontal on the screen.

6.4 The FM audio bandwidth is given by Carson's rule as $2(50+15)=130\,\text{kHz}$, so the bandwidth of a 625-line TV transmission will be $5.5+1.25+0.13=6.88\,\text{MHz}$. This easily fits into the allocated channel bandwidth of 8 MHz.

6.5 The IF spectrum will consist of sidebands with a video carrier at $f_{if}=f_{lo}-f_{vc}=39.5\,\text{MHz}$.

The audio carrier (f_{ac}) will be at $f_{lo}-(f_{vc}+6\,\text{MHz})=33.5\,\text{MHz}$. So, Fig. 6.10 would suggest the IF spectrum shown in Fig. 6.28.

6.6

1. The sampling rate of 32 kHz has been chosen because this gives a rate just above the Nyquist rate of 30 kHz required for a sound frequency up to 15 kHz.

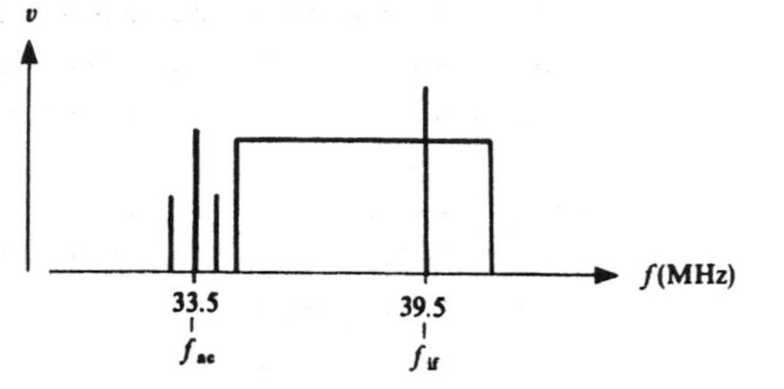

Figure 6.28. A 625-line TV IF spectrum.

2. The number of data bits (n) needed for a sound dynamic range of just over 80 dB is given by $20n \log 2 = 80$. So $n = 4/\log 2 = 14$, giving a range of about 84 dB.

Self-assessment questions

6.1 List the steps required to convert a light image into an electrical signal.

6.2 What is the function of a photoelectric cell?

6.3 Draw the current waveform obtained during a complete scan of the image shown in Fig. 6.29.

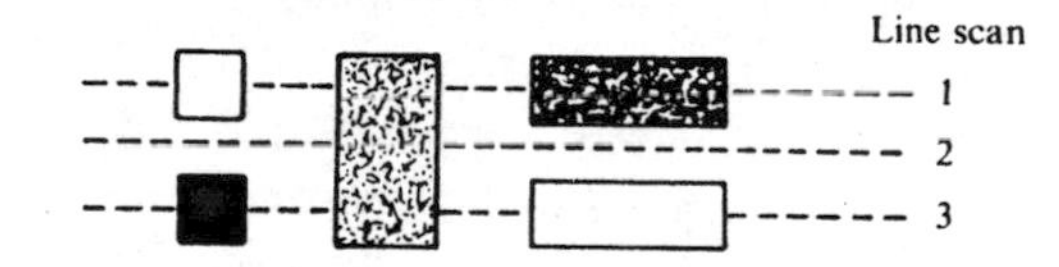

Figure 6.29. A simple image.

6.4 Draw a labelled cross-sectional view of a TV camera tube and the basic circuitry needed to obtain a luminance signal.

6.5 Draw a five-line interlaced raster and the line and frame waveforms required to produce such a scan. Explain why the number of lines used in interlaced scanning is odd.

6.6 Draw the luminance signal for a four-line raster scan so that the whole of the right half of the screen is black and the other half white. Show the line-sync pulses on your waveform.

6.7 Sketch the features of shadow-mask tube that are needed to produce a coloured image.

6.8 Draw the 625-line colour TV UHF and IF spectra.

6.9 How are the chrominance signals formed and carried?

6.10 Why is DSB-SC-AM used to transmit chrominance information whereas VSB-AM is used to transmit the complete TV baseband?

6.11 What feature of the video signal is present to ensure accurate coherent detection?

6.12 How is the chrominance signal accommodated within the luminance frequency band?

6.13 Why does DBS offer more channels of potentially higher quality than terrestrial broadcasting?

7

Pulse carrier and pulse code modulation

Exchanging information using pulse carriers

Objectives

When you have completed this chapter you will be able to:

- Describe the following forms of modulation:
  pulse amplitude
  pulse frequency
  pulse width
  pulse position
- Describe the concept of sampling
- Define the Nyquist frequency and aliasing
- Define quantization error and show how it may be reduced by companding
- Describe how signal quantization may reduce noise in a transmission system
- Outline the action of one form of analogue-to-digital converter
- Outline the action of one form of digital-to-analogue converter
- Describe time-division multiplexing of PCM signals

Chapter 4 described various ways in which a baseband signal may modulate a continuous carrier. The aim of this chapter is to describe how a sequence of pulses may be used to carry signals. The modulating signal varies some property of the carrier pulses, such as their amplitude, frequency, width or position.

A very effective way to convey an audio signal is based on sampling the signal and transmitting binary-coded pulse sequences representing the sampe values. This technique is known as pulse code modulation (PCM). If the sampling rate is greater than double the audio baseband, the audio signal can be recovered from the PCM signal by low-pass filtering. PCM offers great advantages in terms of reduced noise and simple receiver design, and can be time-division multiplexed.

Unlike amplitude or angle modulation, PCM does not produce frequency shifting from baseband to a higher-frequency carrier band nor narrow-banding, both of which are necessary for effective radio transmission. It is therefore used mainly in baseband line communications. It can also be used as an alternative form of baseband signal, which can be frequency-translated by other forms of modulation.

7.1 Pulse time modulation

Pulse time modulation of a rectangular carrier is analogous to angle modulation of a sinusoidal carrier. It may be done in three ways as shown in Fig. 7.1.

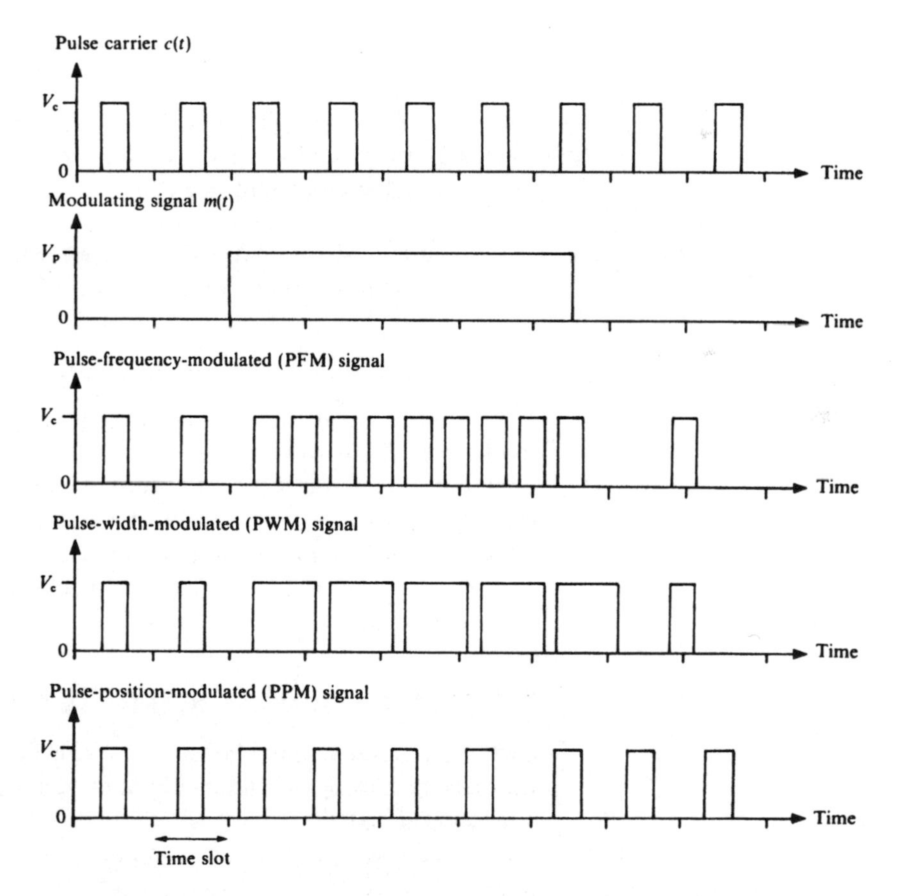

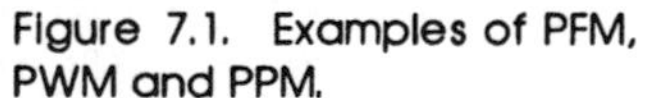
Figure 7.1. Examples of PFM, PWM and PPM.

Pulse frequency modulation (PFM) is, in effect, a hard limited version of sinusoidal-carrier FM. It is used in analogue communication over optical fibres where AM would be degraded by the severe non-linearity in the voltage transfer characteristic of the opto-electric transducers such as LEDs.

In pulse width modulation (PWM) the duration or width of the pulses varies with the amplitude of the modulating signal, and this signal can be recovered by LP filtering. The technique has been used in pulse-carrier TV transmission in optical fibres and in class D audio power amplifiers, where it enables broad-band amplification to take place at very high power conversion efficiency.

In pulse position modulation (PPM) the position of a pulse varies with the amplitude of the modulating signal. In the example of Fig. 7.1, there is no change in pulse position when the modulation is zero, but the pulse positions advance in time when the modulation is positive. A PPM signal may be generated by using a pulse edge of a PWM signal to trigger a monostable multivibrator, which produces a pulse of constant amplitude and width when triggered. Signal recovery is achieved by setting a flip-flop with the unmodulated signal and resetting it with the modulated signal. The result is a PWM signal, which can be converted into amplitude variations with a LP filter. These forms of modulation can be used in TDM providing the time slots are wide enough to allow, for example, the expansion of width in PWM or the pulse shift in PPM.

Exercise 7.1
Check on page 167

Draw the modulated signal waveforms, similar to those of Fig. 7.1, if the pulse modulating waveform is replaced by a single tone.

7.2 Pulse amplitude modulation

Pulse amplitude modulation (PAM) is similar to sinusoidal envelope modulation (DSB-AM), as may be seen from Fig. 7.2.

In Fig. 7.2, a pulse of amplitude V_p changes the magnitude of the carrier amplitude V_c by an amount kV_p, where k is the change in carrier voltage per volt of modulating signal. This form of PAM is analogous to DSB-AM. A more useful form of PAM is the equivalent of DSB-SC-AM.

7.3 Double-sideband suppressed-carrier PAM

Section 4.1 shows that DSB-SC-AM can be achieved by multiplying a modulating tone $v_t = V_t \cos(\omega_t t)$ by a carrier signal $v_c = V_c \cos(\omega_c t)$ to give a modulated signal

$$\tfrac{1}{2} V_t V_c \{\cos[(\omega_c - \omega_t)t] + \cos[(\omega_c + \omega_t)t]\}$$

CW modulation thus causes shifting of a modulating signal (ω_t) to a higher frequency ($\omega_c \pm \omega_t$), assuming $\omega_c \gg \omega_t$.

We now consider the same situation with the sinusoidal carrier replaced with a pulse carrier. That is, v_t is multiplied by v_c as shown in Fig. 7.3. In the system shown in Fig. 7.3 the output sample pulses follow the shape of

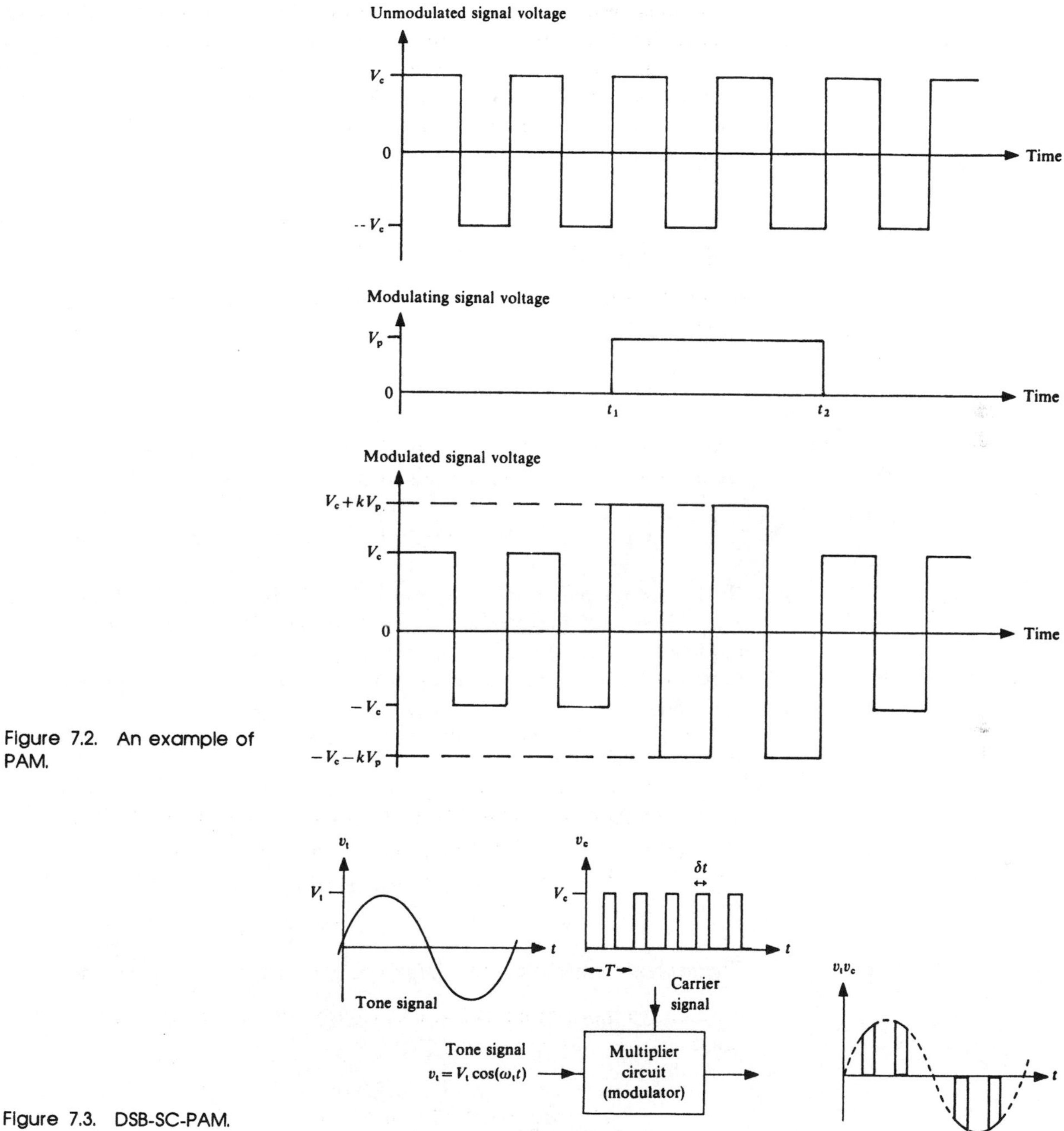

Figure 7.2. An example of PAM.

Figure 7.3. DSB-SC-PAM.

the modulating tone. This is known as *natural sampling*. A simple alternative way to view such sampling is shown in Fig. 7.4. The switch (S) samples the input voltage by making the upper contact. During the remaining time the switch makes the lower contact so that $v_o=0$, assuming that negligible time is spent on changeover.

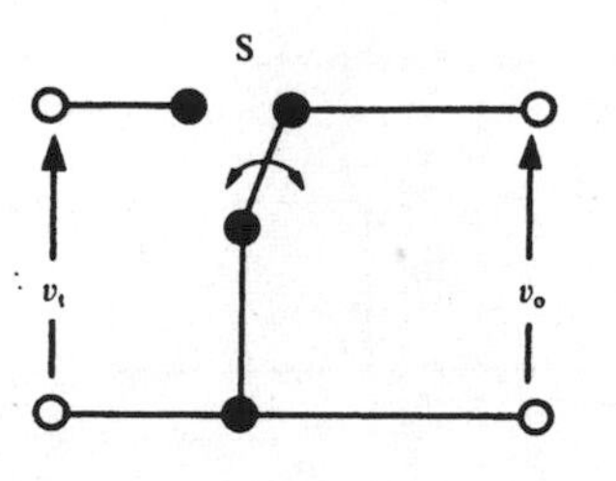

Figure 7.4. Natural sampling.

7.4 A sampling theorem

The Fourier series for a pulse waveform of period T (see Chapter 3) is given by

$$v(t) = c_0 + \sum_{n=1}^{\infty} a_n \cos[2\pi(nt/T)] + \sum_{n=1}^{\infty} b_n \sin[2\pi(nt/T)]$$

where $a_n = H_n \cos\phi_n$, $b_n = -H_n \sin\phi_n$ and c_0 = constant voltage $= \frac{1}{2}H_0$. If we consider the single-sided pulse spectrum shown in Fig. 7.3, we have

$$v(t) = H_0 + \sum_{n=1}^{\infty} H_n \cos(2\pi nt/T) \qquad \text{where } H_n = 2a_n \text{ and } v(t) = v_c$$

If very narrow pulses are used, that is $\delta t \ll T$, then the effect, illustrated in Fig. 7.3, is that the product $v_t v(t)$ produces a sampled version of the modulating tone. It consists of tone amplitude values taken at sampling times of T, $2T$, $3T$ and so on. Over the time interval δt the product is v_t; at all other times the product is zero.

The following analysis yields an important result known as the *sampling theorem*:

$$\begin{aligned} v(t)V_t\cos(\omega_t t) = H_0 V_t \cos(\omega_t t) &+ 2H_1 V_t \cos(\omega_t t)\cos(2\pi t/T) \\ &+ 2H_2 V_t \cos(\omega_t t)\cos[2\pi t/(T/2)] \\ &+ 2H_3 V_t \cos(\omega_t t)\cos[2\pi t/(T/3)] + \cdots \\ &+ 2H_n V_t \cos(\omega_t t)\cos[2\pi t/(T/n)] \end{aligned} \qquad (7.1)$$

The general term of Eq. (7.1), $2H_n V_t \cos(\omega_t t)\cos[2\pi t/(T/n)]$, expands to

$$H_n V_t\{\cos[(2\pi/(T/n) - \omega_t)t] + \cos[(2\pi/(T/n) + \omega_t)t]\}$$

or

$$H_n V_t\{\cos[2\pi(nf_s - f_t)t] + \cos[2\pi(nf_s + f_t)t]\}$$

where f_s is the sampling frequency equal to $1/T$. Applying this expansion to the first three terms in Eq. (7.1) will lead to the normalized ($V_t = 1$) frequency spectrum of Fig. 7.5, which has been drawn assuming that $f_s > 2f_t$ and $\delta t \ll T$.

The spectrum of DSB-SC-PAM shown in Fig. 7.5 contains the original

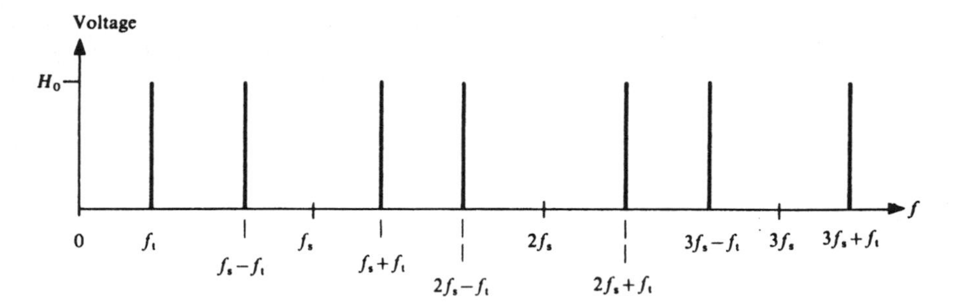

Figure 7.5. Spectrum of DSB-SC-PAM signal carrying a single tone.

tone of frequency f_t, together with upper and lower side frequencies associated with harmonics of the sampling frequency (f_s). If the single tone is replaced by a baseband of frequencies from zero to f_m, all with the same amplitude, the spectrum shown in Fig. 7.5 would be modified to that of Fig. 7.6. Note that in Fig. 7.6 it has been assumed that $(f_s-f_m)>f_m$; that is $f_s>2f_m$.

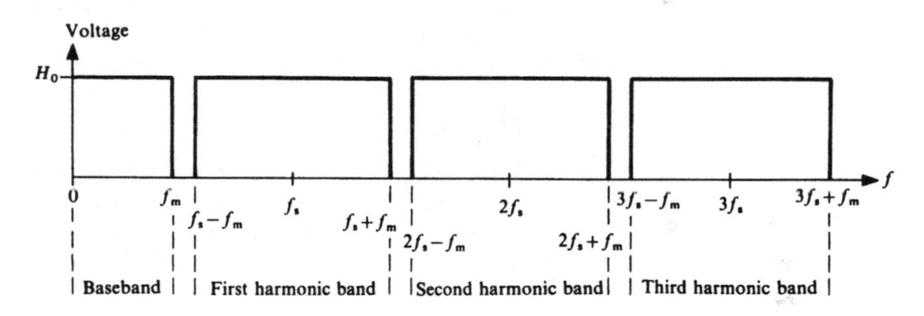

Figure 7.6. Spectrum of DSB-SC-PAM signal carrying baseband 0 to f_m.

The DSB-PAM spectrum consists of the baseband together with upper and lower sidebands centred on the harmonics of the sampling frequency. Notice that the guard band between the baseband and the first harmonic band is equal to f_s-2f_m. If the DSB-SC-PAM signal was passed through an LP filter with a sharp upper cut-off at a frequency within this guard band, the baseband would be recovered and all the harmonic bands rejected. If $f_s \gg 2f_m$ then the guard band will be wide and the filter cut-off need not be so sharp. As $f_s \rightarrow 2f_m$ the guard band tends to zero and practically it becomes impossible to make a filter that will pass the whole baseband while rejecting the harmonics.

The last paragraph describes the concept behind a sampling theorem, which states that the sampling frequency must be greater than twice the highest-frequency spectral component (f_m) of the modulating signal if the modulating signal is to be recovered. That is, $f_s>2f_m$. The critical sampling frequency ($2f_m$) is known as the *Nyquist frequency*.

7.5 Demodulation of PAM signal

Figure 7.6 indicates that the baseband can be recovered by means of a LP filter with a frequency roll-off sufficiently steep to reject effectively all frequencies in the first harmonic band. Just how steep this roll-off must be depends on the separation (the guard band) between the two bands. The guard band increases with sampling frequency, so allowing a cheaper filter to be used.

In the transmission of telephone messages, whose baseband extends from 300 Hz to 3.4 kHz, a sampling frequency of 8 kHz is used, giving a guard band of 1.2 kHz. In compact disc (CD) recording a sampling rate of 44.1 kHz is used for a baseband of 20 kHz, resulting in a guard band of 4.1 kHz. Increasing the guard band to ease the LP filter roll-off requirement means that f_s and, hence, system bandwidth must be increased. Bandwidth is limited in practical lines but in a CD player it is not, so many CD players use *oversampling* to improve performance. In this technique, the sampling frequency of the recording is multiplied, typically four or eight times, in the CD player before LP filtering is applied.

Aliasing

If $f_s < 2f_m$, we have a spectral response with overlapping harmonic bands as exemplified in Fig. 7.7. The LP filter will be unable to reject all the first

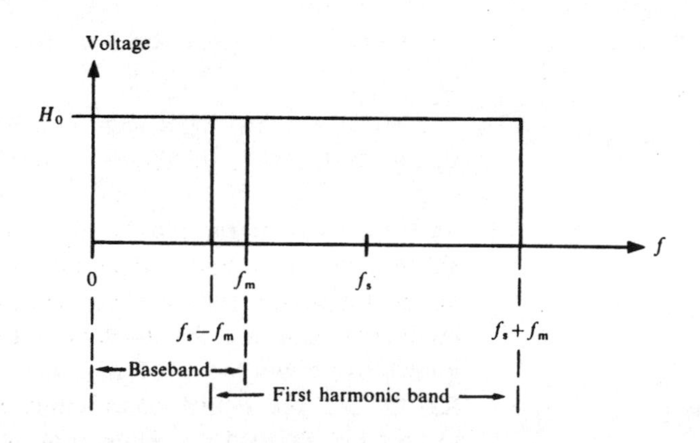

Figure 7.7. Overlapping of baseband and first harmonic bands when $f_s < 2f_m$.

harmonic band. This means that some frequencies, called 'first harmonic band frequencies', will be present with the 'baseband frequencies'. There are thus certain frequencies in the LP filter output that appear twice under different 'names' or aliases, so the recovered baseband will suffer a form of distortion known as *aliasing*.

Exercise 7.2
Check on page 167

Suggest two ways to avoid aliasing.

In order to increase the number of samples that may be interspersed, and hence the number of messages carried over the link by TDM, the sampling period (δt) should be as short as possible. One practical limit on shortness is set by having to have sufficient signal power to obtain an acceptable SNR. A compromise must be made between the SNR and the number of simultaneous signals. In order to reduce δt, while keeping the same signal power, the pulse amplitude may be increased, but there is a practical limit to the size of this voltage.

7.6 Quantization

The modulation methods described so far have been able to represent any value of the modulating signal. In other words, they are forms of analogue modulation. In practice, noise impairs the accuracy of recovery of such signals. If, however, a continuous signal is allowed to assume only certain discrete values, an increase in SNR may be obtained, as will be shown below. This breaking up of an analogue signal voltage into a number of discrete levels (Q_n) is called *quantization*, and an example of how it might be done, for a tone signal, is shown in Fig. 7.8.

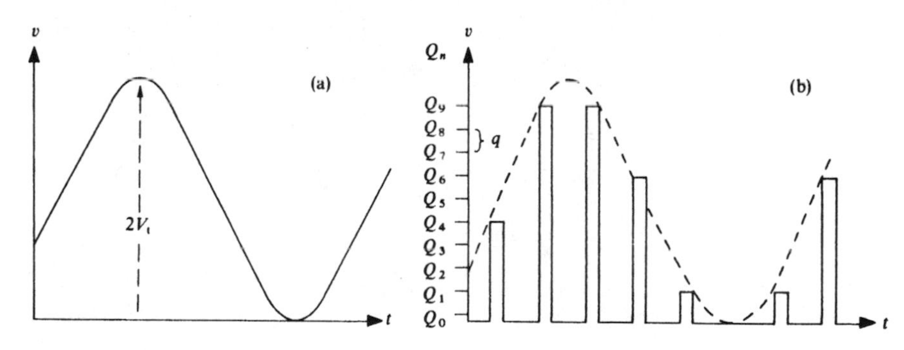

Figure 7.8. Quantization of a sampled tone signal: (a) continuous tone signal; (b) quantized voltage samples of tone signal.

The sampled signal voltage takes a discrete value (Q_n) nearest to its true value during the sampling instant. For example, the tone voltage range is 0 to $2V_t$ and if equal quanta (q) are chosen, then the number of discrete values (n) is equal to $2V_t/q$. In Fig. 7.8, $n=9$. Notice that the samples are now *flat-topped* and not natural as we have assumed until now. However, the sampling theorem still holds for quantized samples because, when $\delta t \to 0$, the difference between natural and flat-topped samples also tends to zero.

Samples provide an approximation to the original signal when a quantized PAM (QPAM) signal is demodulated. The approximation may be improved by reducing q by increasing the number of allowed levels (Q_n). For commercial-quality speech 2^8 levels are used, but 2^{16} levels are needed for Hi-Fi sound. Since the sampled value is being rounded off to the nearest discrete level (Q_n), the sample could be in error by up to $\pm\frac{1}{2}q$. Clearly, this *quantization error* (ε) becomes smaller as more discrete levels are allowed. That is, $\varepsilon \to 0$ as $n \to \infty$.

Exercise 7.3
Check on page 168

1. The audio signals on a CD have their amplitudes quantized at 2^{16} equal intervals of voltage. Calculate the dynamic range in dB obtainable with this number of quanta.
2. If the maximum output voltage of the CD player is 2 V, what voltage is q?
3. What is the maximum possible quantization percentage error for the loudest and quietest sound signals?

Controlling quantization error

We saw in Exercise 7.3 that the quantization error (ε) increases as the signal amplitude decreases. This suggests that, in order to equalize this error over the whole dynamic range (V_d) of a signal, the quanta (q) should be small for weak signals and increase in size with increasing signal amplitude. That is, we want $q=kV_d$. This effect is shown in Fig. 7.9.

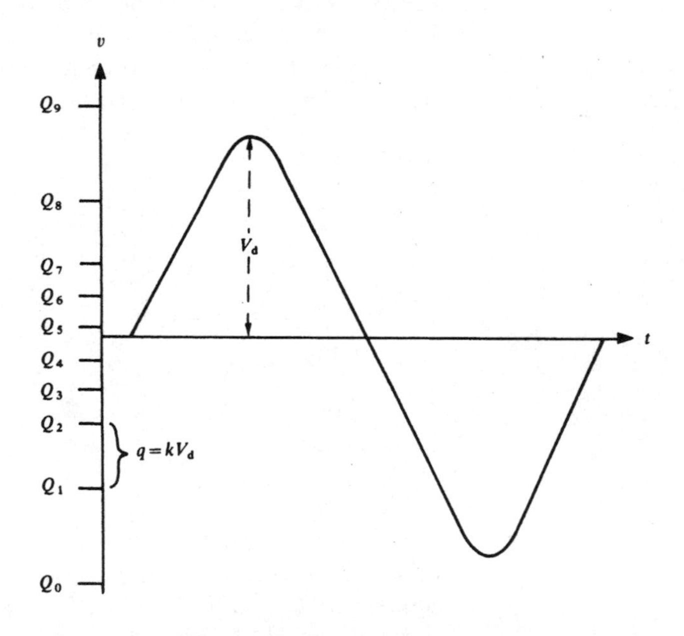

Figure 7.9. Variable quantization of a tone signal.

Exercise 7.4
Check on page 168

Sketch a graph that compares, approximately, the variation in the percentage quantization error obtained with uniform ($q=k$) and non-uniform quantization ($q=kV_d$). Assume that the quanta are the same for the two quantization schemes at the mid-point of the dynamic voltage range (V_d).

An alternative to varying q is to adjust the signal, prior to uniform quantization, so that the larger signals are *compressed* as shown in Fig. 7.10.

Recovery of a compressed signal will involve the inverse process of *expansion* to restore the signal to its original form. Note that it is only the

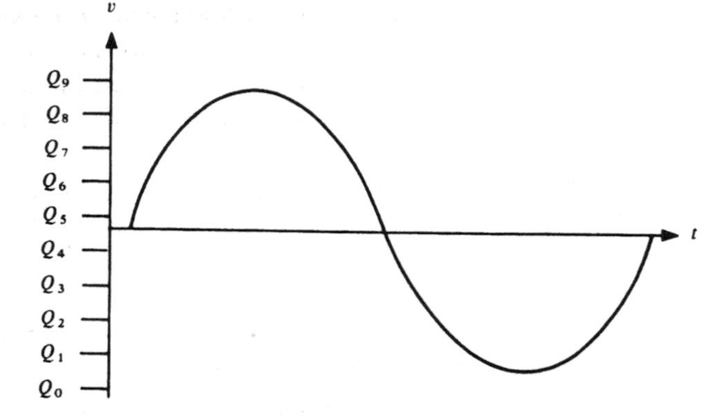

Figure 7.10. Compressed version of the tone signal in Fig. 7.9.

large values of signal voltage that need to be compressed before transmission and *expanded* at reception—the low signal levels are normally unaltered in *companding*. A transmitter voltage transfer characteristic may have the shape shown in Fig. 7.11.

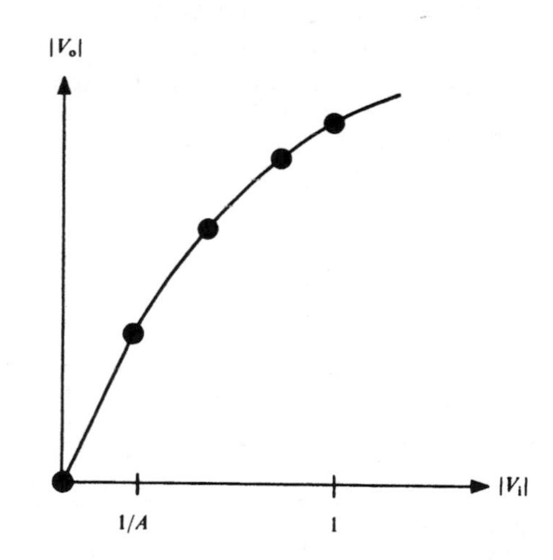

Figure 7.11. Normalized *A*-law characteristic.

The so-called *A*-law characteristic of Fig. 7.11, which is used in Europe, breaks the characteristic into two parts—a linear portion given by

$$|V_o| = (1 + A|V_i|)/(1 + \ln A) \qquad \text{for } 0 \leqslant |V_i| \leqslant 1/A$$

and a logarithmic portion (which is the compressed part) given by

$$|V_o| = [1 + \ln(A|V_i|)]/(1 + \ln A) \qquad \text{for } 1/A < |V_i| \leqslant 1$$

A is the *compression coefficient*. In the USA, Canada and Japan a 'μ-law' characteristic has been adopted. In these characteristics, the small signals are uniformly quantized, so ε increases as the signal gets smaller.

However, this reduces any distortion that might arise, due to a mismatch between the compressor and expander—a problem that would be most serious at low signal level. For telephone communications, the value of A is chosen so that the signal from the statistically quietest talker corresponds to $|V_i| = 1/A$. The remaining logarithmic response is well suited to cope with a wide dynamic range of speech. Data signals, which may also be sent over telephone channels, are restricted in amplitude so that they come within the linear range 0 to $1/A$. The Comité Consultatif International de Téléphonie et de Télégraphie (CCITT) recommend an A of 87.6 as an acceptable working compromise.

Compression has the effect of turning a tone waveform into a complex waveform, that is, of increasing the bandwidth of the signal. This is the price paid for the reduction in quantization error over the speech dynamic range.

A large value of q can bring advantages in terms of noise reduction in a link. The effect of channel noise on a QPAM signal is illustrated in Fig. 7.12. In Fig. 7.12 the noisy QPAM signal has arrived at a repeater with noise superimposed on its quantized levels. The repeater requantizes the signal, and any noise voltage that does not exceed $\frac{1}{2}q$ is thus removed.

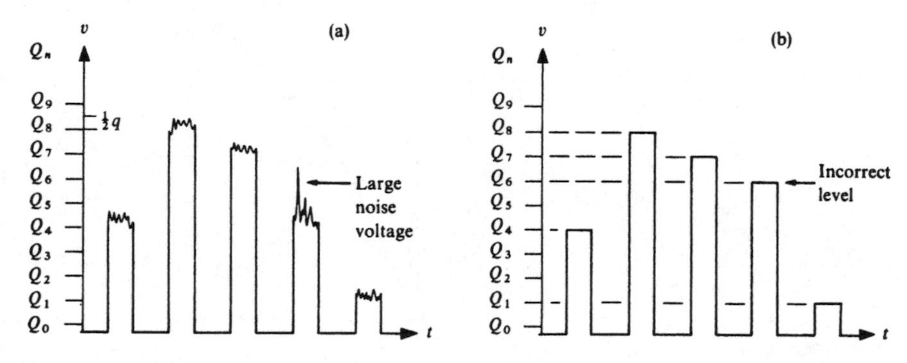

Figure 7.12. Quantized samples with added noise: (a) noisy QPAM signal; (b) signal after requantization.

Occasionally the noise voltage will exceed $\frac{1}{2}q$ and cause an error in a requantized sample. As q is increased and/or the repeaters are moved closer together, the probability of removing all additive noise becomes greater but the requantized signal will always be an approximation to the original due to quantization. In effect, the signal contains *quantization noise*, the effects of which on the SNR are described in Chapter 8. Unpredictable additive noise may be exchanged for predictable quantization noise to achieve an increased SNR.

7.7 Time-division multiplexing

Several signals may be transmitted in the same channel by using the relatively long time intervals between the sampling pulses of one signal to send the samples of other signals. This time-sharing technique is an important way of increasing the number of messages that can be sent along one telecommunication link, and is known as *time-division multiplexing* (TDM). Any form of pulse modulation can be used for TDM, but PAM will be used to illustrate the principle for the two tone signals shown in Fig. 7.13.

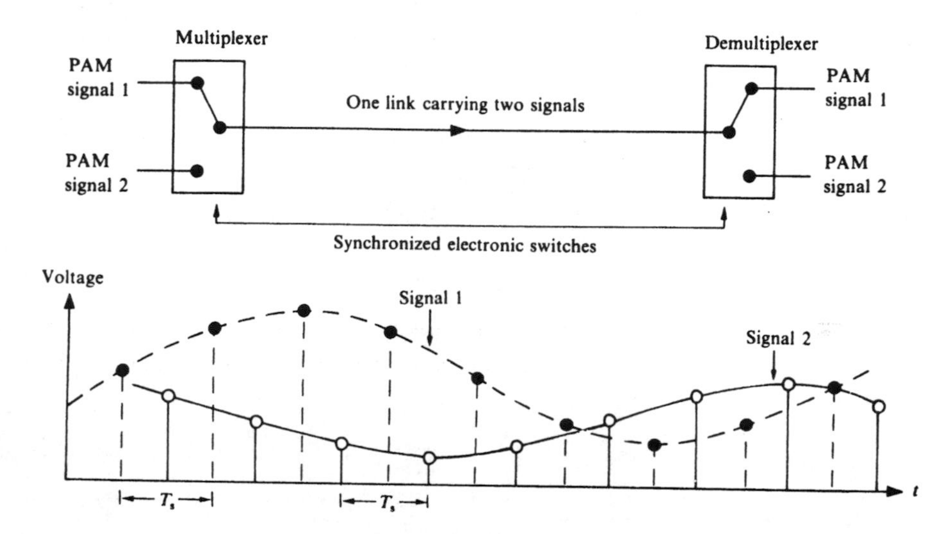

Figure 7.13. The principle of TDM applied to two PAM signals.

The analogue samples from the two signals are taken at the same sampling rate ($f_s = 1/T_s$) and interspersed as shown in the graph of Fig. 7.13, by having the electronic switches alternate between the two inputs and outputs respectively. The graph shows the pulse sequence that would be transmitted down the line.

This idea can be extended to as many separate signals as can be fitted in during the time between samples. Increasing this time will increase the number of messages that can be sent along a single link, but this will be at the expense of reduced sampling frequency and hence reduced message bandwidth. In order to increase the number of messages that can be multiplexed, without sacrificing message bandwidth, the pulse width must be reduced so the bandwidth of the link must be increased. Glass-fibre links would fulfil this requirement ideally.

Exercise 7.5
Check on page 168

Estimate the maximum width (T_w) of the sampling pulse that could be used in the time-division multiplexing of 32 separate PAM commercial speech signals.

7.8 Pulse code modulation

A QPAM signal has two great advantages:

1. It can be time-division multiplexed by transmitting other signals along

the same link during the time intervals between carrier (sampling) pulses.
2. It can restrict the receiver noise to the transmitted quantization noise.

If one more process is carried out on a QPAM signal, the first of the above advantages can be preserved while the SNR can be further improved and signal recovery considerably simplified. This process consists of *binary coding* the quantized value of each sample voltage and transmitting this code as a sequence of binary voltage pulses in the time intervals or time slots between samples. Binary coding, in fact, implies quantization because it is not possible to express all levels of an analogue signal with a finite number of bits. As the number of bits increases, the quantum reduces but can never go to zero.

The PCM transmitter

PCM can be illustrated, for voice signal transmission, by Fig. 7.14. An audio amplifier may precede the sampler, which is a DSB-SC pulse

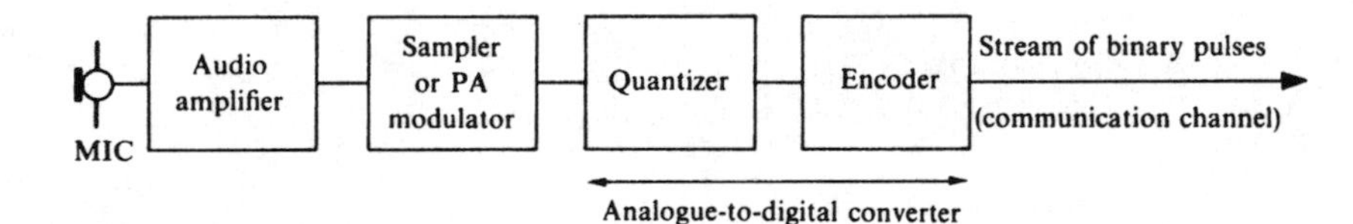

Figure 7.14. A PCM speech transmitter.

amplitude modulator (see Fig. 7.3). The quantizer and encoder form an *analogue-to-digital converter* (ADC), which is commercially available in silicon chip form, in a number of versions of varying speed and complexity. The fastest type, essential for PCM, is the *parallel-comparator* ADC, which uses a chain of resistors to quantize a voltage sample. A two-bit version of this is shown in Fig. 7.15.

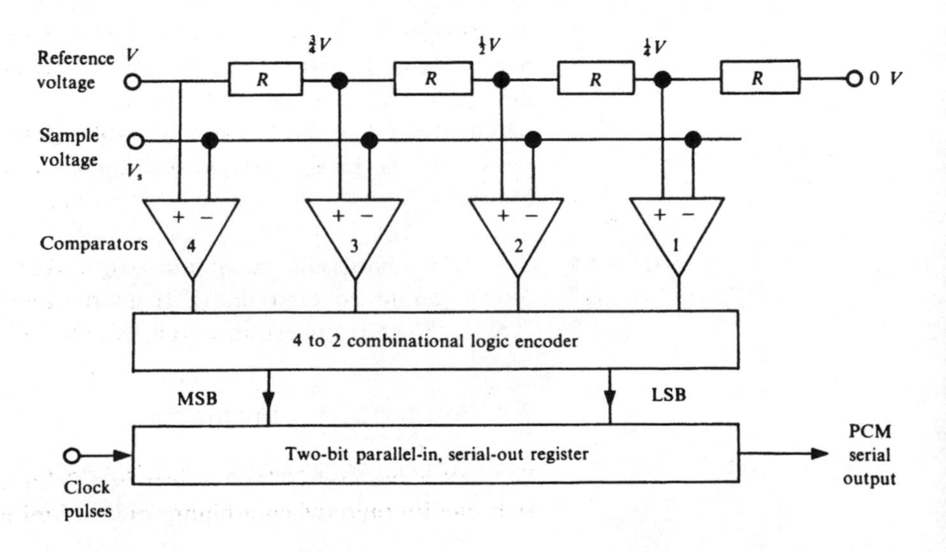

Figure 7.15. A parallel-comparator ADC.

Table 7.1. Outputs for two-bit parallel-comparator ADC

Voltage sample	Comparator outputs				Register output	
V_s	4	3	2	1	MSB	LSB
$0<V_s<\frac{1}{4}V$	0	0	0	0	0	0
$\frac{1}{4}V<V_s<\frac{1}{2}V$	0	0	0	1	0	1
$\frac{1}{2}V<V_s<\frac{3}{4}V$	0	0	1	1	1	0
$\frac{3}{4}V<V_s<V$	0	1	1	1	1	1

As each voltage sample is applied to the ADC, comparator and register outputs are obtained as shown in Table 7.1, with the least significant bit (LSB) entering the output line first followed by the most significant bit (MSB).

Each pulse-coded value of a quantized voltage sample is stored temporarily in a parallel-in, serial-out (PISO) register and is shifted out into the communications channel by two clock pulses, which are applied after each sampling instant (nT) to give the effect illustrated in Fig. 7.16.

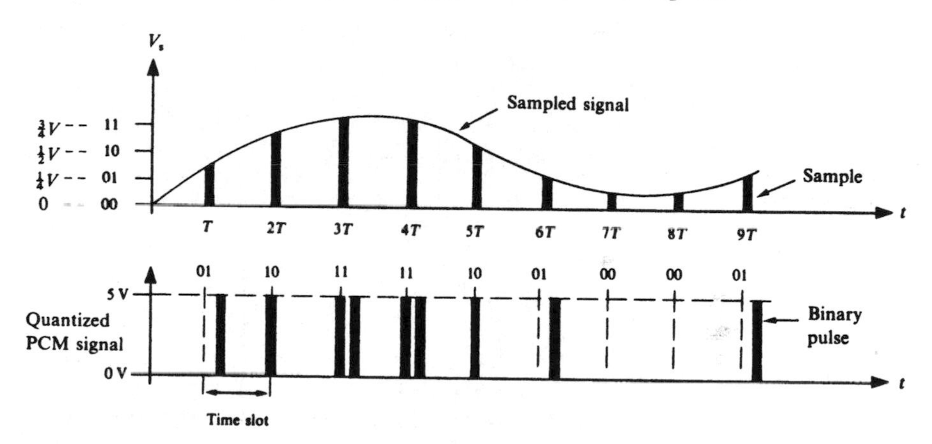

Figure 7.16. PCM of quantized signal samples.

During each time slot, the binary *word* representing the quantized value of the voltage sample is encoded as a series of binary-coded pulses to give a *pulse code modulated* (PCM) signal.

An eight-bit ADC can output a minimum of 1/256 of the full digital input, so it has got 256-digit or eight-bit *resolution*. This means that the smallest detectable change in the input voltage will be $V/(2^8-1)$. The time taken for the input voltage to be converted into binary form is the *conversion time*. The comparator-type ADC is extremely fast, with a conversion time of the order of 1 ns, virtually independent of the number of bits. The time taken to output the pulse train is a more relevant parameter, which depends on the clock frequency.

The PCM receiver

The first stage in a PCM receiver is a Schmitt trigger, which produces a high-quality pulse when a binary pulse of minimum amplitude is received.

This stage restores the PCM signal and its quantization error but removes any additive noise. If repeater operation is intended, the regenerated pulses are merely sent on along the next section of the link. If the original analogue signal is to be recovered, a *digital-to-analogue converter* (DAC) is required to convert the binary voltage pulses back into a QPAM signal, which is then put through a LP filter to remove all frequency components outside the baseband. The recovery of an analogue signal from a PCM signal is summarized in Fig. 7.17.

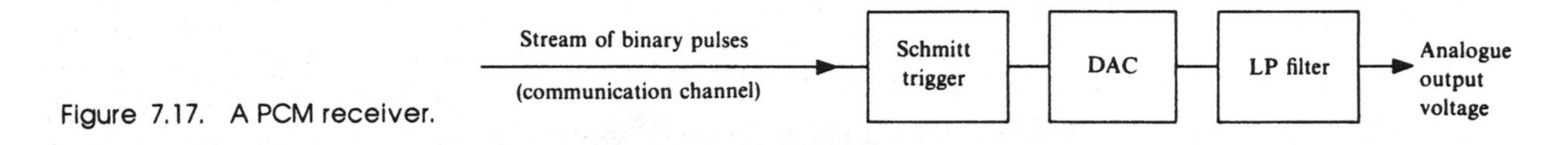

Figure 7.17. A PCM receiver.

Digital-to-analogue conversion

The action of a digital-to-analogue converter depends on summing binary-weighted currents at a node and then converting this current total into a voltage using an operational amplifier (op-amp). One way of doing this is shown in Fig. 7.18.

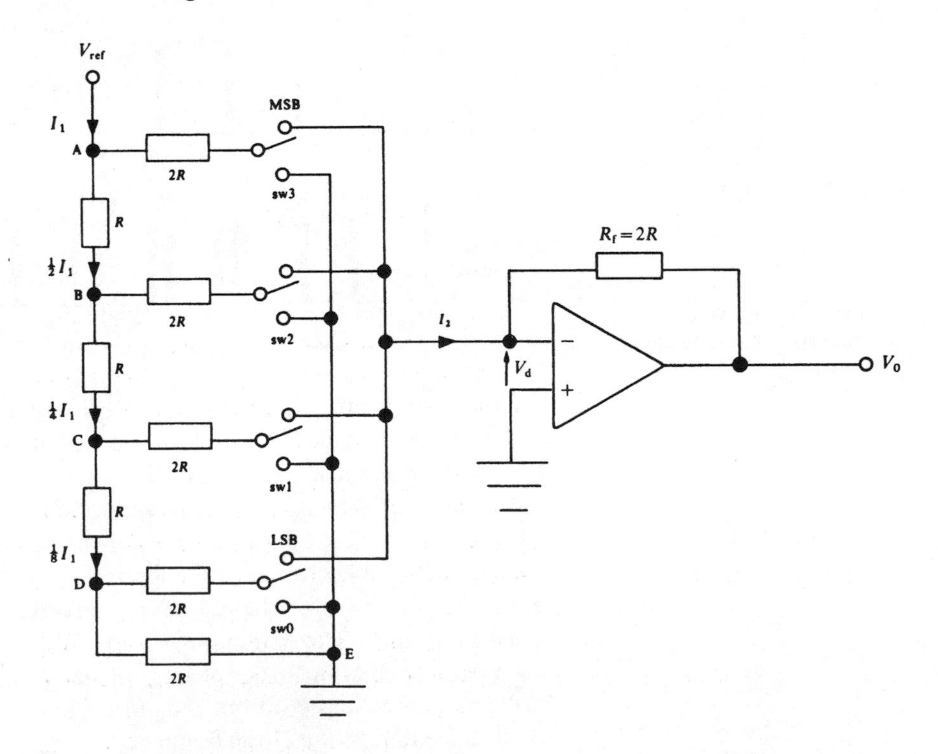

Figure 7.18. A four-bit R–$2R$ DAC.

The digital pulses representing 2^0, 2^1, 2^2 and 2^3 are made to operate the switches are sw0, sw1, sw2 and sw3 respectively. If, for example, the binary input was 0000, all the switches would be connected to earth; whereas if the input was 1111, all the switches would be connected to the inverting input terminal of the op-amp. Since this input is a virtual earth, the R–$2R$ ladder network currents will be unaffected by the position of the switches. Once this point is grasped, it is easy to work out the currents shown in Fig. 7.18 as follows.

The current entering node A is split equally through each $2R$ resistor. The resistance between A and E is $2R$ in parallel with $2R$, that is R. Thus the resistance between C and E is $R+R$ in parallel with $2R$, which is R, so the current entering C splits equally through $R+R$ and $2R$. Similarly, the resistance between B and E is $2R$ in parallel with $2R$ so the current entering B splits equally through another $R+R$ and $2R$. It does not matter how many 'rungs' you have in this ladder, the resistance to earth from each node is $2R$ in parallel with $2R$ so the currents entering all such nodes as A to D are split equally on leaving and the total resistance between each node and earth is R.

The current entering node A will be $V_{ref}/R = I_1$ and this current splits in half at each of the nodes B to D. The current I_2 is the sum of the currents through the switches that have been switched by a binary 1. This current is converted into a voltage V_o by the op-amp. Consider an input of 1111, for example. Then

$$I_2 = \tfrac{1}{2}I_1 + \tfrac{1}{2} \times \tfrac{1}{2}I_1 + \tfrac{1}{2} \times \tfrac{1}{4}I_1 + \tfrac{1}{2} \times \tfrac{1}{8}I_1 = I_1(\tfrac{1}{2} + \tfrac{1}{4} + \tfrac{1}{8} + \tfrac{1}{16})$$

But $I_2 = -V_o/R_f$ and $I_1 = V_{ref}/R$. So

$$-V_o = (1 + \tfrac{1}{2} + \tfrac{1}{4} + \tfrac{1}{8}) V_{ref} R_f / 2R$$

This equation shows that the output of the DAC is proportional to the 8421 binary-weighted input that operates the switches.

The current-to-voltage conversion may be explained as follows. The op-amp gets its energy from the supply voltage $\pm V_s$. It has a voltage gain (G) equal to V_o/V_d when $-V_s < V_o < +V_s$. G is always very large so, for typical values of V_o like 5 V, V_d is only 50 μV. This means that the inverting terminal of the op-amp is virtually earthed, and negligible input current flows. This means that almost all the current (I_2) flows through R_f so $V_o = -I_2 \times R_f$. Thus the input current has been converted into an output voltage on a scale determined by R_f.

Exercise 7.6
Check on page 169

In the DAC shown in Fig. 7.18, $R_f = R = 1\ \text{k}\Omega$ and $V_{ref} = 5$ V. Calculate the switch current and V_o for inputs of 0001, 0010 and 1100.

If the input resistance of the op-amp between the + and − terminals is 1 MΩ and if $R_f = 1\ \text{k}\Omega$ calculate I_2 when $V_o = 5$ V and the op-amp voltage gain is 106 dB.

The DAC described above lends itself to IC construction because only two values of resistor are required (2*R* and *R*) and it is the 2:1 ratio that must be accurately controlled in manufacture, not the actual value of *R*. The switches can be easily formed with (*N*-channel metal oxide semiconductor) (NMOS) gates, for example, as shown in Fig. 7.19.

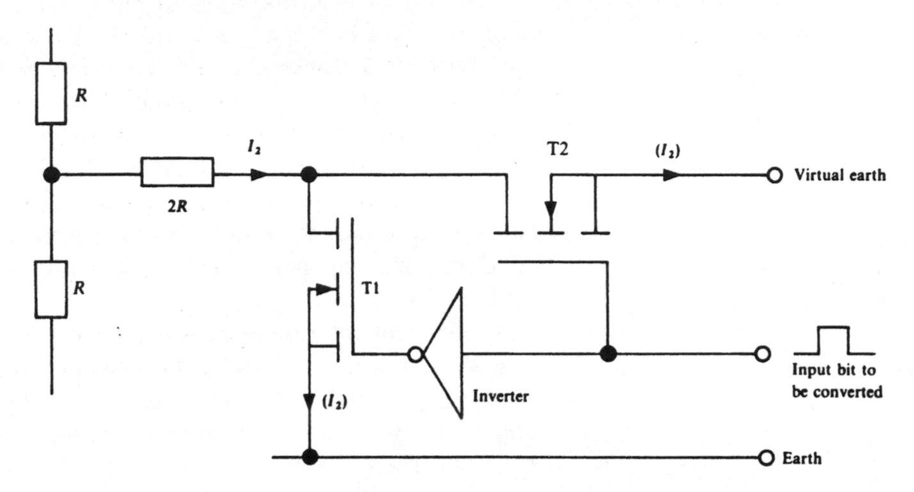

Figure 7.19. An NMOS switch for *R*-2*R* ladder network.

The switch action is as follows. When the input is low, T1 is on and T2 is off, so I_2 flows to earth. When the input is high, T1 is off and T2 is on, so I_2 flows to the virtual earth at the inverting input of the op-amp.

An eight-bit DAC can output a minimum of 1/256 of the full digital input, so it has got 256-digit or eight-bit *resolution*. The difference between the actual output voltage and the correct value at full scale is a measure of *non-linearity* (0.2% is typical). The time taken for the output voltage to reach its final stable value from the instant the digital input is applied is the *conversion time*. The *R*–2*R* DAC is very fast, with a conversion time of about 100 ns for eight-bit resolution.

PCM and noise reduction

The conversion of a QPAM signal into a PCM signal makes the task of the receiver much easier because all it has to do is recognize the presence or absence of a pulse rather than sense a range of perhaps 256 different voltage levels as in QPAM. The noise margin of QPAM is $\frac{1}{2}q$; that of PCM is the voltage difference between the two logic levels. Since the amplitude of a PCM signal carries no information, a PCM receiver simply has to decide whether there is a pulse or not and then decode the pulse sequence. This gives PCM much greater noise immunity than that of QPAM suggested by Fig. 7.12. Additive noise may be removed from a PCM link by simple repeaters, spaced out along the link as shown in Fig. 7.20.

In Fig. 7.20, the PCM transmitter (Tx) feeds a signal (*S*) plus quantization noise (*Q*) into a communications link, which picks up fluctuation

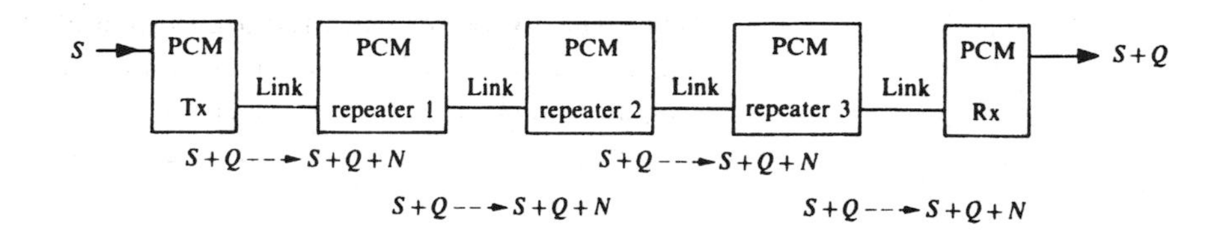

Figure 7.20. A PCM communications link.

noise (N). The length of this link must be restricted so that N is below the threshold of the PCM repeater 1 and then this receiver recovers the original signal $S+Q$. This repeater reshapes the binary pulses—a process that is repeated as many times as it takes to cover the overall distance between the original transmitter and the final receiver. Trunk telephone routes use this kind of arrangement for maintaining low noise levels.

The statistical nature of channel noise makes it inevitable that some errors will be present in the PCM words at the repeater inputs. When such errors occur they can be detected if extra *check bits* are added to the words before transmission along a link section. Error detection is dealt with in Chapter 8.

Forms of pulse coding

A PCM signal need not be of the return-to-zero (RZ) form used so far in this chapter. Prior to transmission, the signal may be put into one of several different forms depending on the channel. The RZ code has a d.c. component that cannot be passed by telephone lines. If alternate pulses (marks) of an RZ signal are inverted, as shown in Fig. 7.21, an *alternate mark inversion* (AMI) code is obtained, which has no d.c. component.

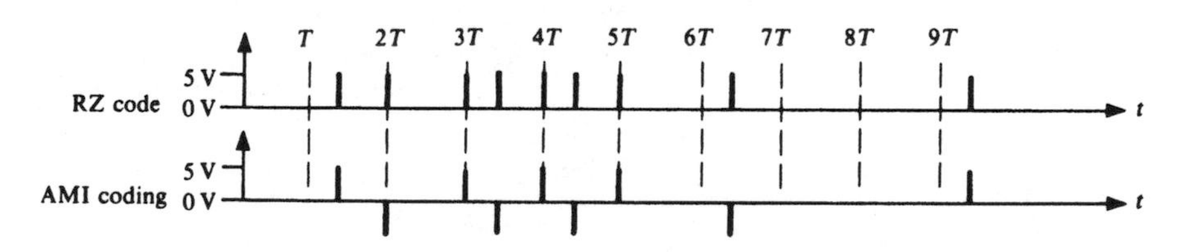

Figure 7.21. RZ and AMI coding.

Long sequences of 0s can result in a loss of alignment control as the clock pulse information vanishes, so, in PCM, a variation of AMI code is used called *high-density bipolar 3* (HDB3) coding. In this code, extra pulses are inserted so that not more than three successive 0s appear. In order that these extra pulses are easy to distinguish from the AMI pulses at the receiver, they must disobey the alternating polarity rule of the AMI code pulses. Further, any extra pulses must alternate in polarity in order to maintain zero mean voltage. An example of how this is done is given in Fig. 7.22.

The decoder has to check for two successive pulses of the same polarity and eliminate the second one, that is, the one marked v in Fig. 7.22. These v pulses are called *violations*. The decoder must also eliminate the *false*

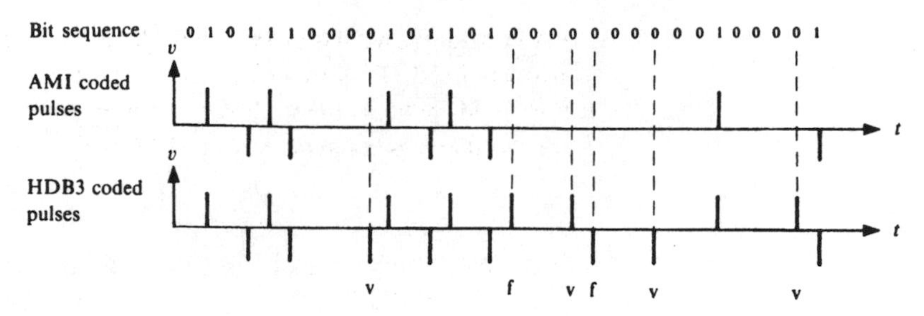

Figure 7.22. Example of HDB3 coding.

ones (f), which it knows will be present after an even number of true ones following a v pulse.

Exercise 7.7
Check on page 169

Draw the AMI and HDB3 code pulse patterns for the bit sequence:

101010000101000011100000001100000

7.9 Time-division multiplexing of PCM signals

It was shown in Fig. 7.13 how two PAM signals could be transmitted along the same link using TDM. Figure 7.23 shows how a similar scheme could be used for two PCM signals by comparing the pulse patterns for

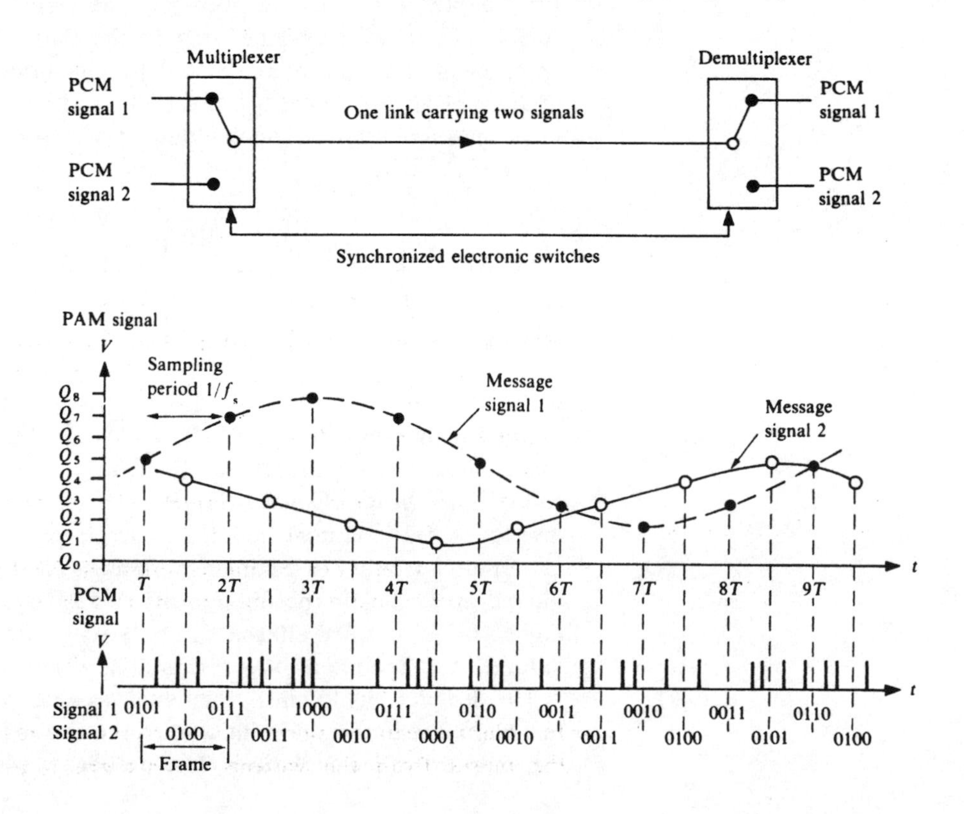

Figure 7.23. A PCM two-channel TDM system.

PAM and PCM when two signals are multiplexed and three-bit quantization is used. This principle may be extended to 32 channels as shown in Fig. 7.24, which also shows how analogue telecommunication can be achieved by using ADCs and DACs.

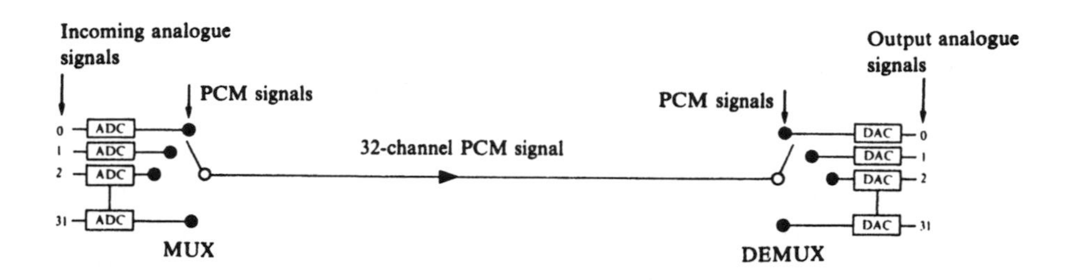

Figure 7.24. TDM of analogue signals using PCM.

When there are two signals, the words (in the form of pulses) are sent during alternate time slots (T) as shown in Fig. 7.23, each pair of words making up a *frame* of two words. If there are 32 signals, the same principle applies: 32 words are sent in a frame of 32 time slots. The way this is done in BT's PCM transmission system, which follows CCITT specifications and uses eight-bit quantization, is shown in Fig. 7.25.

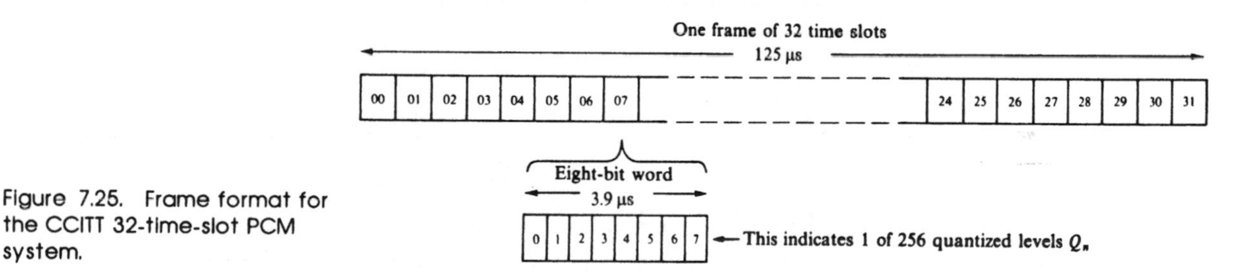

Figure 7.25. Frame format for the CCITT 32-time-slot PCM system.

BT uses an 8 kHz sampling frequency, so a frame, which occupies the period between successive samples of the same speech signal, is 1/(8 kHz) or 125 μs. Each frame consists of 32 time slots of 125 μs/32 = 3.9 μs during which time an eight-bit word is sent down the line. This word conveys a quantized sample voltage.

Exercise 7.8
Check on page 169

1. Calculate the maximum pulse width that could be used in BT's PCM system.
2. What clock frequency does this imply?
3. Calculate the bit transmission rate.

Alignment and signalling

In order to keep the electronic switches in synchronism and to ensure that the pulses from the various signals are routed to the correct DAC, *alignment* information must be sent with the speech signal. In the 32-time-slot system, the first frame in every 32 contains some of this information and the last frame in every 16 contains both alignment and signalling information. Thus, 16 frames (one *multiframe*) must be transmitted to convey all the alignment and signalling information. Only 30 time slots are used to convey the sample values of the speech signals, giving a total of 30 communication channels. The detailed frame format of a 30-channel system is explained in Fig. 7.26.

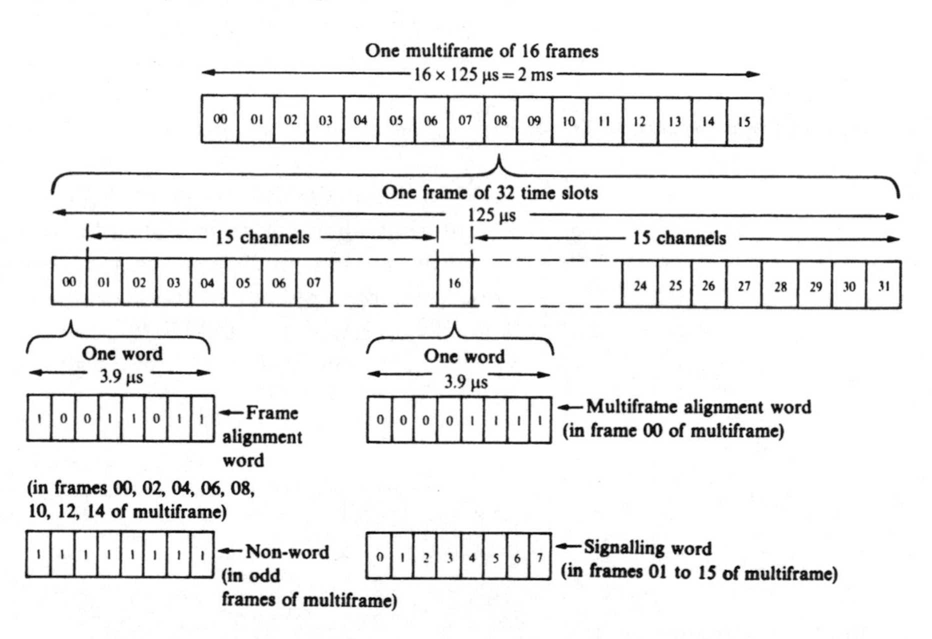

Figure 7.26. Frame format details for a 30-channel PCM system.

The information used to ensure that pulses are routed to the correct receiving channel is distributed between the frame and multiframe alignment words sent in time slots 00 and 16 respectively. The bit contents of these words are as shown while alignment is maintained. If spurious pulses are picked up in the link, the 0s are changed to 1s to recover alignment.

The term 'signalling' indicates the process of sending information that is used to route and monitor messages through a network. The 30-channel PCM system has such provision. Each channel is allocated four signalling bits. The first nibble (bits 0 to 3) of the signalling word is used for channels 01 to 15 and the second nibble (bits 4 to 7) for channels 16 to 30. Frame 01 serves channels 00 and 16, frame 02 serves channels 02 and 17, and so on. A few more details on signalling will appear in Chapter 10.

Summary

This chapter has described various ways in which pulses may be used to carry an analogue signal. Carriers, made up of periodic pulses, may have their amplitude, frequency, width or position varied in order to do this.

If an analogue signal is sampled at a frequency f_s that is at least twice that of its highest frequency component (f_m), it is possible to recover the signal by low-pass filtering. $2f_m$ is known as the Nyquist frequency, and the larger f_s is compared with the Nyquist frequency, the less abrupt need be the high-frequency roll-off of the filter. Aliasing occurs if the sampling frequency is too low. The interval between samples can be used for time-division multiplexing.

If an analogue signal is quantized it is possible to exchange channel interference, distortion and fluctuation noise for quantization noise, which, in turn, can be reduced by increasing the bandwidth to allow a larger number of smaller quanta to be used.

Considerable simplification in receiver design results if binary-coded pulses, representing the quantized amplitude of the analogue samples, are transmitted. This process of pulse code modulation brings further noise reduction in addition to that due to quantization because the receiver only has to distinguish the presence or absence of a pulse against a background of noise.

At the PCM transmitter, quantization and coding are carried out after sampling, by an analogue-to-digital converter. Signal recovery is effected, at the final receiver, using a digital-to-analogue converter followed by a low-pass filter. Intermediate transceivers or repeaters in the line are insensitive to line distortion and fluctuation noise and simply reshape the binary pulses for transmission along the next section of the link. This arrangement prevents cumulative distortion and noise arising in long links.

PCM signals can be multiplexed on a time-division basis, 30-channel TDM with HDB3 coding being the standard adopted in the UK.

PCM is widely used in telephony. It is not used as the sole modulation in radio communications because it does not produce frequency translation from baseband to a higher-frequency carrier band nor narrow-banding, both of which are required for efficient aerial operation. Unlike modulated sinusoidal carrier signals, modulated pulse carriers are really baseband-coded signals. In order to send the latter over radio or optic links, they must modulate a sinusoidal carrier in order to shift their spectra to a suitable higher-frequency band.

► Checks on exercises

7.1 The various pulse-carrier waveforms for a single modulating tone will be as shown in Fig. 7.27.

7.2 Aliasing is avoided by making $f_s > 2f_m$. This may be done by (1) increasing the sampling rate and (2) decreasing the highest baseband frequency (f_m) by passing the message through a LP filter before sampling—this 'anti-aliasing filter' must, of course, have a cut-off frequency less than $\frac{1}{2}f_s$.

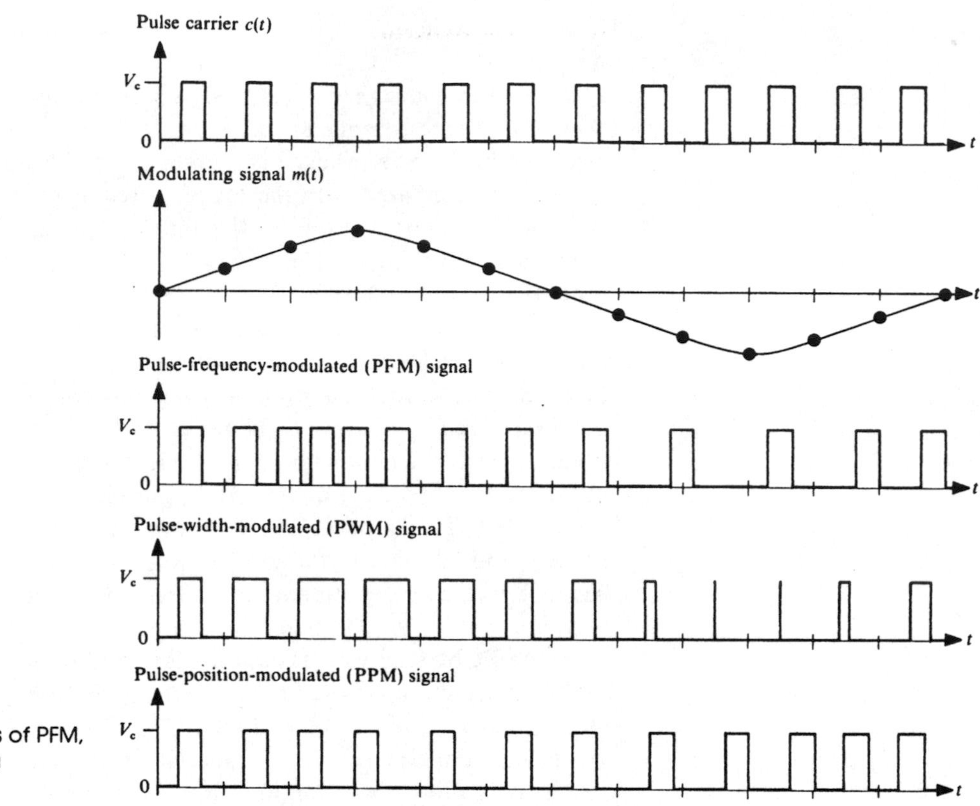

Figure 7.27. Examples of PFM, PWM and PPM for a modulating tone.

7.3 1. The dynamic range obtainable with 2^{16} quanta, using complete silence (0 dB) as the reference, will be $20 \log 2^{16} = 96$ dB. This is an astonishing dynamic range for domestic listening and goes from the threshold of hearing to the sound level corresponding to that of a very heavy lorry. Amplifiers and speakers that can cope with this level of sound are, of course, very expensive.

2. 2^{16} equal quanta over a 2 V range gives $q = 2/2^{16} = 2^{-15} = 31\ \mu\text{V}$.

3. The maximum possible quantization error will be very close to $31\ \mu\text{V}$. This represents a percentage error for the loudest part of the tone of approximately $100 \times 31\ \mu\text{V}/2\ \text{V} = 1.55 \times 10^{-3}\%$. The maximum possible quantization error in the quietest signal will be nearly $100 \times 31\ \mu\text{V}/31\ \text{mV} = 100\%$.

   The quantization error increases as the message signal amplitude decreases. The lowest part of the dynamic range is likely to be below the threshold of hearing, so the huge error in this region does not matter.

7.4 The graph is shown in Fig. 7.28.

7.5 In the transmission of telephone messages, whose baseband extends from 300 Hz to 3.4 kHz, a sampling frequency of 8 kHz is used. This means that

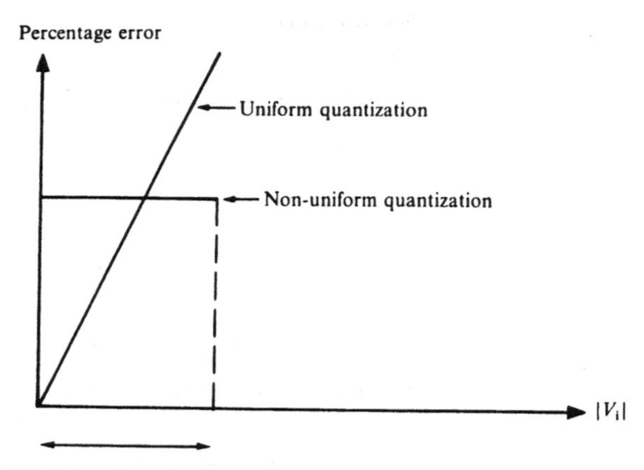

Figure 7.28. Uniform and non-uniform quantization error.

T_s will be about 1/(8 kHz) or 125 μs, assuming that $T_w \ll T_s$. The maximum sampling pulse width (T_w) for 32 signals will be approximately 125 μs/32 = 3.9 μs.

7.6 In the DAC shown in Fig. 7.18, when $R_f = R = 1\ \text{k}\Omega$ and $V_{ref} = 5$ V, the current from the reference voltage will be 5 V/1 kΩ = 5 mA irrespective of the switch positions. When the input is 0001, switch sw0 will connect a current of 5 mA/8 = 0.625 mA and $V_o = -0.625\ \text{mA} \times 1\ \text{k}\Omega = -0.625$ V. Similarly, when the input is 0010 switch sw1 will connect a current of 5 mA/4 = 1.25 mA and $V_o = -1.25\ \text{mA} \times 1\ \text{k}\Omega = -1.25$ V. When the input is 1100, switches sw2 and sw3 will connect currents of 1.25 and 2.5 mA respectively, giving $V_o = -3.75\ \text{mA} \times 1\ \text{k}\Omega = -3.75$ V. An op-amp voltage gain of 106 dB is 2×10^5 so, when $V_o = 5$ V, $V_d = 5\ \text{V}/2 \times 10^5 = 25\ \mu\text{V}$ and I_2 is therefore 5 V/1 kΩ + 25 μV/1 MΩ = 5 mA + 25 pA.

7.7 The patterns are given in Fig. 7.29.

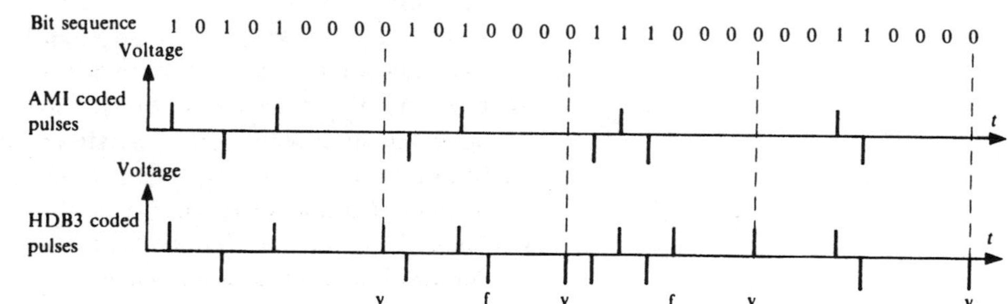

Figure 7.29. Example of HDB3 coding.

7.8

1. Eight bits are transmitted in 3.9 μs so the maximum pulse width that could be used in BT's PCM system will be 3.9 μs/8 = 488 ns.
2. The clock frequency will be 1/(488 ns) = 2 MHz.
3. Eight bits are transmitted every 3.9 μs so the bit rate is 8/(3.9 μs) = 2.051

Mb/s. This is slightly inaccurate because the figure of 3.9 was derived from 125/32, so the true bit rate is $8/(125/32) = 2.048$ Mb/s. An alternative solution is to think of frames, each of which contains 8×32 bits. There are 8000 frames per second, so the bit rate is $8 \times 32 \times 8000 = 2.048$ Mb/s.

Self-assessment questions

7.1 List the various forms of pulse-carrier modulation and state why they are all analogue signals.

7.2 Write out a general expression for a DSB-SC-PAM signal voltage in terms of a sampling frequency and the highest baseband frequency.

7.3 Draw the spectrum of a signal sampled at 8 kHz if the highest baseband frequency is 3.4 kHz.

7.4 What must be the minimum frequency roll-off rate of the LP filter used to recover the message in order to avoid aliasing in question 7.3?

7.5 Distinguish between space- and time-division multiplexing and calculate the maximum width of the sampling pulse if 32 channels are to be available for TDM under the conditions of question 7.4.

7.6 Can quantization be applied to a time-continuous signal as well as to a sampled signal and, if so, what is the point of sampling?

7.7 How can quantization error be reduced and what is the engineering price paid for such reduction?

7.8 How can the quantized signal-to-noise ratio (SNR_Q) be reduced for a given number of quanta?

7.9 What are the main advantages of PCM over DSB-SC-PAM?

7.10 Draw a block diagram showing the main processes that occur in a complete PCM telecommunications system.

7.11 Determine the quantization error in each sample when a cosine signal of maximum amplitude is converted by a two-bit parallel comparator ADC.

7.12 Calculate the resolution of an eight-bit parallel comparator ADC that uses a reference voltage of 5 V.

7.13 Calculate the maximum distance between repeater PCM links if the cable has an attenuation of 20 dB/km and the Schmitt circuits used have outputs of 10 V and the maximum fluctuation noise amplitude is 200 mV.

7.14 State the advantages and disadvantages of NRZ and AMI pulse coding compared with bipolar coding.

7.15 What is the purpose of HDB3 coding?

7.16 Calculate the frame and word periods of a 16-bit PCM transmission using a sampling frequency of 44.1 kHz. What would be the bit rate and dynamic range of this transmission?

8

Noise and errors

Dealing with obscured signals

Objectives

When you have completed this chapter you will be able to:

- Describe, in general terms, how noise affects a telecommunications system
- Distinguish between man-made, atmospheric and thermal noise
- Define white noise
- Describe Johnson and shot noise
- Define signal-to-noise ratio, noise figure and noise temperature
- Derive the noise figure for an amplifier cascade
- Describe the effects of noise in analogue and data demodulation systems
- Describe how the corrupting effects of noise may be minimized
- Derive expressions for some signal-to-quantization-noise ratios
- Describe how data errors may be detected, located and corrected

Noise in telecommunications is any unwanted signal. Noise degrades the performance of a system by reducing the amount of information that a channel can carry in a given time—a feature that will be explained in Chapter 9.

In this chapter various types of noise are described and some noise models are presented. As noise increases, an analogue signal becomes obscured, making it difficult to recover. Data signals are much easier to recover in the presence of noise, but they are likely to contain noise-induced errors. Techniques exist for improving signal-to-noise ratio and for detecting and locating data errors.

Quantization error can be regarded as a special form of noise. The signal-to-quantization-noise ratio may be increased by increasing the number of quantization levels within a given signal dynamic range. This means increasing the number of encoding bits when PCM is used.

8.1 Noise in telecommunications systems

A major problem in speech communication is making yourself understood in noisy surroundings. If the noise level is high, the natural thing to do is to talk louder, that is, increase the *signal-to-noise ratio* (SNR). This is exactly what is done in telecommunications to make a received signal intelligible: a strong signal is sent to enable the receiver to recover the message. What matters in communication is not the noise power itself, but the SNR at the receiver. This must be high enough for the receiving equipment to be able to present an acceptable signal to the viewer or listener.

As the distance between a transmitter and receiver increases, the problem of noise increases for two reasons:

1. The signal gets progressively reduced due to line or atmospheric losses.
2. More unwanted signals get into the link.

These unwanted signals might be other people's signals (cross-talk), spurious electrical signals due to lightning or electrical machinery (atmospheric or man-made interference) and the noise associated with the electrical conduction process (thermal noise).

Much of what goes on in telecommunications is an attempt to maximize the SNR by maximizing the received signal power and minimizing the received noise power. In order to maximize the signal, *amplifiers* are used as shown in Fig. 8.1.

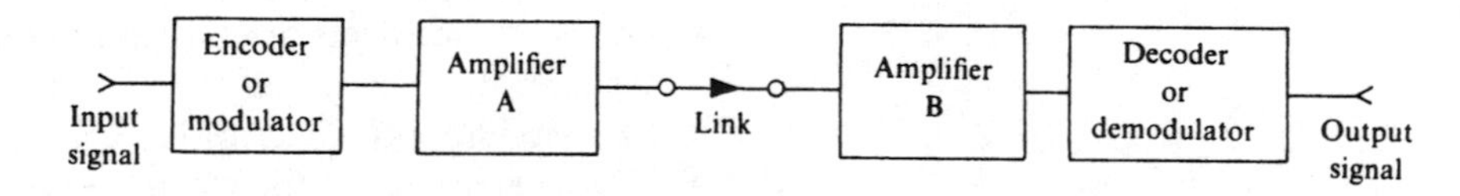

Figure 8.1. Amplifiers in a telecommunications system.

An amplifier increases the amplitude and hence the power of an electrical signal. In the above system we need one amplifier (A) to boost the output, prior to transmission over the link, and another amplifier (B) to make up for the signal power lost to the link.

Exercise 8.1
Check on page 195

Given one amplifier, why is it better to put it at position A rather than B?

In order to minimize cross-talk and interference picked up by a line, coaxial cable may be used. All types of metal cable generate a little thermal noise. By contrast, a radio link does not suffer from thermal noise but is very prone to cross-talk and interference. Glass fibres provide the best form of link in these respects. In practice, noise will appear at the end of any link due to noise generated by electronic sources, and some way must be found of coping with it.

The technique adopted uses signals that are different from the expected noise in such a way that the signal can be picked out, or selected, from the noise by the receiver. This entails encoding the signal information in a particular form before transmission and using decoding circuits in the receiver that can distinguish between the signal and the noise. It is the same technique as that used by humans conversing in noisy environments. Here the human brain (the receiver decoder) picks out the wanted conversation from the unwanted (cross-talk) and ignores interference (other noises).

Exercise 8.2
Check on page 195

One of the earliest forms of radio communications was Morse code, in which messages were spelt out by each letter being coded as a pattern of dots and dashes (short and long tones). Suggest some of the reasons why messages were coded in this way instead of being sent as speech signals.

8.2 The nature of thermal noise

Signal interference from external noise sources can be reduced by shielding. Noise made up of other people's signals has a predictable form and can often be filtered out to leave the wanted signal. Thermal noise is much more difficult to deal with: it is internally generated and random. A noise waveform, when viewed on an oscilloscope, appears to have no regular form: the amplitude varies at random, with voltage spikes appearing haphazardly. If the instantaneous voltage of a given noise component n is v_n, it is found that a graph of v_n against n follows a *normal* or *gaussian* distribution.

The absence of periodicity suggests that thermal noise is made up of a continuous spectrum of frequencies ranging from zero to infinity. If a noise source was applied to a fictitious power meter that had the same response to all frequencies from zero to some variable cut-off frequency (f_c)—an ideal variable low-pass filter response—the result might, for example, be as displayed by the graph in Fig. 8.2.

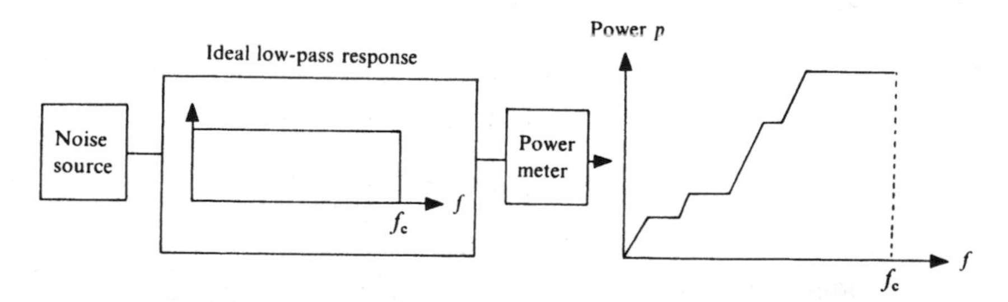

Figure 8.2. Noise measured by an ideal LP power meter.

The power in Fig. 8.2 changes over certain frequency bands. If the noise power increase is proportional to frequency, the result would be as displayed by either of the graphs in Fig. 8.3. In the second graph, dp/df is the noise *power spectral density* (ρ). Noise that has a constant power spectral density (ρ_k) over a very wide frequency range is defined as *white noise* by analogy with white light. Since $dp/df = \rho_k$ then

$$p = \rho_k \int_{f_1}^{f_2} df = \rho_k B$$

where B is the system bandwidth.

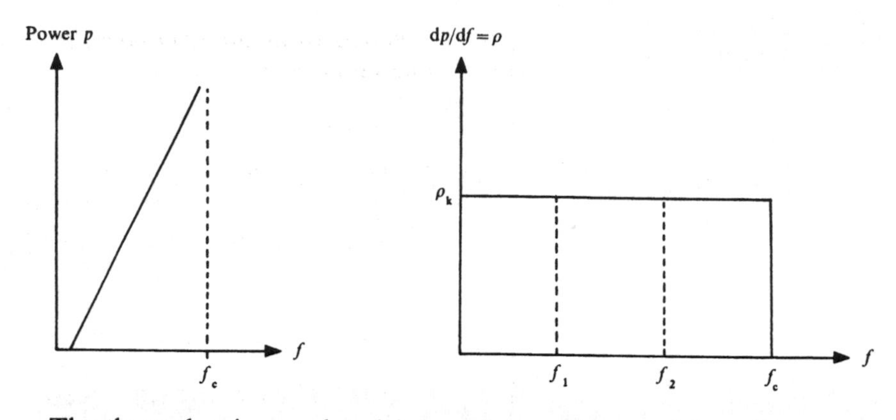

Figure 8.3. Noise power proportional to frequency.

The thermal noise produced by conductors (resistors) is an important example of white noise with guassian voltage distribution among the noise components. In the conduction process, electrons drift due to the voltage. These electrons collide randomly with the lattice ions, which form the structure of the conducting material, and this causes a random variation to be superimposed on the signal current flow. The noise power (N) associated with this variation is proportional to absolute temperature (T) measured in kelvins (K) and the bandwidth (B) of the system, that is $N=kTB$, where k is Boltzmann's constant, equal to 1.37×10^{-23} J/K.

Even when there is no signal current, the thermal agitation of the electrons causes white noise to be produced. In order to extract power from a noisy resistor (R_N), it must be loaded by a resistor (R_L) as shown in Fig. 8.4, R_L will be assumed noiseless for the moment.

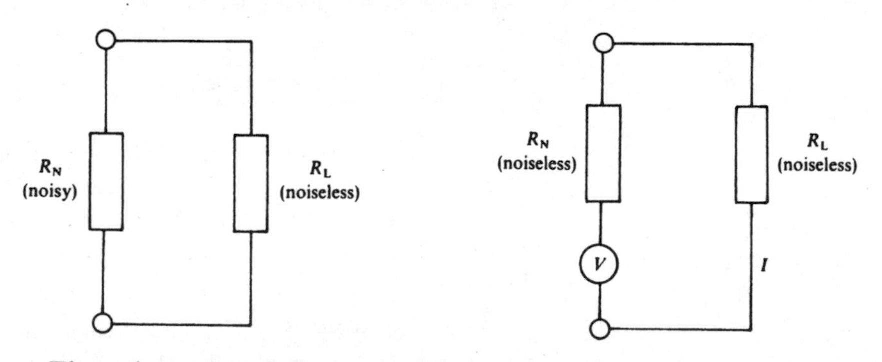

Figure 8.4. Noise voltage due to a resistor.

The noisy resistor may be regarded as a source of noise voltage of V volts r.m.s. that causes a noise current I to flow. The noise power in the load resistor will be

$$kTB=I^2R_L=[V/(R_N+R_L)]^2R_L$$

To extract all the available noise power from R_N, R_L must equal R_N, so the mean-squared noise voltage produced by a resistor R is given by

$$V^2=4kTBR$$

This is referred to as *Johnson noise*. An alternative expression gives the mean-squared noise current

$$I^2 = 4kTBG$$

where $G = 1/R$. A 100 kΩ resistor at 300 K produces about 40 μV r.m.s. of Johnson noise over a bandwidth of 1 MHz. In practice the resistor R_L in Fig. 8.4 will generate noise power, all of which will be absorbed by R_N if $R_L = R_N$. This results in a state of power equilibrium within the system.

Exercise 8.3
Check on page 196

Calculate the r.m.s. noise voltage produced by 10 km of twin cable with a loop resistance of 25 Ω/km at 300 K over a bandwidth of 10 kHz, and comment on the implication of this result.

Another important example of gaussian white noise is *shot noise*, which is produced when holes and electrons flow across the semiconductor pn junctions that proliferate in integrated circuits (ICs). This noise current can be found by thinking of a pn junction in which shot noise and Johnson noise are in equilibrium in the circuit shown in Fig. 8.5.

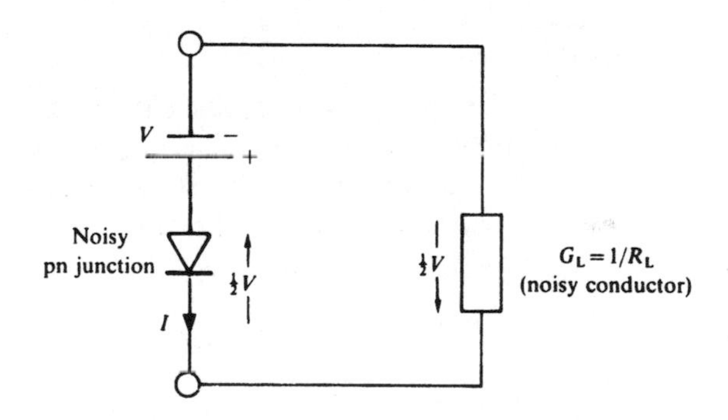

Figure 8.5. Noise voltage due to a pn junction.

The load resistor is made equal to the resistance of the diode semiconductor material. When a voltage V is applied by the battery in the circuit of Fig. 8.5, half of this voltage appears across R_L and half across the resistance of the semiconductor because the voltage across a forward-biased junction is negligible. The current I flowing across the junction is

$$I = I_0 \, e^{qV/2kT}$$

where I_0 is the reverse leakage current across the junction and q is the electronic charge, equal to 1.6×10^{-19}C. At equilibrium G_L absorbs all the shot noise power and must return it all to the junction as Johnson noise power, so

$$G_L = dI/dV = (q/2kT)I$$

The Johnson noise current I_J due to G_L is

$$I_J = \sqrt{(4kTBG_L)} = \sqrt{(2qIB)}$$

and at equilibrium this must be the shot noise current (I_s). Thus each junction of an IC produces an $I_s = \sqrt{(2qIB)}$ r.m.s. of noise current.

The combined effect of Johnson noise and shot noise

Two sources of voltage v_1 and v_2 in series will produce instantaneous power in a resistor R_L of $(v_1+v_2)^2/R_L$. The mean power P will be given by

$$\begin{aligned} P &= \text{mean of } (v_1+v_2)^2/R_L \\ &= \text{mean of } (v_1^2+v_2^2+2v_1v_2)/R_L \\ &= (V_1^2+V_2^2+2\gamma V_1 V_2)/R_L \end{aligned}$$

where V_1 and V_2 are the r.m.s. voltages and

$$\gamma = (1/T)\int_0^T v_1 v_2 \, dt/V_1 V_2$$

The correlation coefficient $\gamma = 1$ for waveforms that differ in magnitude but are in phase, and $\gamma = -1$ for waveforms that differ in magnitude but are in antiphase. For the completely uncorrelated waveforms from different noise sources, the correlation coefficient $\gamma = 0$. In this case

$$P = (V_1^2 + V_2^2)/R_L$$

Signal-to-noise ratios

The signals at the input and output of electronic systems are inevitably combined with thermal noise. The signal-to-noise ratio (SNR) at a pair of terminals is defined as

$$\text{SNR} = \frac{\text{mean signal power at the terminals}}{\text{mean noise power at the terminals}} = \frac{\text{S}}{\text{N}}$$

Thus, the input signal-to-noise ratio (SNR_i) would be written S_i/N_i and the output signal-to-noise ratio (SNR_o) would be written S_o/N_o.

The way that noise affects the SNRs of an amplifier can be seen from Fig. 8.6. The input to an amplifier will typically consist of a signal power S_i and noise power N_i. The amplifier will add thermal noise N_a to the amplified input noise to give a total output N_o equal to $AN_i + N_a$. The input SNR will be greater than the output SNR. The ratio of the former to the latter (the *noise figure*) will therefore

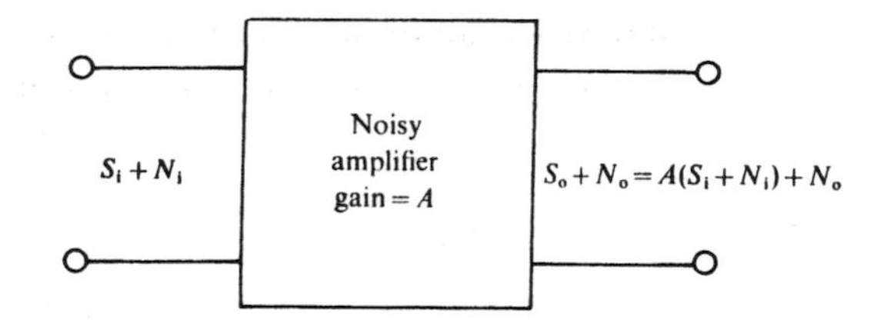

Figure 8.6. SNR of noisy amplifier.

provide a figure of merit for the amplifier:

$$\text{noise figure } F = (\text{SNR at input})/(\text{SNR at output})$$

$$= \frac{(\text{input signal power})/(\text{input noise power})}{(\text{output signal power})/(\text{output noise power})}$$

$$= (S_i/N_i)/(S_o/N_o) = N_o/N_i(S_o/S_i) = N_o/N_iA$$

$$= (AN_i + N_a)/N_iA = 1 + (N_a/A)/N_i$$

N_a/A is the *input-referred noise* and is equal to $(F-1)N_i$. Note that $F=1$ in a noiseless amplifier. Larger values than unity indicate a noisier amplifier.

N_i will be the thermal noise from the source resistance, and N_a will be the sum of the Johnson and Shot noise produced in the amplifier. The noise model and its Thévenin equivalent are shown in Fig. 8.7. In this figure, V_{Ns} is the Johnson noise r.m.s. voltage from the source resistance R_s and is given by $V_{Ns}=\sqrt{(4kTBR_s)}$, V_{Ni} is the input-referred Johnson noise r.m.s. voltage from the amplifier, I_{Ni} is the input-referred Shot noise r.m.s. current from the amplifier, and so

$$F = 1 + (N_a/A)N_i = 1 + [V_{Ni}^2 + (RI_{Ni})^2]/4kTBR_s$$

$$= (N_i + N_a/A)/N_i$$

$$= (\text{total input-referred noise power})/(\text{actual input noise power})$$

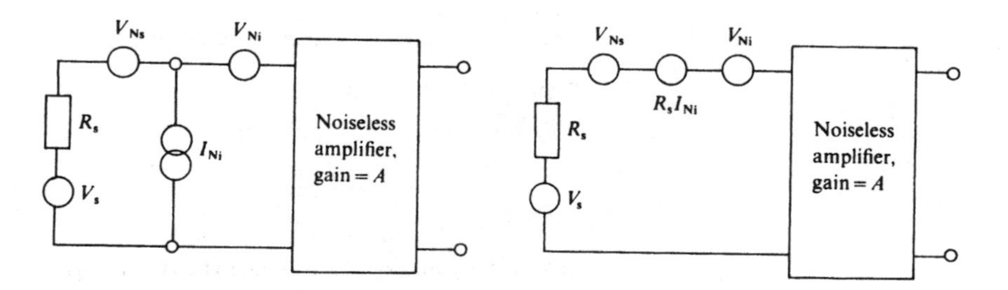

Figure 8.7. Amplifier noise models.

Exercise 8.4
Check on page 196

1. In Fig. 8.7, $V_{Ni}=0.3\,\mu\text{V}$ and $I_{Ni}=1.5\,nA$ and they are uncorrelated. If $V_s=2\,\mu V$ and $R_s=1\,\text{k}\Omega$, calculate F and SNR_o at $T=290$ K and $B=4$ kHz.
2. Show that the noise power at an amplifier input from a source resistance R_s is equal to kTB when R_s equals the input resistance of the amplifier.

Noise temperature

Under matched conditions (when R_s equals the input resistance of the amplifier),

$$F = (N_i + N_a/A)/N_i = (kTB + N_e)/kTB$$

where N_e is the *excess noise power given by* $N_e = N_a/A$. This may be rewritten in the form

$$F = (kTB + kT_eB)/kTB$$

where T_e is the effective noise temperature of the amplifier. We thus get the relationships:

$$N_e = (F-1)kTB \qquad \text{and} \qquad T_e = (F-1)T$$

Noise temperature is a more appropriate figure of merit than noise figure for comparing low-noise amplifiers such as a travelling wave tube. For example, if we have two amplifiers with T_e of 10 K and 20 K it is obvious that one is twice as noisy as the other. However, their respective values of F measured, under matched conditions at 290 K, would would be $(290+10)/290 = 1.03$ and $(290+20)/290 = 1.07$—figures that give a misleading comparison.

Exercise 8.5
Check on page 196

Calculate the noise temperatures of two amplifiers with noise figures of 10 and 20 measured at 290 K.

The noise power available from a resistor over a bandwidth B is given by: $N = kTB$, so the noise temperature of the resistor is $T = N/kB$. Noise power is available from sources other than resistors. For example, an antenna would pick up noise and thus act as a noise source with a noise temperature T_A. The sky also acts as a noise source for the antenna, so there is a sky noise temperature for different parts of the sky.

The noise figure of cascaded amplifiers

If two amplifiers, with noise figures F_1 and F_2, are cascaded as shown in Fig. 8.8, a noise figure for the cascade (F_c) can be found. The input to the

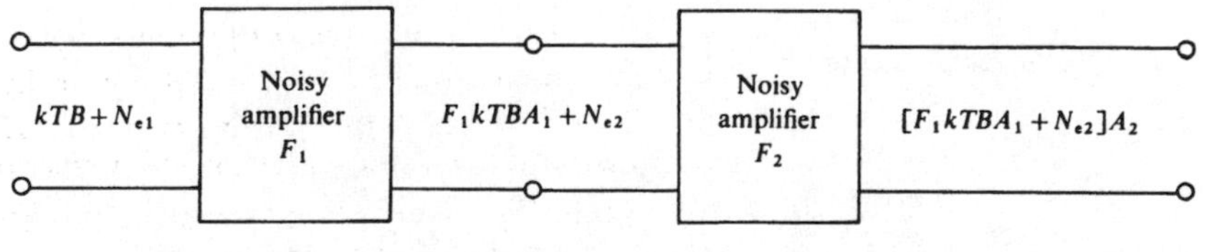

$kTB + N_{e1}$
Noisy amplifier F_1
$F_1kTBA_1 + N_{e2}$
Noisy amplifier F_2
$[F_1kTBA_1 + N_{e2}]A_2$

Figure 8.8. A cascade of noisy amplifiers.

first amplifier stage of Fig. 8.8 consists of actual noise kTB plus excess noise N_{e1}. Since $N_{e1} = (F_1 - 1)kTB$, this gives a total input noise input to the cascade of F_1kTB. The input to the second amplifier stage consists of actual noise F_1kTBA_1 plus excess noise $N_{e2} = (F_2 - 1)kTB$.

This gives a total output noise from the cascade of $[F_1kTBA_1+(F_2-1)kTB]A_2$.

Noise factor may be defined as (input-referred noise)/(actual input noise) so the noise factor of the cascade of two amplifiers is given by:

$$F_c=\frac{[F_1kTBA_1+(F_2-1)kTB]A_2/(A_1A_2)}{kTB}$$

$$=F_1+(F_2-1)/A_1 \tag{8.1}$$

Equation (8.1) makes the point that low-noise amplifier cascades rely on a high-gain, low-noise first stage. Equation (8.1) can be extended to n stages:

$$F_c=F_1+(F_2-1)/A_1+(F_3-1)/A_1A_2+\cdots+(F_n-1)/A_1A_2\ldots A_{n-1}$$

Although amplifiers have been used to illustrate the effect of noise on the input and output SNRs and the noise figure, the same principles apply to other electronic devices such as demodulators. In this case, however, the bandwidth (B) at the input will normally be different from that at the output.

8.3 Noise effects in analogue modulation systems

We will now compare the performance of various analogue demodulation systems in terms of their ability to minimize the system output SNR. They all suffer, in the same way as amplifiers, the effects of Johnson and shot noise. Unlike amplifiers, however, demodulators may have different input and output bandwidths and this also affects the noise power. The input bandwidth will be the IF amplifier bandwidth but the output signal will occupy the base bandwidth, and these may not be the same. The signal carrier powers may not be the same in different systems, and this will also affect the signal-to-noise ratios. These are the features we take into account in this section.

AM systems

A single tone input signal of frequency ω_t will be assumed for simplicity. It will also be assumed that the noise entering the demodulator from the IF amplifier is spread uniformly over the IF bandwidth (B) and that the output noise is baseband (W) limited. To effect a fair comparison, coherent demodulation, using a synchronous signal of unit amplitude as shown in Fig. 8.9, will be used in all cases.

For DSB-AM

$$v_{if}=V_{if}[1+m\cos(\omega_t t)]\cos(\omega_{if}t)+v_{Nif}$$

Hence, the normalized input signal power (based on v_{if} applied across a

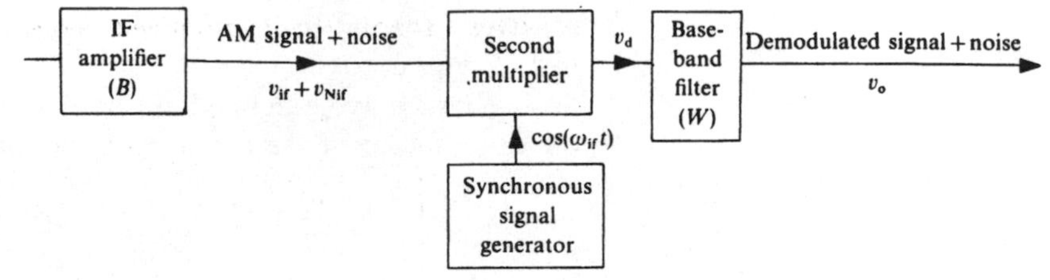

Figure 8.9. Coherent demodulation in the presence of noise.

1 Ω resistor) is

$$S_i = \tfrac{1}{2}V_{if}^2 + \tfrac{1}{2}(\tfrac{1}{2}mV_{if})^2 + \tfrac{1}{2}(\tfrac{1}{2}mV_{if})^2 = \tfrac{1}{2}V_{if}^2(1 + \tfrac{1}{2}m^2)$$

The noise will be made up of a continuous spectrum of voltages of general frequency ω_N. That is, each noise voltage component will be given by $v_{Nif} = V_N \cos(\omega_N t)$ with each component covering $\delta\omega$ of the IF band (B). Since this is white noise, the noise power input to the multiplier will be

$$N_i = \int_0^B \rho\, \delta\omega = \rho B$$

where ρ is the noise spectral density. Hence

$$\mathrm{SNR}_i = \tfrac{1}{2}V_{if}^2(1 + \tfrac{1}{2}m^2)/\rho B$$

The multiplier shifts the signal from the IF down to baseband as follows:

$$\begin{aligned} v_d &= V_{if}[1 + m\cos(\omega_t t)]\cos^2(\omega_{if} t) \\ &= V_{if}[1 + m\cos(\omega_t t)]\tfrac{1}{2}[1 + \cos(2\omega_{if} t)] \\ &= \tfrac{1}{2}V_{if}[1 + m\cos(\omega_t t)] + \text{higher-frequency terms} \end{aligned}$$

The output from the baseband filter is

$$v_O = \tfrac{1}{2}mV_{if}\cos(\omega_t t)$$

so

$$S_o = \tfrac{1}{4}(mV_{if})^2$$

Each noise component, $V_N \cos(\omega_N t)$ from the IF amplifier will be multiplied by $\cos(\omega_{if} t)$ to give a noise signal:

$$\begin{aligned} v_N &= V_N \cos(\omega_N t)\cos(\omega_{if} t) \\ &= \tfrac{1}{2}V_N\{\cos[(\omega_{if} - \omega_N)t] + \cos[(\omega_{if} + \omega_N)t]\} \end{aligned}$$

This outputs from the baseband filter as $\tfrac{1}{2}V_N \cos[(\omega_N \omega_{if} - \omega_N)t]$, so the normalized noise power (δN) over bandwidth ($\delta\omega$) at frequency ω_N is

$\frac{1}{4}V_N^2 = \rho\delta\omega$. Hence, the normalized output noise power over the whole baseband is given by

$$N_o = \int_0^W \delta N = \int_0^W \rho \, d\omega = \rho W$$

$$SNR_o = \tfrac{1}{4}(mV_{if})^2/\rho W = \tfrac{1}{2}(mV_{if})^2/\rho B$$

(since $B = 2W$). So

$$SNR_i/SNR_o = [\tfrac{1}{2}V_{if}^2(1+\tfrac{1}{2}m^2)/\rho B]/[\tfrac{1}{2}(mV_{if})^2/\rho B] = (1+\tfrac{1}{2}m^2)/m^2$$

When $m = 1$ we get the minimum value of $SNR_i/SNR_o = 1.5$. This ratio is not the noise figure because the Johnson and shot noise generated by the electronic circuits of the demodulator have been ignored. However, the ratio enables a simple comparison to be made between the various demodulation systems.

For DSB-SC-AM the same expressions hold good as for DSB-AM except for the carrier signal term, which disappears to give

$$SNR_i = \tfrac{1}{2}V_{if}^2(0+\tfrac{1}{2}m^2)/\rho B$$

so

$$SNR_i/SNR_o = (0+\tfrac{1}{2}m^2)/m^2 = 0.5$$

In SSB-AM the carrier and one sideband are missing, so the DSB-AM input signal-to-noise ratio expression becomes

$$SNR_i = \tfrac{1}{2}V_{if}^2(0+\tfrac{1}{4}m^2)/\rho B$$

The output signal-to-noise ratio expression becomes

$$SNR_o = \tfrac{1}{4}(mV_{if})^2/\rho W = \tfrac{1}{4}(mV_{if})^2/\rho B$$

(since now $B = W$) so

$$SNR_i/SNR_o = \tfrac{1}{2}V_{if}^2[0+\tfrac{1}{4}m^2/\tfrac{1}{4}(mV_{if})^2] = 0.5$$

SC-AM systems appear to have the same overall noise performance. The reason is that, in this comparison, the SSB demodulator is presented with half the signal power that appears in the DSB-SC case, but it is also presented with half the noise power, because $B = W$ instead of $2W$. If the same signal power is input to the demodulator, the SSB system is twice as good as the DSB-SC system.

The noise performance of a DSB-AM system is always inferior to that of its DSB-SC or SSB equivalent because much of the transmitted power is in the carrier, which contains no information. The same drawback applies, of course, to PAM systems. DSB systems may have a lower SNR than the equivalent SSB system because there may be interference when sideband path differences occur.

FM systems

A single tone input signal will be used for simplicity and again it will be assumed that the noise is spread uniformly over the IF bandwidth (B) and baseband limited by an output filter. A frequency demodulator will be used instead of the synchronous demodulator of Fig. 8.9. This will take the form shown in Fig. 8.10.

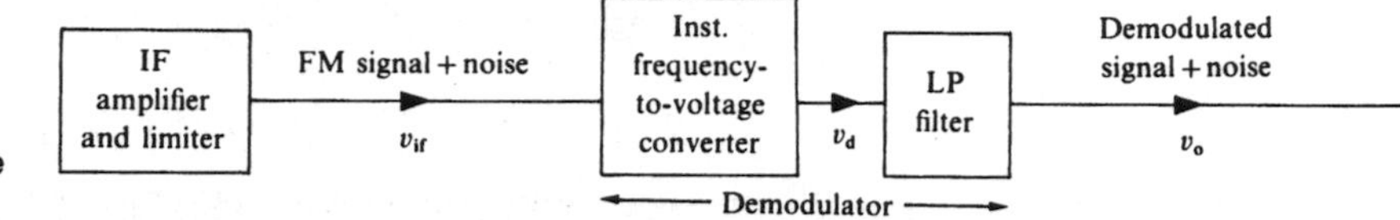

Figure 8.10. Frequency demodulation in the presence of noise.

The output (at maximum signal amplitude) of the IF amplifier will be

$$V_{if} = V_{if} \cos[\omega_{if} t + \beta \sin(\omega_t t)] + v_{Nif}$$

where $\beta = d\omega_m/\omega_t$. Hence, the normalized demodulator input powers are

$$S_i = \tfrac{1}{2} V_{if}^2 \quad \text{and} \quad N_i = \rho B$$

where, according to Carson's rule,

$$B = 2(\beta + 1)\omega_t = 2(d\omega_m + \omega_t)$$

If $\beta \gg 1$, then B becomes $2d\omega_m$ (that is 150 kHz in UK FM broadcasting). Hence

$$\text{SNR}_i = \tfrac{1}{2} V_{if}^2/\rho B = V_{if}^2/4\rho\, d\omega_m$$

The frequency-to-voltage converter produces an output voltage $v_d = d\theta/dt$ where $\theta = \omega_{if} t + \beta \sin(\omega_t t)$. Therefore $v_d = \beta\omega_t \cos(\omega_t t) = d\omega_m \cos(\omega_t t)$, where $d\omega_m$ is the maximum frequency deviation. This produces a normalized signal power of $\tfrac{1}{2} d\omega_m^2$. When this is translated from IF to baseband, half this signal voltage is passed to the LP filter, that is $S_o = \tfrac{1}{4} d\omega_m^2$.

Any noise component $v_N = V_N \cos(\omega_N t)$ input to the frequency converter will be multiplied with the IF carrier by the frequency converter (for example, see diagram of PLL converter (Fig. 5.13)). This results in both amplitude and phase modulation of the IF carrier by the noise component as shown in Fig. 8.11.

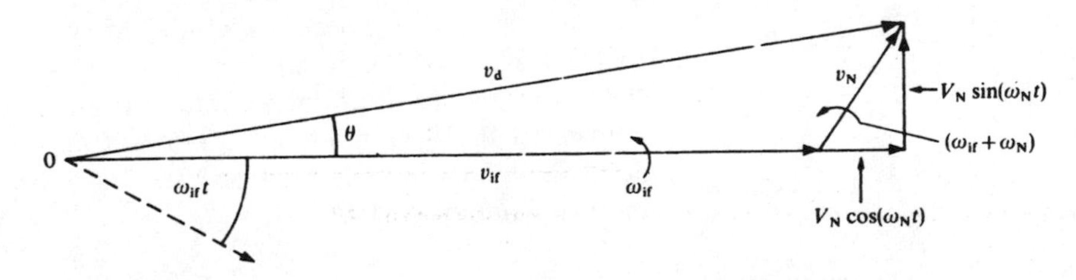

Figure 8.11. Phasor diagram showing how noise combines with the IF carrier.

The multiplying action of the frequency-to-voltage converter will give

$$v_d = v_{if} v_N$$
$$= \{V_{if} \cos[\omega_{if} t + \beta \sin(\omega_t t)]\}[V_N \cos(\omega_N t)]$$

From Fig. 8.11, the noise component in v_d will be

$$v_N = V_{if} + V_N \cos(\omega_N t) + jV_N \sin(\omega_N t)$$

and the instantaneous phase of the noise-modulated IF carrier relative to the unmodulated carrier is given by

$$\theta = \tan^{-1}\{V_N \sin(\omega_N t)/[V_{if} + V_N \cos(\omega_N t)]\}$$
$$= (V_N/V_{if}) \sin(\omega_N t)$$

(since normally $V_{if} \gg V_N$, but see subsection on the comparison of SNRs in DSB-AM and FM below).

Thus, the converter output noise voltage at any spot frequency (ω) will be

$$v_d = d\theta/dt = (V_N/V_{if})\omega_N \cos(\omega_N t)$$

giving a normalized noise power over bandwidth ($\delta\omega$) at frequency ω of

$$\delta N = \tfrac{1}{2}(\omega V_N/V_{if})^2$$

But $\frac{1}{2}V_N^2 = \rho\delta\omega$ so $\delta N = \rho\delta\omega\omega^2/V_{if}^2$. Hence, the normalized output noise power (N_o) which passes from the converter through the LP filter is given by integrating the power spectral density (ρ) of the noise δN over the baseband (W). Thus

$$N_o = \int_0^{+W} dN = \int_0^{+W} (\rho\omega^2/V_{if}^2)\, d\omega = \rho W^3/3V_{if}^2$$

The summation of component noise powers is illustrated in Fig. 8.12. Since the equation of the curves in Fig. 8.12 is $dN/d\omega = \rho\omega^2/V_{if}^2$, then

$$\text{SNR}_o = \tfrac{1}{4}\, d\omega_m^2/(\rho W^3/3V_{if}^2) = 3V_{if}^2(d\omega_m)^2/4\rho W^3$$

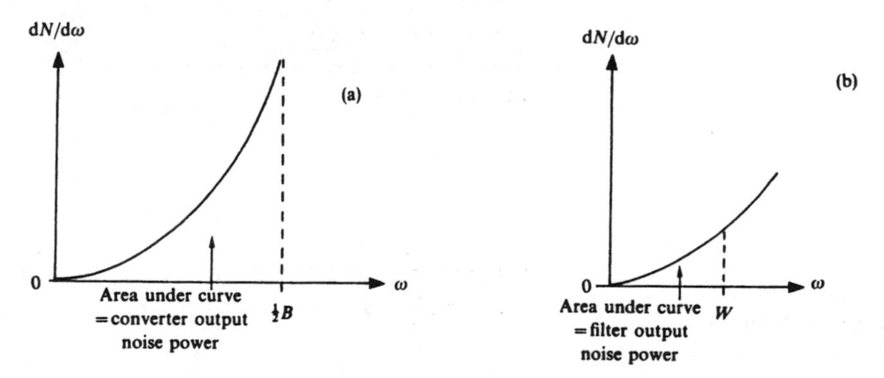

Figure 8.12. Spectral density of noise at (a) converter and (b) filter outputs.

and

$$\mathrm{SNR_i/SNR_o} = [V_{\mathrm{if}}^2/4\rho\, \mathrm{d}\omega_m]/[3V_{\mathrm{if}}^2(\mathrm{d}\omega_m)^2/4\rho W^3] = (W/\mathrm{d}\omega_m)^3/3$$

As would be expected, the smaller the baseband and the larger the frequency deviation, the better the noise performance of a demodulator.

Comparison of SNRs in DSB-AM and FM

We have deduced that $(\mathrm{SNR_o})_{\mathrm{FM}} = 3V_{\mathrm{if}}^2(\mathrm{d}\omega_m)^2/4\rho W^3$ and $(\mathrm{SNR_o})_{\mathrm{AM}} = \frac{1}{2}(mV_{\mathrm{if}})^2/\rho B$. Therefore

$$\begin{aligned}(\mathrm{SNR_o})_{\mathrm{FM}}/(\mathrm{SNR_o})_{\mathrm{AM}} &= [3V_{\mathrm{if}}^2(\mathrm{d}\omega_m)^2/4\rho W^3]/[\tfrac{1}{2}(mV_{\mathrm{if}})^2/\rho B] \\ &= 3B\,\mathrm{d}\omega_m^2/2m^2W^3\end{aligned}$$

As a basis of fair comparison we might make the AM bandwidth (B) equal to the FM output bandwidth of 15 kHz. The maximum frequency deviation in UK FM broadcasting is 75 kHz, so $(\mathrm{SNR_o})_{\mathrm{FM}}/(\mathrm{SNR_o})_{\mathrm{AM}} = 1.5 \times 25/m^2$, implying a great improvement in SNR when FM is used instead of AM since $m < 1$.

Exercise 8.6
Check on page 196

1. Calculate $(\mathrm{SNR_O})_{\mathrm{FM}}/(\mathrm{SNR_o})_{\mathrm{AM}}$ for a DSB-AM broadcast system with $m = 0.25$ and a output bandwidth of 5 kHz, and an FM broadcast system with an output bandwidth of 15 kHz and a maximum frequency deviation of 75 kHz.
2. Calculate the transmission bandwidth ratios for these two systems.

The improvement in $\mathrm{SNR_o}$ obtained with FM was made on the assumption that $V_{\mathrm{if}} \gg V_{\mathrm{N}}$. The value of V_{if} at which this assumption breaks down is the *FM threshold*. Below this value, the FM $\mathrm{SNR_o}$ reduces rapidly, becoming much less than that of AM. Above the FM threshold, we get the reverse effect in which the $\mathrm{SNR_o}$ increases considerably. If the main noise happens to be a signal from another station, we can see that this signal will be rejected and the FM receiver locks on to the stronger signal. This useful FM feature is known as the *capture effect*.

An increase in the modulation index (m) will increase the SNR of an AM system. This effectively increases the modulating signal amplitude. An increase in the modulation index ($\beta = \mathrm{d}\omega_m/\omega_t$) will increase the SNR of an FM system. In this case the increase in the modulating signal amplitude is achieved by an increase in the maximum frequency deviation ($\mathrm{d}\omega_m$). The improvement in $\mathrm{SNR_{FM}}$ is achieved at the expense of transmission bandwidth instead of at the expense of transmitted power, as is the case with AM.

Figure 8.12 shows that the noise power spectral density (ρ) of an FM receiver increases with the square of frequency. For AM receivers, ρ is constant. In order to increase the SNR_{FM} at higher frequencies, *pre-emphasis* is used. This is obtained with a transmitter pre-amplifier which has treble boost as shown in Fig. 8.13. The signal is restored to normal at the receiver by a de-emphasis circuit with a treble-cut response of 20 dB/decade from 3.182 kHz.

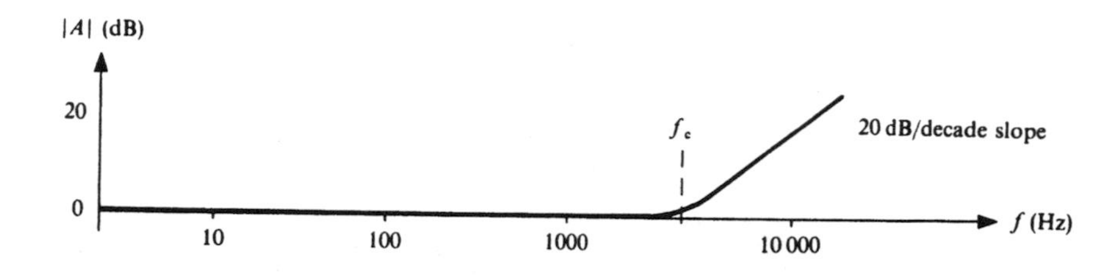

Figure 8.13. Pre-emphasis gain/frequency response used in UK FM broadcasting.

Pre-emphasis is also used in analogue gramophone recording to reduce the effect of groove surface noise and a sophisticated version, patented by Ray Dolby, is used in magnetic tape recording to reduce tape hiss. The principle of this noise reduction system is to apply pre-emphasis as shown in Fig. 8.13, but with the cut-off frequency (f_c) increasing as the signal level increases to ensure that noise reduction is greatest at low signal level, where it is most needed, and is progressively removed at high signal level. The reason for doing this is that a non-linear system gives most distortion at high signal level, so the removal of the non-linear frequency response at these levels gives better dynamic performance and a SNR improvement of about 10 dB. The principle of the system is shown in Fig. 8.14.

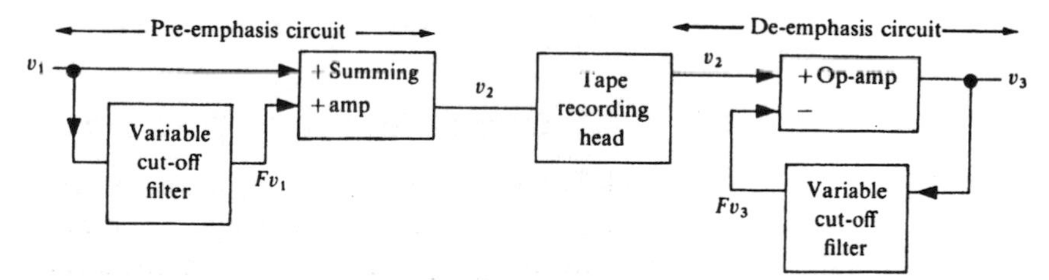

Figure 8.14. Layout of basic noise reduction system.

A variable filter, with a transfer function F, gives pre-emphasis to low-level input signals prior to recording. On play-back, the filter controls the negative feedback to an op-amp de-emphasis circuit. Since $v_3 = v_2 - Fv_3$, then the output $v_3 = v_2/(1+F)$. We want $v_3 = v_1$, so we must record $v_2 = (1+F)v_1$ as shown.

8.4 Noise effects in pulse modulation systems

Analogue PAM systems have a noise performance similar to those of DSB-AM. In quantized PAM and PCM systems the main noise is due to quantization. When an analogue signal is allowed to take on only discrete values, each sample can be in error by $\frac{1}{2}q$ (see Sec. 7.6). If we assume that, over a long time interval, all values of error voltage in the range Q_{n-1} to Q_n are likely to occur the same number of times, then the mean-square quantization error will be given by

$$\varepsilon_{ms} = (1/q) \int_{-q/2}^{+q/2} Q^2 \, dQ = q^2/12$$

The mean-square value of a tone voltage is

$$\tfrac{1}{2}V_t^2 = \tfrac{1}{2}(\tfrac{1}{2}nq)^2 = (nq)^2/8$$

So, the signal-to-quantization-noise ratio for a tone is given by

$$\text{SNR}_Q = \tfrac{1}{8}(nq)^2/(q^2/12) = 1.5n^2 = (10 \log 1.5 + 20 \log n)\ \text{dB}$$
$$= (1.8 + 20 \log n)\ \text{dB}$$

Thus, for example, the SNR_Q of a CD system that uses 2^{16} levels would be

$$10 \log 1.5 + 320 \log 2 = 98\ \text{dB}$$

Exercise 8.7
Check on page 196

Calculate SNR_Q for a triangular voltage signal using eight-bit uniform quantization.

Quantization noise is quite different in character to interference and thermal noise, whose magnitudes can fluctuate randomly over a wide dynamic range. It can be made arbitrarily small by reducing the quantum voltage (q) at the expense of increasing the transmission bandwidth. It can also be made non-cumulative by breaking the telecommunications link into sections with repeaters as explained in Sec. 7.8.

8.5 Error reduction in data signals

A feature of data signals is the way noise can completely alter the data by putting a 1 in place of a 0 and vice versa. An error in only one digit of a word completely alters it. A similar 'error' or noise voltage would have a much less radical effect on an analogue signal. The error probability (p_e) in data transmission increases up to a maximum value of 0.5 as the SNR increases as shown in Fig. 8.15.

An error probability of 0.5 occurs when noise dominates because the receiver is mainly receiving gaussian noise of zero mean value, which has a probability of 0.5 of being either positive or negative. At $S/N > 20$ the error

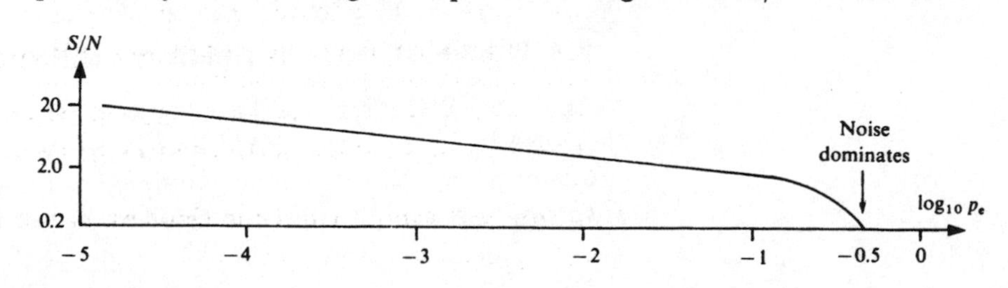

Figure 8.15. Variation of SNR with error probability.

rate becomes less than 10^{-5}, which can be coped with by including redundant information in the message. This extra information can be used either to *detect* that an error has occurred so that a request may be made for retransmission, or to *locate* the position of an error so that it may be *corrected* at the receiver without having to go through the time-wasting procedure of retransmission.

Redundant codes

A simple way to locate errors is to transmit the same signal twice and compare the two at the receiver. If the two signals are the same, then it is assumed that transmission has been error-free because it is improbable that a random error will occur twice at the same place in the bit stream. The second signal is completely redundant because all the information resides in the first signal; the second signal provides no more. Double transmission gives a high likelihood of error detection but is very inefficient. Code efficiency is defined as the ratio of information bits/total bits, so double transmission is only 50% efficient. A signal with no redundancy is 100% efficient. Most error detection schemes have efficiencies between these two values, with efficiency being traded for improved error detection.

Parity checks

A simple and effective error detection scheme for high SNR systems is the parity check in which an extra bit is added to a message to make the total number of 1s even. An example of this is shown for the ASCII symbol 'E' in Fig. 8.16.

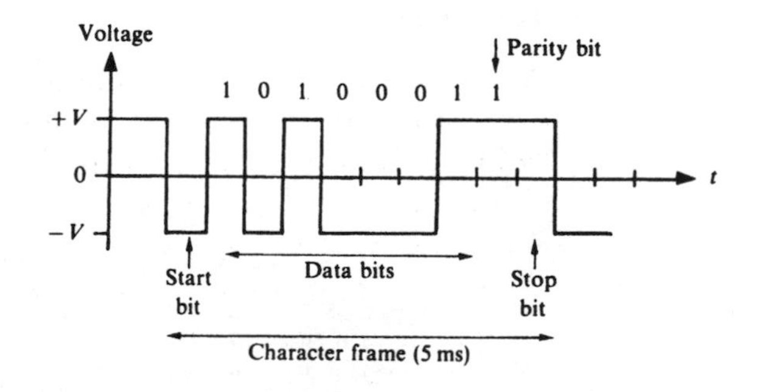

Figure 8.16. ASCII symbol with even parity bit.

The even parity check detects only one error; two errors will go unnoticed because a double error will give an even number of 1s.

When an error is detected the receiver must signal back to the transmitter a request to retransmit the data—this is called *automatic repeat request* (ARQ). There are three forms of ARQ, which will now be briefly described.

1. *Stop and wait ARQ* is the simplest. If the receiver detects no error it sends back a positive acknowledgement (ACK) signal and the transmit-

ter then sends the next word; otherwise a negative acknowldgement (NAK) signal is returned causing the transmitter to re-send. The process is repeated until no error is detected.

2. *Go-back N ARQ* allows the transmitter to send words without waiting for an acknowledgement. When an error is detected a NAK signal is returned, which causes the transmitter to re-send all the *N* words from the one in error. This system is easy to implement and allows for speedier data transmission than method 1.
3. *Selective-repeat ARQ* again allows the transmitter to send words without waiting for an acknowledgement after each word. If an error is detected the receiver requests retransmission of that word only and then continues to transmit from where it left off. This technique has the highest transmission rate but is the most expensive to implement.

An extension of the parity check idea is the *sum check* in which blocks of data may be checked by sending a series of bits representing their binary sum. Transmission errors are highly likely to cause the sum check to fail.

The simple parity and sum checks described will only detect single errors. Double errors go undetected. To locate an error, so that it can be corrected at the receiver and thus avoid the need for retransmission, more elaborate codes must be devised. Multiple-error detection and single-error location require the addition of more redundancy to a message. Many sophisticated codes have been devised, which not only provide multiple-error checks, but pinpoint errors so that they can be corrected at the receiver. These codes use up channel capacity.

Block codes

An important form of error locating code is the *block code*. A block of symbols may be checked if symbol and block check bits are used as shown in Fig. 8.17. Each symbol contains four information bits (*b*) and one check

	b_5	b_4	b_3	b_2	c_1
First symbol	0	0	1	1	0
Second symbol	1	0	1	1	1
Third symbol	0	1	0	0	1
Fourth symbol	0	0	0	0	0
Block check bits	1	1	0	0	0

Figure 8.17. Parity checks on a four-symbol block of data.

bit (*c*). A check bit checks the parity of its row. Each column is parity checked by a block check bit in that column. So, if b_5 of the second symbol was received as 0, this would show as parity error in the second symbol row and the b_5 column, thus locating the position of the erroneous bit at the 'grid reference' shown in Fig. 8.18.

First symbol	0	0	1	1	0
Second symbol	0 ···	0 ···	1 ···	1 ···	1
Third symbol	0	1	0	0	1
Fourth symbol	0	0	0	0	0
Block check bits	1	1	0	0	0

Figure 8.18. Parity checks locating erroneous bit.

Exercise 8.8
Check on page 197

Locate the incorrect digit in the following bit sequence, which represents a block of four ASCII symbols followed by a block check byte (BCB):

00011011 10011111 10111111 10011111 10000100

– – → data flow

Error location in the block code example of Fig. 8.18 requires nine check bits for 16 information bits, giving an efficiency of 9/16=56%, little better than double transmission. Hamming codes can improve on this.

Hamming codes

These use an extension of the block code principle with the parity bits distributed in certain positions throughout a symbol instead of being concentrated at the end of a block of symbols. They way Hamming code works can be appreciated with an example. Consider a symbol or word 1011 with its digits arranged in the bit positions (*b*) 3, 5, 6 and 7 shown in Fig. 8.19(a).

(a)

b_7	b_6	b_5	c_4	b_3	c_2	c_1
1	1	0		1		

(b)

b_7	b_6	b_5	c_4	b_3	c_2	c_1
1	1	0	0	1	1	0

Figure 8.19. Example of a 7,4 Hamming code.

The bit positions (*c*) 1, 2 and 4 are reserved for even parity check bits to satisfy the following equations:

$$b_7(+)b_6(+)b_5(+)c_4=0 \tag{8.2}$$

$$b_7(+)b_6(+)b_3(+)c_2=0 \tag{8.3}$$

$$b_7(+)b_5(+)b_3(+)c_1=0 \tag{8.4}$$

The symbol (+) means modulo-2 addition, a process carried out by an XOR gate (see Fig. 4.27). This gives the encoded word 1100110, shown in Fig. 8.19(b). Now suppose the received word is 1000110. Then Eqs (8.2) to (8.4) give

$$\begin{aligned} b_7(+)b_6(+)b_5(+)c_4 &= \\ b_7(+)b_5(+)b_3(+)c_2 &= \\ b_7(+)b_6(+)b_3(+)c_1 &= \end{aligned} \begin{bmatrix} 1 \\ 1 \\ 0 \end{bmatrix} \leftarrow \text{syndrome indicates error in } b_6$$

The column of bits on the right-hand side of the equals signs is the *syndrome* (S). S gives the position address (MSB uppermost) of the digit received in error. In this example $S = 110$, which indicates b_6 to be in error. An error may be corrected at the receiver using a system such as that of Fig. 8.20.

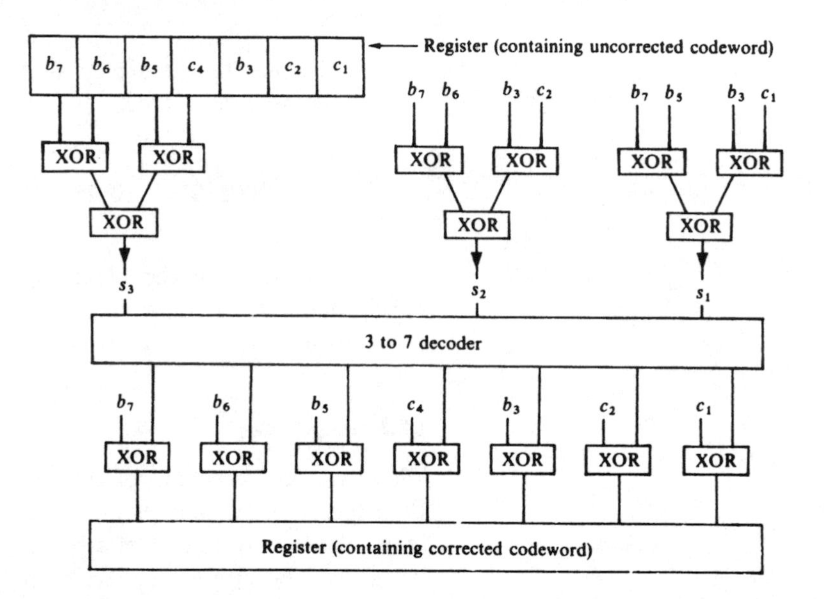

Figure 8.20. Error-correcting system.

Exercise 8.9
Check on page 197

Deduce which bit is in error in the Hamming encoded sequence 1101110.

Exercise 8.9 raises the question of why positions 1, 2 and 4 are chosen for the parity bits. Consider the list of binary numbers in Table 8.1. This table shows that if the LSB of the error address is 1, the error must be in one of the positions 1 (001), 3 (011), 5 (101) or 7 (111); if the next bit is 1, the error must be in one of the positions 2 (010), 3 (011), 6 (110) or 7 (111); if the MSB is 1 the error must be in one of the positions 4 (100), 5 (101), 6 (110) or 7 (111). The bit positions 1, 2 and 4 appear only once in each of these three sets of positions, so these are chosen for the parity bits. This gives a 7,4 codeword with seven bits, four of which carry information.

Table 8.1. Binary numbers

Denary	Binary	Denary	Binary
0	000	4	100
1	001	5	101
2	010	6	110
3	011	7	111

Exercise 8.10
Check on page 197

By considering a list of binary numbers from 1 to 15, write out the codeword that will locate a single error in the sequence 10101010101.

Block structure of Hamming code

The structure of a 7,4 Hamming code is based on forming a single-row, seven-column codeword matrix $[C]$ where

$$[C]=[b_7 \quad b_6 \quad b_5 \quad b_4 \quad b_3 \quad c_2 \quad c_1]$$

A Hamming matrix $[H]$ gives the bit positions in binary form:

$$\begin{matrix} 7 & 6 & 5 & 4 & 3 & 2 & 1 \end{matrix} \quad \leftarrow \text{bit positions}$$

$$[H]=\begin{bmatrix} 1 & 1 & 1 & 1 & 0 & 0 & 0 \\ 1 & 1 & 0 & 0 & 1 & 1 & 0 \\ 1 & 0 & 1 & 0 & 1 & 0 & 1 \end{bmatrix}$$

Applying modulo-2 addition to the bit elements of these matrices as follows:

$$[H][C]^{\mathrm{T}}=[S]$$

That is

$$\begin{bmatrix} 1 & 1 & 1 & 1 & 0 & 0 & 0 \\ 1 & 1 & 0 & 0 & 1 & 1 & 0 \\ 1 & 0 & 1 & 0 & 1 & 0 & 1 \end{bmatrix} \begin{bmatrix} b_7 \\ b_6 \\ b_5 \\ c_4 \\ b_3 \\ c_2 \\ c_1 \end{bmatrix} = \begin{bmatrix} s_3 \\ s_2 \\ s_1 \end{bmatrix}$$

gives

$$b_7(+)b_6(+)b_5(+)c_4=s_3$$
$$b_7(+)b_6(+)b_3(+)c_2=s_2$$
$$b_7(+)b_5(+)b_3(+)c_1=s_1$$

Hamming distance

The various codewords, which are made up of information bits (b) and parity check bits (c), will be different by one or more digits. Consider $w_i=1001110$ and $w_j=0000111$. These differ in the MSB, mid-bit and LSB columns. That is, they differ in three respects. This difference or *distance* is expressed as $d_{ij}=3$. Each pair of a set of codewords will have a value for d_{ij}. The smallest value of d_{ij} for the set of words is the *minimum distance*

(d_{min}) or *Hamming distance*. The greater d_{min}, the easier it will be to distinguish words received over a noisy channel.

To detect that a received codeword is not valid, d_{min} must be at least one more than the number of digit errors. To locate an error d_{min} must be at least one more than twice the number of digit errors. On this basis, the 7,4 Hamming code must have $d_{min}=3$ because we have seen that it will locate one error.

Exercise 8.11
Check on page 198

Verify that the 7,4 Hamming code has $d_{min}=3$ by listing all the codewords.

A Hamming code is one in which $d_{min}=3$. In general the number of bits in a Hamming coded and uncoded word, n and b respectively, are related to the number of check bits c by; $c=n-b$ and $n=2^c-1$.

In a seven-bit Hamming code these equations are satisfied by $n=7$, $b=4$ and $c=3$, giving a 7,4 Hamming code. If $c=4$, as it does in the code of Exercise 8.10, a single error in one of $n=15$ bit positions can be located. In this example, $b=11$, so we have a 15,11 Hamming code.

The price paid for error location is high—in the seven-bit Hamming code there are three redundant bits for every four information bits, giving a code rate of 4/7. This code will work only if the error rate is not greater than one in seven bits. This is quite often the case over a long period of transmission, but errors may occur in bursts. Then interleaving is employed to spread the errors evenly. For example, instead of sending the bits in the order 1, 2, 3, 4, 5, 6, 7, 8 and 9, they might be sent in the order 1, 4, 7, 2, 5, 8, 3, 6, 9 to ensure that a burst of three successive errors would result in one error in three bits.

Exercise 8.12
Check on page 198

Write out the bit sequence showing the interleaving required in order that one of three successive digit errors could be located with a 7,4 Hamming code.

An extension of the block code idea, for high-error systems, treats groups of bits or *symbols* instead of individual bits to give *Reed–Solomon* (RS) block codes, which locate symbol errors rather than bit errors. This is a sensible way to locate *error bursts*. RS codes are used in CD technology to correct for errors due to disc defects. These errors tend to occur in bursts—where there is a scratch on the disc surface for example. Also, the bursts themselves tend to be grouped together so, in order to spread the error bursts over the symbol sequences, interleaving is used. Computer disk systems use simple parity checks, so floppy and hard disks require much greater physical protection than CDs.

Hamming and RS codes are special cases of a general set of codes known as *Bose, Chaudhuri, Hocgenghem* (BCH) codes, which all employ b information bits with c check bits to give a word of $n=b+c$ bits. These differ from block codes in that the data is sent in continuous streams rather than blocks, with the parity bits interspersed or *convolved* according to the code rules.

Cyclic codes form another group of parity check codes, which are important because short check words are easily produced with a feedback shift register as shown in Fig. 8.21 where an eight-bit word generates a four-bit cyclic redundancy check (CRC) codeword.

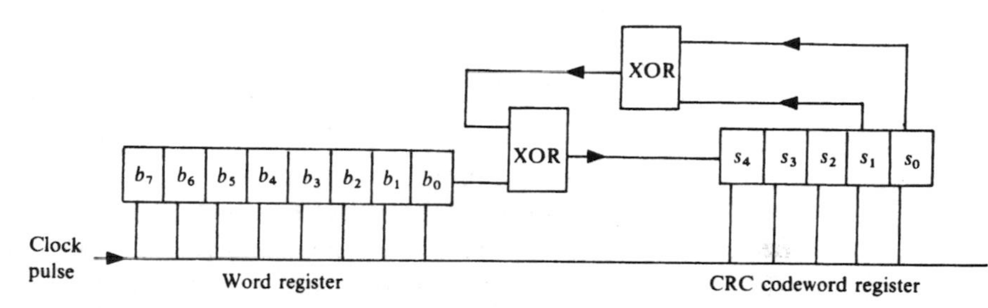

Figure 8.21. Generation of a CRC signature.

The registers are cleared initially and the information word is loaded serially into the eight-bit register by a sequence of eight clock pulses. A further eight clock pulses will clear the word register and generate a CRC codeword, which is then transmitted after the information word. A CRC codeword is generated at the receiver in exactly the same way and compared to the received CRC codeword. Any difference between these CRCs indicates transmission errors. The technique is really equivalent to double transmission but only one-and-a-half bytes are transmitted. Thus cyclic codes can give almost as much error detection as double transmission but have a greater efficiency—67% in the example of Fig. 8.21.

A cyclic code is a form of block code, in which successive rows in the block are lateral shifts of the previous row. Consider, for example, the word 100. Then

$$[G]=\begin{bmatrix}1 & 0 & 0\\ 0 & 1 & 0\\ 0 & 0 & 1\end{bmatrix}$$

Division of any row of the matrix G by the next row below gives a quotient 2. This can be checked at the receiver in order to detect errors.

Channel coding and source coding

In this section we have been concerned with the way extra bits can be added to a coded signal at source, so that any errors incurred in the communications channel can be detected and, perhaps, located. This *channel coding* is quite different to the *source coding* that is necessary to carry information. For example, PCM is source coding and a single-parity bit channel coding is used with each PCM word for error detection purposes. In channel coding redundancy is increased; in source coding, the aim is to reduce the number of bits needed to encode a message. A signal, such as human speech, contains much natural source redundancy, which makes it understandable against a noisy background, but more efficient forms of redundancy could be devised for hearing-error correction. In telecommunications systems coding economy is needed to conserve bandwidth.

Summary

Unwanted signals, known as noise, constitute a basic limitation on the performance of a communications system. Noise may be external, such as atmospheric or man-made, or internal to the system. Two forms of internal noise, known as Johnson noise and shot noise, are white and gaussian. The former arises from the random motion of conducting electrons making collisions with the thermally vibrating lattice ions of the conductor. Shot noise occurs in current flow across pn junctions and vacuum tubes.

External noise may be reduced by screening and using highly selective circuits, but internal noise can only be reduced by lowering temperature. What matters at the receiver is not the level of the noise but the signal-to-noise ratio. This can be maximized by ensuring that the signal level is high. This is done by transmitting at a sufficiently high power or bandwidth. Increasing the latter, however, also increases the noise power (see Chapter 9).

Several other figures of merit are used to describe the performance of a system in the presence of noise apart from SNR. Noise figure (F) can be useful and is the ratio (SNR at input)/(SNR at output). For low-noise systems, noise temperature is more appropriate. The input-referred noise power, equal to $(F-1)$ times the actual input noise power, can be a convenient figure for calculating the performance of cascaded systems. When $F=1$ we have a noiseless system. Large values of F indicate a noisy system—which makes F a figure of de-merit.

The performance of a DSB-AM system in the presence of noise is inferior to that of its DSB-SC or SSB equivalents because much of the transmitted power is in the carrier, which contains no information. An increase in the modulation index (m) will increase the SNR of an AM system. This effectively increases the modulating signal amplitude. An increase in the modulation index (β) will increase the SNR of an FM system. In this case the increase in the modulation signal amplitude is achieved by an increase in the maximum frequency deviation. The improvement is achieved at the expense of transmission bandwidth instead of at the expense of transmitted power as is the case with AM. Further output SNR improvement is achieved in FM broadcasting by pre-emphasis of the treble frequencies.

Analogue PAM systems have a noise performance similar to those of DSB-AM. In quantized PAM and PCM systems the main noise is due to quantization because repeaters can be used along the transmission link to remove the interference and atmospheric noise induced into each section between the repeaters. The use of repeaters ensures that the noise along a complete link is non-cumulative. Quantization noise can be made arbitrarily small by reducing the quantum voltage at the expense of increasing the transmission bandwidth.

Random noise can completely alter data by putting a 1 in place of a 0 and vice versa. An error in only one digit of a word completely alters it. A similar 'error' or noise voltage would have much less effect on an analogue signal. The error probability in data transmission increases up to a maximum value of 0.5 as the SNR increases.

Parity digits can be used either to detect that an error has occurred (leading to a request for retransmission) or to detect the position of an error so that it can be corrected at the receiver. This may be done by adding parity bits on a block or serial basis. Such codes work only if the error rate is fairly well distributed. Where errors are likely to occur in bursts, interleaving must be employed to spread the errors evenly throughout the transmission period. The Reed–Solomon (RS) block codes are used for this purpose. The Hamming and RS codes are special cases of a general class of Bose, Chaudhuri, Hocqenghem (BCH) codes.

► Checks on exercises

8.1 An amplifier at position B would boost the signal and the noise picked up in the link equally, leaving the SNR unaltered. An amplifier at position A would boost the signal only, thus giving a higher SNR at the receiver input. The amplifier at B cannot increase this SNR. The reason for putting an amplifier at B is to increase the signal power before the decoder or demodulator adds more noise. The simplest way to maximize the SNR is to have a powerful input signal.

8.2 In the early days of radio communication, the crude equipment used resulted in the received signals being weak, distorted and obscured by noise. The dots and dashes of Morse code were distinctive enough to be heard against levels of background noise that would have made speech signals unintelligible. Binary coded signals are easier to distinguish from noise than analogue signals.

It takes longer to convey a message by Morse code than by speech, although certain code conventions, like SOS, were used to overcome this drawback. The electronic equipment used today can encode a signal much faster than the Morse operator, so binary coded signals can be used for speech and even for the high frequencies required for video transmission. Morse code has, therefore, become obsolete.

8.3 The r.m.s. noise voltage produced by 10 km of twin cable with a loop resistance of 25 Ω/km at 300 K and 10 kHz bandwidth is given by

$$V^2 = 4kTBR = 4 \times 1.37 \times 10^{-23} \times 300 \times 10^4 \times 250$$

so

$$V = 0.2\,\mu\text{V}$$

This small voltage suggests that thermal noise is not the main noise in narrow-band telephone lines. Cross-talk and interference are much more important.

8.4 1. $F = 1 + [V_{\text{Ni}}^2 + (RI_{\text{Ni}})^2]/4kTBR_s$

$$= 1 + [(0.3\,\mu\text{V})^2 + (1.5\,\mu\text{V})^2]/4 \times 1.37 \times 10^{-23} \times 290 \times 4000 \times 1000$$
$$= 1 + [2.34 \times 10^{-12}]/6.36 \times 10^{-14} = 37.8$$

Now $F = \text{SNR}_i/\text{SNR}_o$ so

$$\text{SNR}_o = \text{SNR}_i/F = V_s^2/4kTBR_sF$$
$$= (4 \times 10^{-12})/6.36 \times 10^{-14} \times 37.8 = 1.6$$

2. If the source resistance is matched to the input resistance of an amplifier, the r.m.s. input noise current due to the source resistance will be given by

$$I_N = \sqrt{(4kTBR_s/2R_s)}$$

Hence, the noise power at an amplifier input will be $I_N^2R_{\text{in}}$. This will equal kTB when $R_s = R_{\text{in}}$.

8.5 $T_e = (F-1)T = 9 \times 290 = 2610$ K and $19 \times 290 = 5510$ K. High noise temperatures reflect the relative noise of the two amplifiers just as well as the noise figures, but the noise figure gives a more manageable number and would be used in preference to noise temperature at these noise levels.

8.6 1. $(\text{SNR}_o)_{\text{FM}}/(\text{SNR}_o)_{\text{AM}} = 3B\,\text{d}\omega_m^2/2m^2W^3$ where $m = 0.25$, $B = 5$ kHz, $W =$ 15 kHz and $\text{d}\omega_m = 75$ kHz. So the FM to AM SNR figure is 200 or 23 dB.

2. The transmission bandwidth ratio for the systems (FM to AM) = $150/10 = 15$.

The cost of a 200 times improvement in SNR is thus a 15 times increase in transmission bandwidth. A further improvement in the FM SNR is achieved in practice by pre-emphasizing the treble at the transmitter and de-emphasizing the treble at the receiver.

8.7 The mean-square value of a triangular voltage is $V_t^2/3 = (nq)^2/3$. Hence, the SNR_Q for a triangular voltage signal is $[(nq)^2/3]/[q^2/3] = n^2 =$ $20 \log 8 = 18$ dB.

8.8 Rearranging the bit sequence into a block highlights parity violations in the third column and the third byte:

```
10000100  1st byte
10011111  2nd byte
10111111  3rd byte  ←parity violation
10011111  4th byte
00011011  BCB
  ↑
parity violation
```

The third byte should be 10011111.

8.9 In the received Hamming encoded sequence 1101110,

$$\begin{aligned} b_7(+)b_6(+)b_5(+)c_4 &= \\ b_7(+)b_6(+)b_3(+)c_2 &= \\ b_7(+)b_5(+)b_3(+)c_1 &= \end{aligned} \begin{bmatrix} 1 \\ 0 \\ 0 \end{bmatrix} \leftarrow \text{syndrome } [S]$$

$[S]=[100]$, indicating that the bit in position 4 is in error. Notice that the code locates an erroneous check bit!

8.10 Inspection of a list of binary numbers from 1 to 15 in Table 8.2 reveals

Table 8.2. Binary numbers from 0 to 15

	4321		4321
0	0000	8	1000
1	0001	9	1001
2	0010	10	1010
3	0011	11	1011
4	0100	12	1100
5	0101	13	1101
6	0110	14	1110
7	0111	15	1111

that: if the LSB of the error address is 1, the error must be in one of the positions 1, 3, 5, 7, 9, 11, 13 or 15; if the next bit is 1, the error must be in one of the positions 2, 3, 6, 7, 10, 11, 14 or 15; if the next bit is 1 the error must be in one of the positions 4, 5, 6, 7, 12, 13, 14 or 15; and if the MSB is 1, the error must be in one of the positions 8, 9, 10, 11, 12, 13, 14 or 15. The bit positions 1, 2, 4 and 8 appear only once in each of these four sets of positions, so these are the positions chosen for the four parity bits.

The Hamming codeword for the information bits 10101010101 must satisfy the equations:

$$b_{15}(+)b_{14}(+)b_{13}(+)b_{12}(+)b_{11}(+)b_{10}(+)b_9(+)c_8 = 0$$

$$b_{15}(+)b_{14}(+)b_{13}(+)b_{12}(+)b_7\ (+)b_6\ (+)b_5(+)c_4 = 0$$

$$b_{15}(+)b_{14}(+)b_{11}(+)b_{10}(+)b_7\ (+)b_6\ (+)b_3(+)c_2 = 0$$

$$b_{15}(+)b_{13}(+)b_{11}(+)b_9\ (+)b_7\ (+)b_5\ (+)b_3(+)c_1 = 0$$

So 10101010101 becomes

$$\begin{array}{ccccccccccccccc} b_{15} & b_{14} & b_{13} & b_{12} & b_{11} & b_{10} & b_9 & c_8 & b_7 & b_6 & b_5 & c_4 & b_3 & c_2 & c_1 \\ 1 & 0 & 1 & 0 & 1 & 0 & 1 & 0 & 0 & 1 & 0 & 1 & 1 & 0 & 1 \end{array}$$

This a 15,11 Hamming codeword.

8.11

7654321	7654321
bbbcbcc	*bbbcbcc*
0000000	1001011
0000111	1001100
0011001	1010010
0011110	1010101
0101010	1100001
0101101	1100110
0110011	1111000
0110100	1111111

$$b_7(+)b_6(+)b_5(+)c_4 = 0$$
$$b_7(+)b_6(+)b_3(+)c_2 = 0$$
$$b_7(+)b_5(+)b_3(+)c_1 = 0$$

8.12 A 7,4 Hamming code can locate one error in seven bits, so the bit sending sequence required will be

1, 4, 7, 10, 13, 16, 19, 2, 5, 8, 11, 14, 17, 20, 3, 6, 9, 12, 15, 18, 21

In general interleaving a 7,4 Hamming code requires storage at the transmitter and receiver for $7n$ digits, where n is the statistically expected burst length.

Self-assessment questions

8.1 Define 'noise' in a telecommunications system.
8.2 List the types of noise in terms of their origin.
8.3 Explain the terms 'white' and 'gaussian' as applied to noise.
8.4 How may the effects of external noise be minimized?
8.5 Define signal-to-noise ratio.
8.6 How may the SNR be maximized?

8.7 Define noise figure.

8.8 Why might noise temperature be more appropriate than noise figure?

8.9 Derive an expression for SNR_i/SNR_o for DSB-AM and DSB-SC-AM.

8.10 Show that the noise power spectral density (ρ) of an FM receiver increases with the square of frequency.

8.11 Explain why pre-emphasis is used in FM broadcasting.

8.12 How can quantization noise be minimized?

8.13 Explain what is meant by Hamming distance.

8.14 Which bit is in error in the Hamming encoded sequence; 1111100?

8.15 Locate the incorrect digit in the following sequence of four bytes followed by a BCB:

00010001 10010011 10011111 10011101 10000100

– – → data flow

8.16 State the purpose of interleaving in codes.

9

Information theory

Describing what is being telecommunicated

Objectives

When you have completed this chapter you will be able to;

- Describe the processes that occur in an electronic telecommunications system
- Distinguish between data and information
- Describe, in general terms, what constitutes information
- Describe how telecommunications information may be quantified
- Establish a relation between information and probability
- Define information rate
- Distinguish between information and signalling rates
- Define channel capacity
- State the laws of Shannon and Hartley
- Describe how bandwidth and SNR may be traded

The signal obtained from the perfect transducer contains all the information concerning the input variable, which may be sound, light or data. If all this information could arrive intact at the human receiver, we would have the ideal communications system. All practical systems suffer from the distorting effects of non-linearity, interference and thermal noise. In this chapter we consider the behaviour of the ideal system so that the concepts inherent in information transfer are not obscured by such practicalities.

Information may be simply defined as the difference between our knowledge at one instant and our knowledge at the next instant. This definition will be refined in this chapter so that information may be quantified. Information rate and channel capacity will then be defined.

9.1 Electronic information systems

Most electronic systems are data processors that perform operations on analogue and digital signals. When these processors get involved with exchanging information over a distance, they become part of a telecommunications system. This idea is summarized in Fig. 9.1.

The functions outlined in Fig. 9.1 are as follows. Thoughts are put into words or pictures by the thinker (the information source) and are then converted into electrical form by a transducer. Further encoding may be

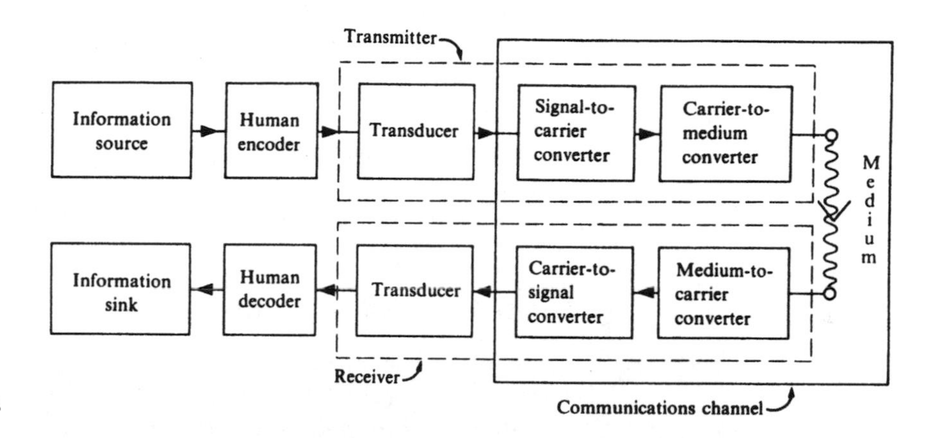

Figure 9.1. A telecommunications system.

carried out, according to the nature of the telecommunications link, to facilitate transmission. At the receiving end of the link, which is made up of some medium such as copper, glass or space, the reverse process occurs, finally leaving another thinker (the information link) to recreate the original thoughts. The system can be broken down into three subsystems: the transmitter, the receiver and the channel.

The human codes used to convey thoughts are the spoken word, pictures or symbols. Sound is converted into an analogous electric signal by a microphone transducer and pictures by a TV camera tube. Symbols are converted into digital signals by a teletypewriter or document scanner. Telecommunications is basically concerned with the exchange of information by means of data transmission through an electrical channel.

9.2 Data and information

The terms 'data' and 'information' are often used interchangeably, but information theory (IT) distinguishes them. In Fig. 9.1 it is data that traverses the channel and information is what we are left with after this data has been sensed and processed in some way. A basic simplex data telecommunications system is shown in Fig. 9.2.

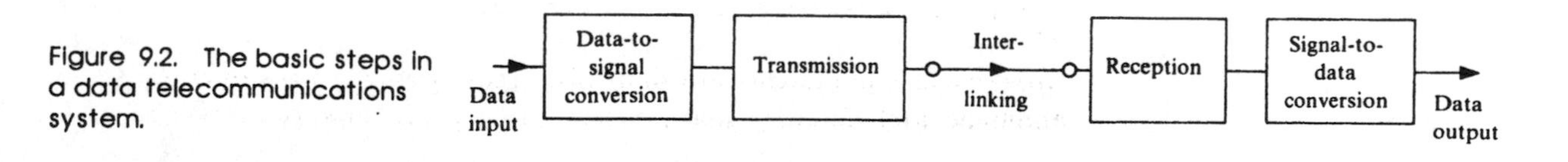

Figure 9.2. The basic steps in a data telecommunications system.

Electrical input data from a transducer is converted into a form suitable for the transmission medium. The transmission process may include modulation of the baseband signal. In a practical system, unwanted signals and random noise will be received along with wanted data. At reception the wanted signal will be extracted and converted back into the original data. The reception process may include demodulation.

An important part of the sending process is the putting of the data into some form of order. The more ordering done in the input process, the easier it is for the output processor to extract the data from the received signal in the presence of noise. As far as telecommunications is concerned, the electrical energy flowing between a transmitter and receiver carries a pattern, a set of ordered events such as a pulse train, that constitutes data from which information may be extracted.

9.3 The nature of information

To appreciate further how we might distinguish between data and information, consider the following example. If you received a telephone directory with the names, addresses and numbers each arranged in random order, it would be impossible to link the names, addresses and numbers, so the directory would be useless as a source of information. All the data is there, but it does not tell us anything we did not know before we perused the directory, except the tiny scrap of information that the directory was of no use in finding the required information! Data conveys information only if it is ordered. Randomness cannot be informative.

Randomness cannot provide information because it implies uncertainty. As soon as randomness is reduced, by making one set of events more likely than another, a pattern may become discernible. Patterns convey information. A receiver starts off with no information—just a set of random possibilities. As information is sent, these possibilities become certainties. A teleprinter, for example, knows that it will receive a finite number of specified alphanumerics (data), but it does not know in what order it will receive them. The more ordered the data received, the less processing (ordering) there is to do at the receiver. We could say that the disorder at the receiver decreases as more information (energy) is absorbed. This is the second law of thermodynamics applied to telecommunications.

If information is to be telecommunicated, the receiver must have no previous knowledge of that information. You cannot inform someone of what they already know. In the case of DSB-AM, for example, there is no point in transmitting the carrier, as far as conveying information is concerned, because the carrier frequency is already known by the receiver. If an information source is predictable (that is, you are certain what it is going to do) it cannot convey information. This suggests that we might look upon information as that which reduces uncertainty.

Let us take this idea a little further. If you asked the question: 'Who won the three-thirty at Newmarket?', you would get the name of a horse. If you knew that there were only two starters, this bit of news would not give you much information because you already knew that it had to be one of them (assuming that the probability of both not finishing or being disqualified was negligible). If you had also known the favourite and the favourite had won, you would have heard what you were expecting to hear, so the answer told your even less in this case. If there had been a field of 10 runners, the answer to the above question would convert greater initial uncertainty into certainty, so much more information would be obtained with the answer to this question. Even more information would have been obtained had a rank outsider won.

9.4 Quantifying information

Information appears to be measured by its unexpectedness. An unexpected event is one that has a low probability of occurring. In other words, information (I) is inversely proportional to some function (f) of the probability of an event:

$$I \propto f(1/\text{probability of an event})$$

We now apply this idea to telecommunications, where an event would be the receiving of a message. Suppose an information source has Q symbols at its disposal and sends messages N symbols in length as illustrated in Fig. 9.3. If any one of the Q symbols is equally likely to occur in any position of the message, the number of possible messages (M) will be Q^N.

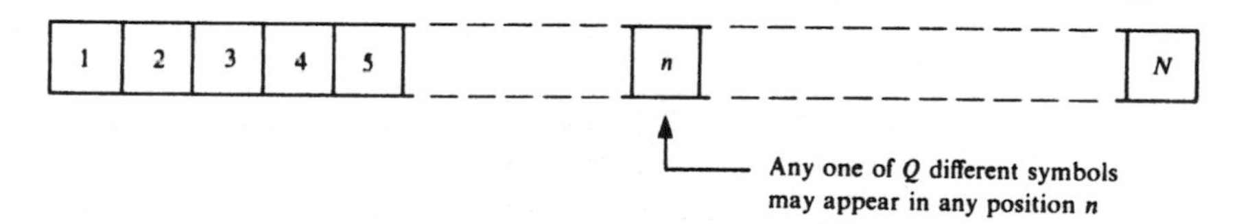

Figure 9.3. A message format.

The more possible messages there are, the more information will be obtained when one of them is received, so information (I) will be depend on M, that is

$$I = kf(M) = kf(Q^N)$$

where k is a constant of proportionality and f is some function of M. In order to find this function, time must be considered. This seems reasonable because the amount of information received must have something to do with the time available for transmission and how fast the messages are sent. Suppose that R symbols are transmitted every second, then the total number transmitted in time T will be RT, so the information transmitted in times T_1 and T_2 will be

$$I_1 = kf(Q^{RT_1}) \qquad \text{and} \qquad I_2 = kf(Q^{RT_2})$$

Thus

$$I_1/I_2 = f(Q^{RT_1})/f(Q^{RT_2})$$

If we assume that the information transmitted is proportional to the time of transmission for the same values of Q and R, then $I_1/I_2 = T_1/T_2$. So

$$T_1/T_2 = f(Q^{RT_1})/f(Q^{RT_2})$$

This relationship is satisfied if

$$f(Q^{RT}) = \log(Q^{RT}) = RT \log Q$$

Hence

$$I = kRT \log Q = kN \log Q \tag{9.1}$$

Equation (9.1) is stating the obvious in one respect: that information is proportional to the number of symbols (N) in the message. What is not so obvious, until the information rate (R) is considered, is that information is proportional to the logarithm of the number of equiprobable symbols (Q) that could be selected to fill the N places.

The unit of information can be defined so that $k = 1$, but this still gives a choice of which kind of log function to adopt. If base 2 is used the unit is the *binit*, for base 10 the unit is the *decit*, and if natural logs are used the unit is the *nat*. So

$$I = N \log_2 Q \text{ binits} \quad \text{or} \quad N \log_{10} Q \text{ decits} \quad \text{or} \quad N \log_e Q \text{ nats}$$

In general

$$I = N \log Q \tag{9.2}$$

Exercise 9.1
Check on page 212

Calculate the information content, in binits and decits, of the following:

1. A message with only one possible symbol.
2. A message containing two equally probable symbols.
3. A message containing one of 128 equally probable symbols.
4. A PCM word of eight equally probable binary symbols.
5. An A4 sheet of 50 lines each of 64 equally probable characters.

9.5 Information and probability

We have seen that the information (I) in a message N symbols long, consisting of Q equally probable symbols, is given by $I = N \log Q$. As Q increases, the probability (p) of occurrence of a particular symbol decreases, that is $p = 1/Q$. Hence

$$I = N \log(1/p) \tag{9.3}$$

An example of how this definition of information may be used is the way in which a chess board position may be indicated. A chess board has its squares numbered in ascending order from top left to bottom right as in Fig. 9.4. A chess piece must occupy one of the 64 squares. In this example, we are not concerned with which piece it is in a particular square, so 'piece' represents a single-symbol message. This means that $N = 1$. Since a piece is equally likely to be on any square, the probability of it being on a particular square is 1/64. If you were told that 'a piece was on square 20', for example, the information (I) received would be $\log_2(64) = \log_2(2^6) = 6$ binits.

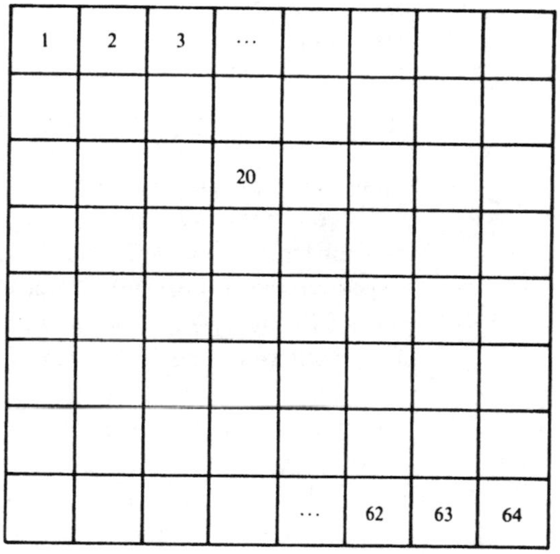

1	2	3	...				
			20				
				...	62	63	64

Figure 9.4. A chess board.

This example may be used to illustrate other ways of giving information about the position of a piece. Instead of the statement 'a piece is on square 20', the same information may be given in the form 'a piece is in column 4, row 3'. In this case, since the probabilities of a given row or column are both 1/8,

$$I = \log_2 8 + \log_2 8 = 3 + 3 = 6 \text{ binits}$$

A third method of learning which square was occupied would be to ask a series of questions to which there are equally probable 'yes' or 'no' answers. Each answer would give 1 binit of information. A minimum of six such questions would be needed to locate the square, a typical starting one being 'is the occupied square in the top half of the board?' This would be followed by 'is the occupied square in the left-hand half of the board?' Each question in this binary search reduces the uncertainty of the square's location by a half (1 binit), so these six questions elicit a total of 6 binits of information.

Exercise 9.2
Check on page 212

1. How much information has been given if someone states that they obtained the score of 7 when they rolled two dice?
2. How would this information value be affected if the score had been 12?

Binits and bits

Exercises 9.1 and 9.2 show that the information transmitted by either one of two equally probable symbols is 0.3 decits or 1 binit. The information content of one binary unit (bit) is 1 binit. This numerical equivalence between binary and information units, which arises because $\log_2 2 = 1$, has meant that the unit of binary signal information is usually called 'bit' rather than 'binit'.

Messages of unequal probability

Equation (9.3) gives a definition of information in terms of equally probable events, such as the reception of a particular symbol. But what if the probabilities are not equal? The definition inherent in Eq. (9.3) can be refined to cover unequal probabilities of occurrence by considering the concept of the *average information* in a message as follows. Suppose there are M different and independent messages m_1, m_2, m_3 and so on, with probabilities of occurrence p_1, p_2, p_3 etc. Suppose also that a sequence of L messages is sent during a long transmission period. For very large L, it would be reasonable to expect that in this message sequence there would occur p_1L of m_1 messages, p_2L of m_2 messages, p_3L of m_3 messages and so on. The total information in such a message sequence would be given by

$$I_{\text{total}} = p_1L\log(1/p_1) + p_2L\log(1/p_2) + p_3L\log(1/p_3) + \cdots$$

Each message contains different amounts of information, but the average information (I_{av}) of a sequence of M such messages would be given by

$$I_{\text{av}} = I_{\text{total}}/L = p_1\log(1/p_1) + p_2\log(1/p_2) + \cdots + p_M\log(1/p_M)$$

That is

$$I_{\text{av}} = \sum_{n=1}^{M} p_n\log(1/p_n) \tag{9.4}$$

I_{av} is called *entropy*—a term borrowed from thermodynamics, where a similar equation to (9.4) appears and a number of parallels with information theory exist. The second law of thermodynamics states that a closed system (one without any further energy input) will always end up in the condition that can be realized in the maximum number of ways. In other words, the disorder or entropy in the system will increase. If energy is put into a thermodynamic system, order increases. When information, which is a form of energy, is received, randomness or entropy or uncertainty are reduced. Average information is the entropy in a telecommunications system.

Equation (9.4) puts no restriction on the number of *symbols* in a message—a message may consist of one symbol only, selected from a set of symbols of unequal probability. So, for two messages m_1 and m_2, each with one symbol whose probability of appearing in that message is p_1 and p_2 respectively, the average information is given by

$$I_{av} = \sum_{n=1}^{2} p_n \log(1/p_n) = p_1 \log(1/p_1) + p_2 \log(1/p_2)$$

Exercise 9.3
Check on page 212

1. A source can send only two symbols 0 to 1. If the probability of transmitting a 0 is 0.3, what is the average information transmission?
2. If the symbols above were equally probable, show that $I_{av} =$ 1 bit.
3. The letters in a particular four-letter word have probabilities of occurrence: $\frac{1}{2}$, $\frac{1}{4}$, $\frac{1}{8}$ and $\frac{1}{8}$. Calculate the average information in the word.
4. Show that the average information from two messages is a maximum when both messages are equally probable.

In a communications channel the average information in a message of two or more symbols is a maximum when each symbol is equally probable. In general, information comes in binits, but the maximum average information is the number of bits if they are all equally probable, such as they are in a binary signal.

9.6 Information rate

The rate at which information is transmitted is given by

$$R = I_{av}/(\text{time taken to transmit } I_{av})$$

Exercise 9.4
Check on page 212

Calculate the information rate of a PCM word that consists of eight bits occupying a time slot of 3.9 μs.

The maximum information rate will be achieved when a binary (two equiprobable symbols) signal is used and this rate may be measured in bits per second (b/s). When the symbols are not equally probable, the information rate will be less than maximum. A complete message is made up of symbols, such as in the dots and dashes of Morse code or the alphanumeric characters of ASCII. The rate at which such symbols are transmitted determines the *signalling rate*. Each symbol may not take the same time to transmit. In Morse code, for example, the most frequently used letter 'e' is simply one dot—which might be as short as 0.1 s; while the letter 'j' is one dot and three dashes—which might take about 1 s. The signalling rate is defined as:

1/(time taken to transmit the shortest symbol)

The signalling rate in the above Morse code example would be 1/0.1 s = 10 *baud*. The signalling rate, measured in baud, cannot be numerically greater than the information rate, measured in b/s. The unit of information rate might, one day, take the name of some famous engineer. Perhaps a change from b/s to shannon (Sh) or hartley (Hy) might be appropriate!

Exercise 9.5
Check on page 213

The NRZ form of the symbol 'M' in ASCII is shown in Fig. 9.5. Calculate the signalling rate and the data rate. How would these rates be affected if there were four logic levels instead of two (that is, the signal was quaternary instead of binary)?

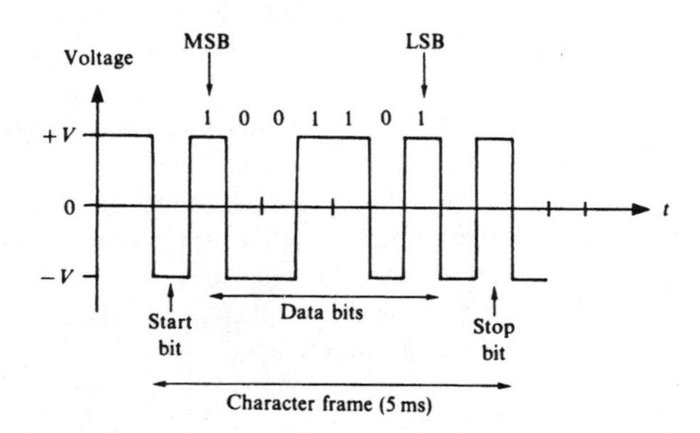

Figure 9.5. Data bits for ASCII symbol.

9.7 Channel capacity

An ideal low-pass channel of bandwidth B can transmit pulses at a maximum rate of $2B$ pulses per second. If there are Q equiprobable pulse levels (Q symbols in effect) discernible at the channel output, then the rate at which information is transmitted through the channel—the *channel capacity* (C)—will be given by substituting $N=2B$ in Eq. (9.2). This gives

$$C = 2B \log_2 Q \text{ binits/s} \qquad (9.5)$$

Equation (9.5) is often referred to as Hartley's law. For binary signals, $Q=2$ so the channel capacity will be $2B$ b/s.

Exercise 9.6
Check on page 213

1. What is the capacity of a telephone line carrying PCM transmission, assuming a bandwidth of 4 kHz?

2. The transducer of a CD player sends out 2^{16} equiprobable pulse levels at a sampling rate of 44.1 kHz. What minimum bandwidth would this need and what would be the channel capacity required?

The effect of noise on information exchange

When information is received in the presence of noise, we cannot be sure that we have got the right information. In an analogue system 4 V might be received instead of 5 V because 1 V of noise was present in the communications channel. That might be an acceptable error. The same error in a digital system, using 1 V to represent binary 1 and 0 V for binary 0, would be unacceptable.

An interesting view of Eq. (9.3) appears when we realize that, in a noiseless system, the probability of an event at a receiver after a message has been received is unity. Equation (9.3) may then be restated in the form:

$$I = N \log\left(\frac{\text{probability of event at receiver after a message is received}}{\text{probability of event at receiver before a message is received}}\right)$$

Noise reduces channel capacity—it is as though a channel can carry a given amount of power and, the more noise power there is, the less the signal power. Equation (9.5) gives the channel capacity when there are M signal levels. The average power of a received signal in the presence of noise is $S+N=v_{\mathrm{R}}^2/R$ where v_{R} is the root-mean-square voltage at the receiver input and R is the receiver input resistance. Thus $v_{\mathrm{R}}=\sqrt{[R(S+N)]}$. Errors will occur if the signal voltage levels are separated by the noise voltage $[v_{\mathrm{N}}=\sqrt{(RN)}$ or less, so the number of signal levels (M) cannot then exceed $v_{\mathrm{R}}/v_{\mathrm{N}}=\sqrt{[R(S+N)/RN]}$. Substituting this value of M in Eq. (9.5) gives the channel capacity in the presence of noise as

$$\begin{aligned} C &= 2B\log_2\{\sqrt{[R(S+N)/RN]}\} \\ &= B\log_2[(S/N)+1]\ \text{b/s} \end{aligned} \tag{9.6}$$

Equation (9.6) is the Shannon–Hartley law. It applies to channels in which the noise power is evenly spread throughout the whole bandwidth, that is gaussian white noise. For non-gaussian channels the value of C turns out to be greater than that given by the law, so Eq. (9.6) indicates the minimum (worst-case) capacity.

Exercise 9.7
Check on page 213

1. What is the information capacity of a telephone line for analogue transmission if the SNR is 30 dB and B = 4 kHz?
2. What is the information capacity of a noiseless gaussian channel?

Channel capacity does not increase without limit as the bandwidth increases because white noise power increases with bandwidth. For a given signal power (S) C reaches a maximum value C_m as B increases. This value may be found as follows:

$$C = B\log_2[1+(S/N)]$$

Recall that $N=\rho B$ where ρ is the density of the noise power spectrum in W/Hz. Hence

$$C=(S/\rho)(\rho B/S)\log_2[1+(S/\rho B)]=(S/\rho)\log_2\{[1+(S/\rho B)]^{\rho B/S}\}$$

Since the limit as $x \to 0$ of $(1+x)^{1/x} = \text{e}$ (the natural log base), then, thinking of x as $S/\rho B$, we get the maximum channel capacity of

$$C_m=(S/\rho)\log_2 \text{e}=(S/\rho)(\log_{10}\text{e})/(\log_{10}2)=1.44S/\rho$$

Trading bandwidth and signal-to-noise ratio

Equation (9.6) shows that, for a given value of C, an increase in B results in a decrease in S/N. The variation of S/N with B for a normalized value of $C=1$ b/s is found by using the relation $B=1/\log_2[(S/N)+1]=0.3/\log_{10}[(S/N)+1]$. The result of this is shown in Fig. 9.6.

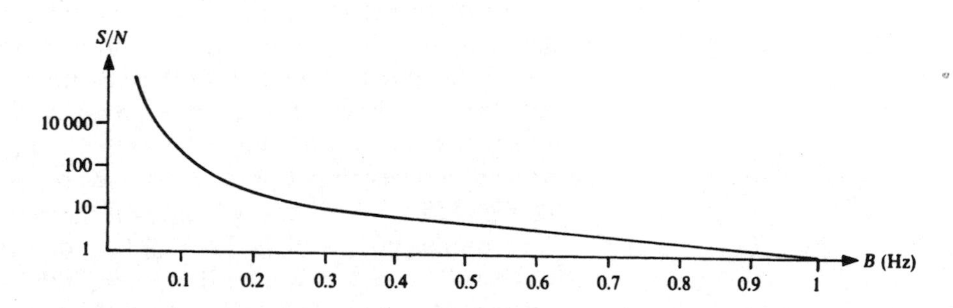

Figure 9.6. Variation of SNR with bandwidth for $C=1$ b/s.

As would be expected, the rate (R) at which information can be sent along a channel is limited by the channel capacity (C). If $R \leqslant C$, the received information will be error-free, if error location coding is used. However, if $R > C$, then the probability of error is almost unity. These last two sentences state *Shannon's theorem.* Examination of Eq. (9.6) (the Shannon–Hartley law) shows that it is possible to receive a signal with an arbitrarily small probability of error providing $C > R$. This may seem surprising because, as B falls below the frequency range of the signal (f_r), the signal received will be attenuated and distorted. However, it has to be realized that as $B \to 0$, $N \to 0$, so the signal can be made good with a noiseless repeater.

Perfectly reliable communication is possible over a noisy channel if the information rate is less than the channel capacity. So long as the signal is made sufficiently different from the noise, a receiver can be devised that can recover the signal. This requires adding *redundancy* to the message, which the receiver can use in order to sort out the signal from the noise. For very noisy or error-prone channels, much redundancy is needed, and this requires a large channel capacity. On the other hand, low-error operations, such as transferring data between memories within computer systems, require quite simple redundancy schemes, consisting of single parity bits to detect errors (refer to Chapter 8). We can understand poor speech or handwriting because there is much redundancy in everyday language.

Summary

The receipt of information implies that some ignorance or uncertainty, existing in the mind of the recipient, is reduced after the information has been received. The information content of a message may thus be measured by the reduction in uncertainty that occurs after a message has been received.

When an information source has Q symbols at its disposal and sends messages N symbols in length, the number of possible messages will be Q^N. The more possible messages there are, the more information will be obtained when one of them is received, so information (I) will be depend on M, that is $I = kf(Q^N)$. If we assume that the information transmitted is proportional to the time of transmission for the same values of Q and R, then $I = kN \log Q = N \log_2 Q$ binits. Thus, the information content of one bit is one binit. Hence, for a binary signal, the same information may be expressed in bits or binits.

Maximum information rate is achieved when a binary signal is used. A complete message is made up of symbols. The rate at which such symbols are transmitted determines the signalling rate. Since each symbol may not take the same time to transmit, the signalling rate, measured in baud, is defined as the reciprocal of the time taken to transmit the shortest of the Q symbols. For binary signals, $Q = 2$, so the bit rate and baud rate are the same.

The rate at which information is transmitted through a channel is the channel capacity $C = 2B \log_2 Q$ b/s. This is Hartley's law. For $Q = 2$, the channel capacity will be $2B$ b/s.

The Shannon–Hartley law gives the channel capacity in the presence of noise as $C = B \log_2 [(S/N) + 1]$ b/s. The law applies to gaussian white noise. The maximum channel capacity for a given signal power (S) is $1.44S/\rho$ where ρ is the density of the noise power spectrum (W/Hz). For a given maximum channel capacity set by $1.44S/\rho$, bandwidth and SNR may be traded.

Shannon's theorem states that the rate (R) at which information can be sent along a channel is limited by the channel capacity (C) such that if $R \leqslant C$ then the received information will be error-free, and if $R > C$ the probability of error is almost unity.

► Checks on exercises

9.1

1. $I = \log 1 = 0$. This confirms that you cannot get information if you already know what you are going to receive—and you certainly know what you are going to receive if there is only one possible symbol.
2. $I = \log_2 2 = 1$ binit or $\log_{10} 2 = 0.3$ decit.
3. $I = \log_2 128 = \log_2 2^7 = 7$ binits or $\log_{10} 2^7 = 2.1$ decits.
4. $I = 8 \log_2 2 = 8$ binits or 2.4 decits.
5. Each line of the page is a message of 64 characters so $I = 50 \log_2 64 = 50 \log_2 2^6 = 300$ binits or 90 decits. In practice not all characters are equally probable, so we will have to find a way of dealing with this reality.

9.2

1. A score of 7 can be thrown in six ways, each one of which has a probability of 1/36, so the total probability of throwing 7 will $6 \times 1/36 = 1/6$. The information is $\log_{10} 6 = 0.8$ decits or $\log_2 6 = \log_{10} 6 / \log_{10} 2 = 0.78/0.3 = 2.6$ binits.
2. A score of 12 can be thrown if only one way, so the information content of this rare score being announced is $\log_{10} 36$ decits $= 1.56$ decits or $\log_2 36 = 1.56/0.3 = 5.2$ binits.

9.3

1. The average information would be given by

$$I_{av} = p_1 \log(1/p_1) + p_2 \log(1/p_2)$$

If p_1 is taken as 0.3, then $p_2 = 0.7$, since $p_1 + p_2 = 1$. Hence

$$I_{av} = 0.3 \log_2 3.3 + 0.7 \log_2 1.4 = 0.520 + 0.340 = 0.86 \text{ binits}$$

2. $I_{av} = 0.5 \log_2 2 + 0.5 \log_2 2 = 1$ binit

3. $I_{av} = \frac{1}{2} \log_2 2 + \frac{1}{4} \log_2 4 + \frac{1}{8} \log_2 8 + \frac{1}{8} \log_2 8$ binits
$= \frac{1}{2} + \frac{1}{2} + \frac{3}{8} + \frac{3}{8} = 1.75$ binits

4. $I_{av} = p_1 \log(1/p_1) + p_2 \log(1/p_2)$. If $p_1 = p$ then $p_2 = 1 - p$, since for only two occurrences $p_1 + p_2 = 1$. So

$$I_{av} = p \log(1/p) + (1-p) \log[1/(1-p)]$$

For maximum I_{av},

$$dI_{av}/dp = 0 = p^2 + \log_e(1/p) - (1-p)^2 - \log_e[1/(1-p)]$$

So $p = \frac{1}{2}$ and maximum $I_{av} = \log_2 2 = 1$ binit/message. The general result for M messages is $p = 1/M$, with maximum $I_{av} = \log M$.

9.4 A PCM word consists of eight equally probable symbols (0 or 1) so the information rate is $8 \log_2 2$ binits/3.9 μs $= 2.048$ Mb/s.

9.5 One ASCII symbol is transmitted in 5 ms so the signalling rate is 1/(5 ms) = 200 baud. The bit rate is 9/(5 ms) = 1800 b/s. If there were four logic levels instead of two, these rates would be doubled. Since the bandwidth limits the bit rate the use of M-ary signals enables the signalling and data rates to be increased within a bandwidth constraint. This is accomplished at the expense of greater system complexity.

9.6
1. Assuming a bandwidth of 4 kHz, the telephone line capacity for binary transmission is $2 \times 4\,\text{kHz} \times \log_2 2 = 8$ kb/s.
2. The minimum bandwidth required would be 89.2 kHz, giving a channel capacity of $2 \times 89.2\,\text{kHz} \times \log_2 2^{16} = 2.822$ Mb/s.

9.7
1. SNR(30 dB) = $10 \log_{10}(S/N)$ so $S/N = \text{antilog}\, 3 = 1000$. Therefore

$$C = B \log_2[(S/N) + 1] = 4000 \log_2 1001$$
$$= 4000(\log_{10} 1001)/(\log_{10} 2) = 40.0\ \text{kb/s}$$

2. In a noiseless gaussian channel N is zero, so the information capacity will be infinite according to the Shannon–Hartley law.

Self-assessment questions

9.1 Outline the processes in the exchanging of information over a distance.

9.2 Distinguish between the terms ‘data’ and ‘information’.

9.3 Define the binit.

9.4 Calculate the information in a message of 32 equally probable symbols.

9.5 Why can the binit and the bit be used interchangeably in binary signals?

9.6 If the letters in a particular four-letter word have probabilities of occurrence 0.4, 0.3, 0.2 and 0.1, calculate the average information in the word.

9.7 Define information rate.

9.8 A symbol of one byte is sent in 100 μs. Calculate the bit and baud rates.

9.9 Define channel capacity.

9.10 State Hartley’s law.

9.11 State how channel capacity is affected by the SNR.

10

The telephone network

Making connections between subscribers

Objectives

When you have completed this chapter you will be able to:

- Trace the development of telephony
- Describe the UK public switched telephone network (PSTN)
- Describe how telephone traffic is measured and how it varies
- Outline the operation of space and time switches
- State the principles of signalling
- Describe some of the features of System X
- Describe the main facilities offered by private branch networks
- Describe how a cordless telephone works
- Outline the principle of telepoint
- Describe how pagers work
- Describe the operation of cellular telephone systems
- Outline the principles of mobile radio telephones

The telephone network is, by far, the largest engineering system in existence. It is basically a network of cables that enclose the globe like a string bag around a large beach ball. It developed from small local networks that carried conversations, and expanded as these networks grew and joined together. The demand for telephones has always been on the increase and this demand has been satisfied over the years by telephone companies investing heavily in new technology. The huge present-day development of the network has been made possible by the availability of cheap and powerful micro-circuits, high-quality optical fibres and numerous space satellites. The present state of the global network reflects not only these developments but also the way in which the network grew. A knowledge of the history of the UK public switched telephone network (PSTN) gives some insight into why the present system is the way it is.

The telephone network must use switching to connect the millions of subscribers. Each subscriber is connected to a local switching centre (exchange) via a twin-wire cable. The local exchanges are interconnected to a hierarchy of larger switching centres via trunk links, which span the UK. The UK PSTN is linked to other similar national networks via submarine cables and satellite microwave links. Trunk links carry many communication channels by using frequency- or time-division multiplexing. PCM has improved the quality of the received signal and TDM has enabled digital switching to replace space switching. By using the same signal formats for switching and transmission, great advantages of economy, compatibility and flexibility are achieved.

Private branch networks with their own exchanges have developed alongside the public network. These are usually connected to the PSTN and can offer many sophisticated facilities, which are now being gradually introduced into the PSTN as it goes digital. Private networks have led the way technologically because it is much easier to modernize a small network than it is to alter the vast PSTN.

Data telecommunication has developed in parallel with the telephone system. The old telegraph network has now given way, with the help of satellite links, to a global system for data exchange. The scope of data and satellite communication has become so great that it will be dealt with separately in Chapter 11.

The 1990s will see a great increase in the use of radio telephones, which remove, to varying degrees, the restrictions imposed by wire links. This will be the era of the personal telephone with such mobile communications facilities as cordless handsets, pagers, telepoint and cellular phones available for use with the PSTN and private mobile radio.

10.1 The development of telephony

The first successful attempt at telecommunication was made in 1831 when an electric *telegraph*, based on an electromagnetic semaphore, was used to send messages between railway signalmen. Samuel Morse later invented a system in which an electromagnet operated a pen over a moving paper when the current at the sending end of the line was interrupted. It was the beginning of a crude teleprinter service. At the same time, Morse code communication was being developed. In this system, a key or switch tapped out a binary code of long and short electric currents and used the sound of the pen over the paper to give an audible signal. In 1839 Morse demonstrated his system over a line 40 miles long, and within about 10 years there were about 50 telegraph companies in the USA alone using his system. The first electric telegraph in Britain connected Euston and Camden Town railway stations, a distance of only one mile. This was followed, in 1839, by a 20-mile link between Paddington and Slough railway stations along the Great Western Railway line.

Not long after, the inland telegraph service was established with a network of cables running to most parts of the country. Telegraph lines were laid under the sea: across the Dover Straits in 1851 and across the Atlantic in 1866. By the first few years of this century, the British telegraph network, run by the Post Office, encircled the globe. In 1861 Wheatstone invented the punched paper tape system for recording and receiving messages automatically over telegraph lines. Then came the first teleprinters and the telegram service.

While all this telegraph development was going on, the invention of the microphone came in 1875. With it came the first attempts at telephone communications using two phones connected together via a dedicated line, with a hand-operated magneto crank to ring the bell at the called end. When more than two phones were connected to form a network, a coded bell ring was used to signal which phone was being called. As the number of phones increased, the codes became too complicated, so the next development was a switchboard to which all phones were connected. Lifting the phone and winding the crank called a human operator who would make the required connection to the number given by the caller. A light would indicate to the operator when the call was finished. In highly populated areas, several switchboards were grouped together to form a telephone exchange, whose function was to route calls between telephones in that area and to establish trunk connections to other local exchanges.

The number of calls that can be handled by a human operator is limited, so automatic exchanges now do this connecting job. It has been estimated that to handle today's telephone traffic manually would need about half the World's population being employed as operators! The connecting apparatus in the automatic exchange was controlled electromechanically by a dial on the telephone handset, but this is now being superseded by push-buttons, which operate electronically.

The UK telephone network quickly developed, as many companies were set up to serve local communities. Users of the service paid a subscription (rental) and a call charge and became known as subscribers. The organization of the network on a national scale came with the 1868 Telegraph Act, which gave the Postmaster General the sole right to operate public telegraph systems by operating the system himself or licensing others. When tne PSTN was established in 1879 a legal ruling declared that telegraphy included telephony—that is how the Post Office got involved with telephones.

For a long time, the Post Office operated its own system and also gave licences to local independent companies up and down the country. These were linked together by trunk lines provided by the Post Office. Most of these companies were gradually brought together into the National Telephone Company. In 1912 the Post Office decided not to renew the licence of this company and itself took over the network. This must have been the first example of nationalization of a private company. Up to 1969, public telecommunications was the exclusive privilege of the Post Office, with the curious exception of the Kingston-upon-Hull Corporation, which remained private. The Post Office Act 1969 established the Post Office Corporation (POC) as an independent public body with discretionary powers to license others to operate telecommunication services.

The British Telecommunications Act 1981 removed telecommunications from the POC and transferred it to a new public corporation: British Telecommunications (BT). Under this same Act, the Secretary of State for Trade and Industry licensed the following public telephone operators (PTOs): British Telecommunications plc, Mercury Communications Ltd (Mercury), Kingston Communications plc (Hull), Telecom Securicor Cellular Radio Ltd (Cellnet) and Racal Vodaphone Ltd (Vodaphone). The British Approvals Board for Telecommunications (BABT) was set up, under the same Act, to approve apparatus. The advent of radio telephones brought such apparatus within the regulations of the Wireless Telegraphy Acts 1949 and 1967.

The Telecommunications Act 1984 privatized BT and created the Office of Telecommunications (Oftel). Section 1 of the Act requires the Secretary of Trade to appoint a Director General of Telecommunications to run Oftel, the objectives of which are as follows:

1. To promote the interests of consumers, purchasers and other users in the UK in relation to prices and the quality and variety of telecommunications services provided and the apparatus supplied.
2. To promote and maintain effective competition between those engaged in commercial activities connected with telecommunications.
3. To enable persons producing telecommunications apparatus in the UK to compete effectively in the supply of such apparatus within and outside the UK.

Section 4 of the 1984 Act defines the term 'telecommunications system' as any system using electric, magnetic, electromagnetic, electrochemical or electromechanical means to convey speech and other sounds, visual images, other information and signals for the actuation or control of apparatus.

The PSTN is operated by BT, which also offers a range of other services in competition with the other main PTO, Mercury, a subsidiary of Cable and Wireless plc. Mobile telephone networks are now being provided by independent operators using cellular radio links and telepoint.

In addition to public networks, there are private ones, which can provide a more economical service by using *tie-lines*. A tie-line is a direct link between sites belonging to the same organization such as a London head office and a regional sales office. Generally, the higher the volume of traffic, the cheaper the tie-line calls are compared with those on the PSTN.

The PSTN has developed considerably since the first UK telephone exchange was opened by the National Telephone Company in London in 1879. Subscriber trunk dialling (STD), for example, was introduced in 1958, and covered all the UK by 1979. In 1966, international direct dialling

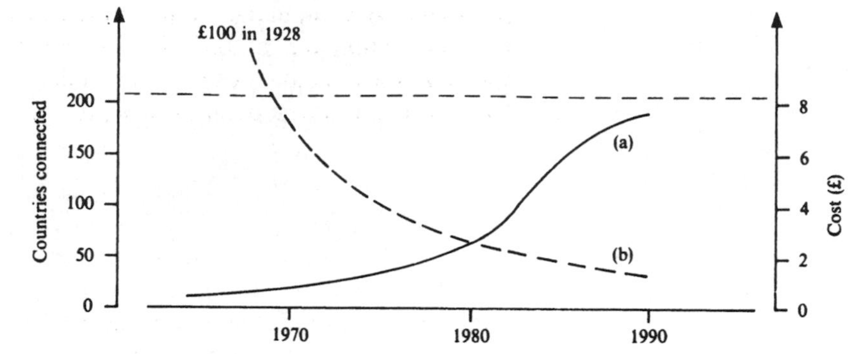

Figure 10.1. Growth of UK IDD and decreasing cost of a transatlantic call: (a) number of countries connected by IDD; (b) inflation-adjusted cost of 3 min transatlantic call.

(IDD) on digital links became available, and the extension of IDD has seen a dramatic fall in global call costs as shown by Fig. 10.1.

BT now offers a variety of data transmission services over the PSTN as well as private lines. Mercury offers high-speed digital networks primarily for the business user. Its trunk network is based on two optical fibre loops: a Birmingham–London–Bristol–Birmingham loop and a Birmingham–Manchester–Leeds–Sheffield–Birmingham loop. A microwave link is used to connect Liverpool to the network from a node in Warrington. Mercury's cables have been laid alongside British Rail's Intercity routes. In 1986 a switched network service for business-to-business dialling was provided by the Mercury network and agreement has been reached betwen BT and Mercury for interconnection of their networks. Both PTOs offer access to satellite communications through their networks and both offer network services such as viewdata, electronic funds transfer and electronic point-of-sale transactions. Some of these will be described in Chapter 11.

The interconnection of telephones is done by the PSTN. As demand for phones has increased, this network has been expanded so that it can carry more signal traffic and cope with all the extra switching that this entails. The obvious way of doing this is to lay down more cables and build extra exchanges, but more subtle engineering solutions to the problem have been found. Instead of using more cable over the long-distance routes between-trunk exchanges, microwave links and optical fibres are used. Such links have much greater bandwidth than copper cable and so can carry much more information. The extra switching has been accomplished by taking advantage of PCM voice transmission, which was introduced in the late 1950s on the trunk routes in order to improve the SNR. Digital switching requires huge memories with short write/read access times. These became available in the late 1960s in the form of very-large-scale integrated (VLSI) silicon chips, and digital trunk exchanges were installed from the early 1970s. BT now connects all trunk traffic by System X digital switching units and is gradually converting all its local exchanges to this system.

When the PSTN began to carry digitally encoded voice signals, this expedited the transmission of other digital information such as that generated by computers. An integrated services digital network (ISDN) is due for completion in the UK sometime in the 1990s. In this system, exchanges will handle both voice and data in digital form. This ISDN will be described in more detail in Chapter 11.

10.2 The structure of the public switched telephone network

The PSTN is a huge network of copper and fibre cables, radio links and switching centres (exchanges), which carry audio, video and data signals. The signals between the subscribers and the local exchange are at baseband and those along the trunk routes are pulse code modulated according to the CCITT 30-channel standard described in Chapter 7.

Locally, a twin-wire link is permanently connected from each subscriber's terminal to a local analogue exchange or its modern replacement, a digital *remote concentrator unit* (RCU). This line link is rented and is terminated at the subscriber's end by a standard BT socket into which the terminal apparatus (a telephone or computer modem for example) may be plugged. Most adjacent local exchanges or units are linked by connections to a *digital cell centre exchange* (DCCE) as illustrated in Fig. 10.2.

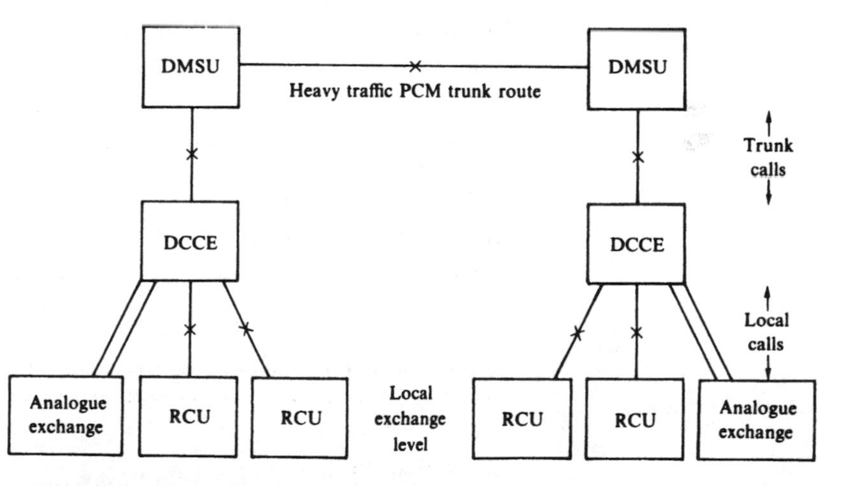

Figure 10.2. Routing and switching centre hierarchy of the PSTN.

The links marked '×' carry PCM signals, the others use analogue signals. The DCCEs are linked to *digital main switching units* (DMSU), of which there are about 60, all interlinked to give a national network. The DMSUs are all System X. The DCCEs are mainly System X, but a few use the Ericsson System AXE10—a system introduced to promote competition between equipment suppliers.

Note that local calls are switched by an analogue exchange but an RCU merely concentrates local traffic, which is normally switched at a DCCE.

Interlinking of all DMSUs gives the possibility of *automatic alternative routing* (AAR), which gives some security against network failure and helps to maintain the grade of service during busy periods. AAR works as shown in Fig. 10.3. If unit B tries to route traffic through the shortest route to unit C but finds this route congested, it will automatically use an alternative route, which is not congested, by going via unit A in this example. Each of the alternative routes will cost the same as far as the subscriber is concerned.

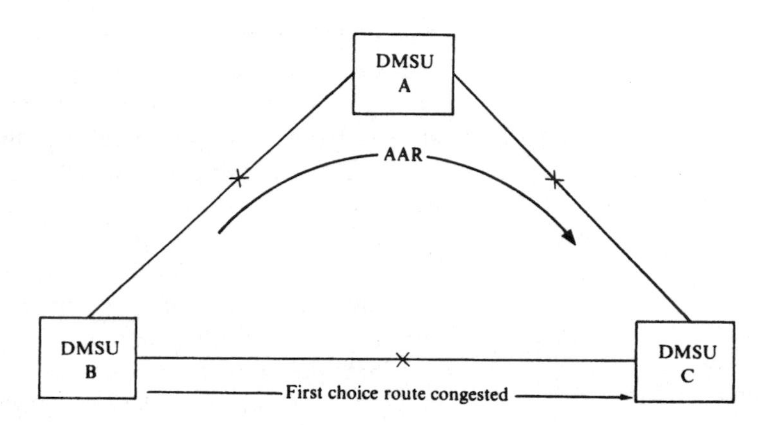

Figure 10.3. Automatic alternative routing.

The telephone numbering scheme and subscriber trunk dialling

Table 10.1 gives some examples of UK telephone numbers that reflect the hierarchical system of Fig. 10.2. From the examples of Table 10.1, it can be seen that a telephone number may be made up according to the two alternatives in Fig. 10.4. Telephone numbers include a local area (LA) code, which may be serviced by a local exchange or a DCCE, and a *subscriber trunk dialling* (STD) code.

To see how the numbering scheme might work, consider what occurs when a trunk call is dialled, for example, from Portsmouth football

Table 10.1. Examples of UK telephone numbers

DMSU number (name)	Local area number (name)	Subscriber's number (name)
071 (Inner London)	385 (Fulham)	5545 (Chelsea AFC)
091 (Tyneside)	514 (Roker)	0332 (Sunderland AFC)
0705 (Portsmouth)	73 (Fratton)	1204 (Portsmouth AFC)

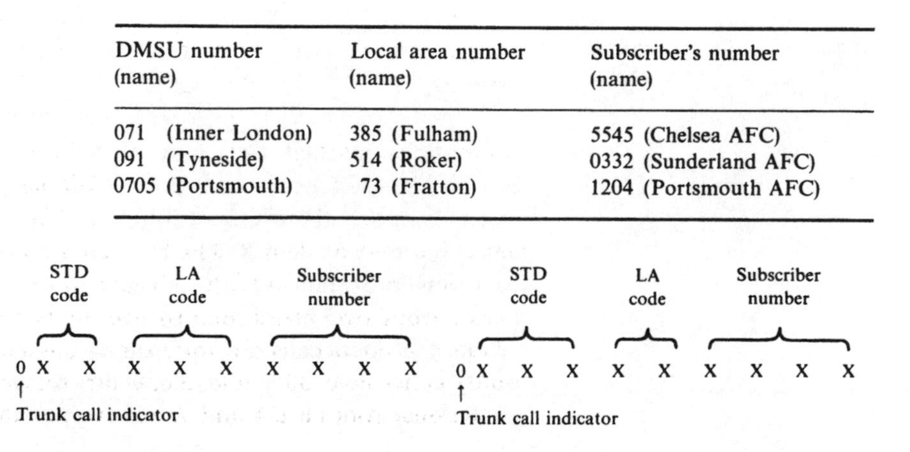

Figure 10.4. National telephone number schemes.

club to Chelsea football club. Immediately on recognition of 0 in the STD code number 071, the Fratton local exchange searches for and connects to a free trunk route. The connection to a DCCE has now been made. The remainder of the STD code number (71) will be used by the DCCE to establish, via a DMSU, a connection to the inner London DCCE. Meanwhile, the digits 385 are being sent along the route and 385 will enable the London DCCE to select the Fulham exchange, which, using the digits 5545, will connect to the switchboard of the Chelsea club.

Exercise 10.1
Check on page 249

Outline the routing operation that would occur if the Chelsea club manager called the manager of the Sunderland club.

International telephone communications

STD has been extended to most of the world as *international direct dialling* (IDD). This has required international agreement in order that each subscriber has a unique number. The CCITT (See Chapter 11) has allocated country codes: for example, the UK is 44, the USA and Canada are both 1 and the Virgin Islands are 1 809 49. An IDD telephone number is made up according to the scheme in Fig. 10.5.

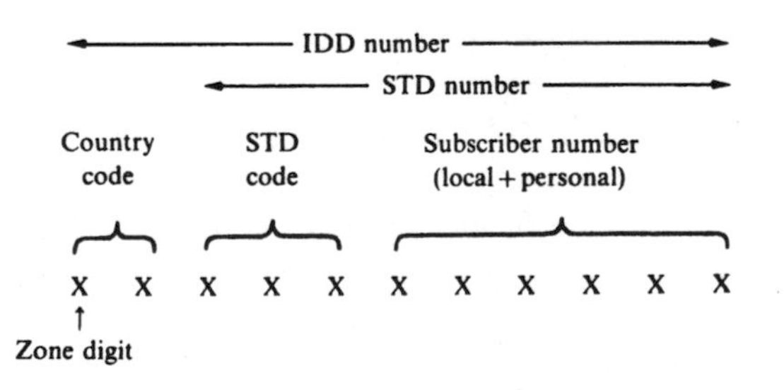

Figure 10.5. Example of an international telephone number structure.

The CCITT has recommended that the total length of an IDD number should be 11 digits which may be distributed between the country, STD and subscriber codes as appropriate (the example of Fig. 10.5 applies to the UK). In the USA and Canada, where the country code is only one digit, an STD number of up to 10 digits may be allocated, but in the Virgin Islands only five digits would be available for the STD number. The more populous countries must be allocated the shorter codes. The zone digit in a country code is allocated according to Table 10.2.

In all but two zones, one to five digits are added to the zone digit to produce a country code: for example, 4 for the UK to give 44, 59 for Bulgaria to give 359, and 3 628 for Andorra to give 33 628. The two exceptional zones are 1 and 7. This means that Canada and the USA can

Table 10.2 International zone digit allocation

Digit	Zone	Digit	Zone
1	North America	6	Australasia
2	Africa	7	USSR
3	Europe	8	Eastern Asia
4	Europe	9	Far and Mid East
5	South America	0	Not used

have 10-digit subscriber numbers, three for the area and seven for the personal number, and no one in the USA can have the same number as another in Canada.

Network linking

Having looked at telephone network switching structure, we now consider the network linking techniques, starting with an example of the kind of links that might be established between a subscriber in London and another in Canberra, Australia. In sequence, these might be:

1. A wire pair from the caller's handset to the local exchange
2. A glass fibre to the satellite Earth station at Goonhilly Downs in Cornwall
3. A 6 GHz radio beam directed at the satellite over the Atlantic Ocean
4. A radio beam of 4 GHz from the satellite to an eastern Canadian Earth station
5. A land line from Mill Village in eastern Canada to Vancouver
6. A submarine cable from Vancouver to Sydney
7. A microwave link between Sydney and Canberra
8. A wire pair from the Canberra local exchange to called subscriber's phone

There are other alternatives, but it would not be practicable, for instance, to use the Pacific Ocean satellite as well as the Atlantic one in the above link-up, because it takes a little over a quarter of a second for the radio waves to travel between the Earth and a satellite. This would give a total time of over one second between speaking and getting an answer when two satellite links are used, and this delay has proved unacceptable. The wire links will carry an analogue voice signal and the remaining links will carry a PCM voice signal.

The subscriber to local exchange link

The premises of each subscriber are linked, by a short underground or overhead twin wire, to a distribution point (DP). The DPs are linked to large junction boxes or cabinets which, in turn, are linked to the exchange, usually some distance away, by thick cables containing hundreds of twin wires—one twin for each subscriber.

Inter-exchange links

The use of wire cables over trunk routes is expensive, so other means of trunk linking are now used together with the land and submarine cables that were laid down in the early days of telephony. These take the form of the glass fibre and microwave links described in Chapter 2. The microwave links may be short, land links or long-distance satellite links. Glass fibre cables are being used more and more for terrestrial links.

Exercise 10.2
Check on page 250

'Telecommunication systems transmit information by electric current flow along wires or by radio waves which are either guided or beamed into space.' Comment on how this statement accords with the ever-increasing use of optic fibres.

A layered view of a telephone link

When a telephone link is established as exemplified in the London to Canberra example above, a series of steps is required to accomplish the connection. These steps are governed by rules or *protocols*, which standardize the procedures and equipment. Each of the steps is equally important and dependent on the previous step. Figure 10.6 shows how the procedures can be modelled as a series of layers.

In a simple communications system, consisting of direct conversation, only level 7 would be involved. The protocols would be the

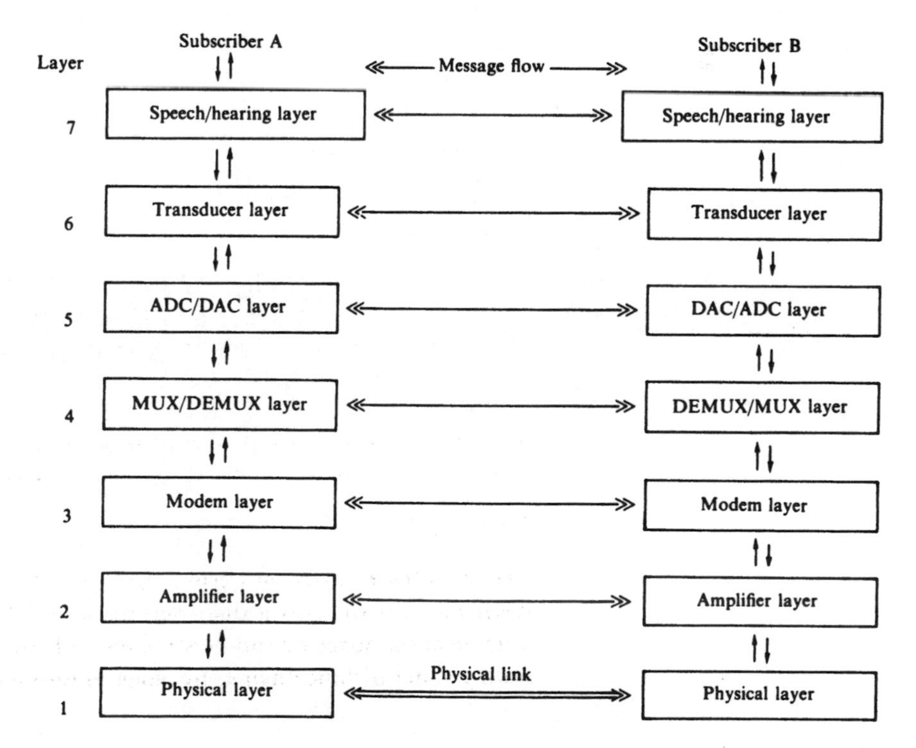

Figure 10.6. A telephone link seven-layer protocol model.

rules of grammar and the link would be the air or whatever between speaker and listener. If smoke signals were used, transducers in the form of fires would enter the model. The protocols of level 7 would change to the smoke signal code and the protocols of level 6 would specify the way the smoke was produced. In a PCM telephone system all seven levels would be involved, each with its own set of rules, specifications or protocols.

Exercise 10.3
Check on page 250

Which of the seven layers would be involved in an analogue FDM microwave link?

10.3 Telephone traffic

The amount of use made of a telephone network depends on the number of calls and their duration or holding time. The holding time is random and has a mean value between 2 and 3 min in practice. Most calls tend to be made at certain periods in the day, giving the kind of traffic variation shown in Fig. 10.7.

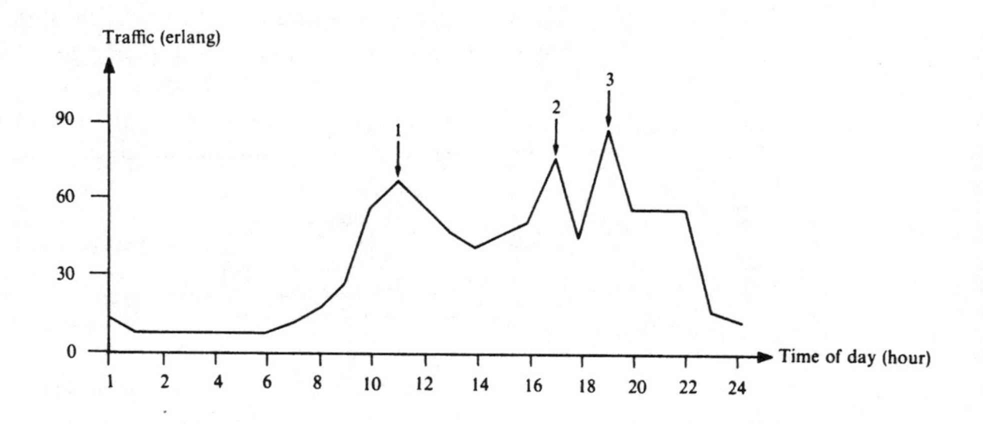

Figure 10.7. Telephone traffic distribution.

Exercise 10.4
Check on page 250

The graph of Fig. 10.7 shows the traffic through a typical System X local exchange in a town with an even spread of residential and business customers. Suggest the cause of the peaks marked 1, 2 and 3.

The 'busy peak' is a most important feature to the telephone service provider, since it is used to determine the traffic-handling capacity needed in exchanges in a given area. If the equipment can deal with the number of calls made during the busy peak, then it will be more than adequate for the rest of the time. Since the telephone system is seldom fully loaded, BT and other providers offer non-voice services to take up some of this idle capacity and boost revenue. These ever-increasing data services are known as *value added network services* (VANS) because they add value to that present due to voice communications. The VANS are furnished by independent providers who lease the necessary lines from BT. Examples of such VANS will be described in Chapter 11.

The measurement of telephone traffic

Telephone traffic (A) is equal to the product of the mean number of calls/unit time (N) and the mean holding time (H), that is $A = NH$. If $N = 1$ call per second and H is 1 s then $A = 1$ erlang. This unit of traffic is equivalent to one continuous call.

Exercise 10.5
Check on page 250

If the average domestic subscriber makes four calls per day averaging 2 min a call, calculate the traffic in an exchange serving 8000 such subscribers.

The low traffic level of the average subscriber suggests that, although each must be individually linked to an exchange, there is no need to provide this number of links within or between exchanges. We saw in Chapter 2 (Fig. 2.29) how two exchanges, each serving four subscribers, could be connected with only two links using a 4×2 cross-point matrix. This principle of concentration and expansion (illustrated in Fig. 2.30) will now be taken further. Figure 10.8 shows how 100 subscribers might be connected by only 25 trunk links. This arrangement saves links but requires 5000 cross-point switches. Further economies can be made by using the arrangement shown in Fig. 10.9.

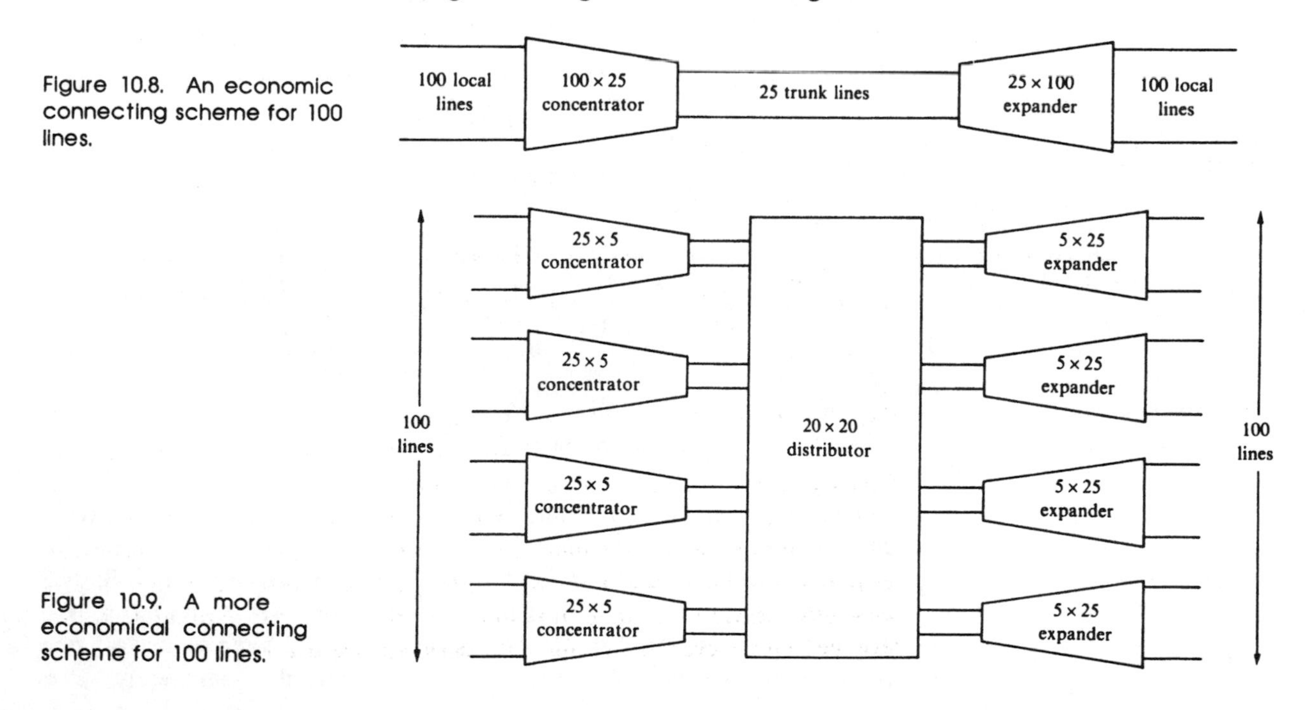

Figure 10.8. An economic connecting scheme for 100 lines.

Figure 10.9. A more economical connecting scheme for 100 lines.

The scheme of Fig. 10.9 requires 500 concentrator switches, 400 distributor switches and 500 expander switches: a saving of $5000-1400=3600$ switches on the scheme of Fig. 10.8. Economies are obtained at the price of a greater probability of blocking, since only 20 of the 100 subscribers can be linked to another 20 simultaneously. The way the scheme alters the line traffic is shown in Fig. 10.10.

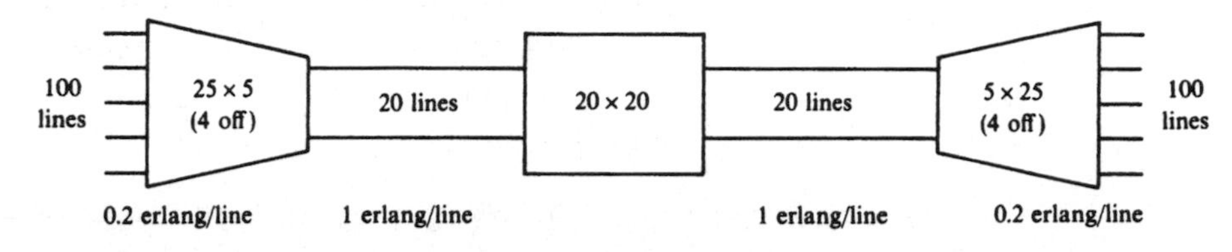

Figure 10.10. Concentration, distribution and expansion of traffic.

10.4 Switching pulsed voice signals

TDM enables *time-slot interchanging* to be used for exchange switching, which is different from the *space* switching used with time-continuous signals described in Chapter 2. The principle is illustrated, for two PAM input signals, in Fig. 10.11.

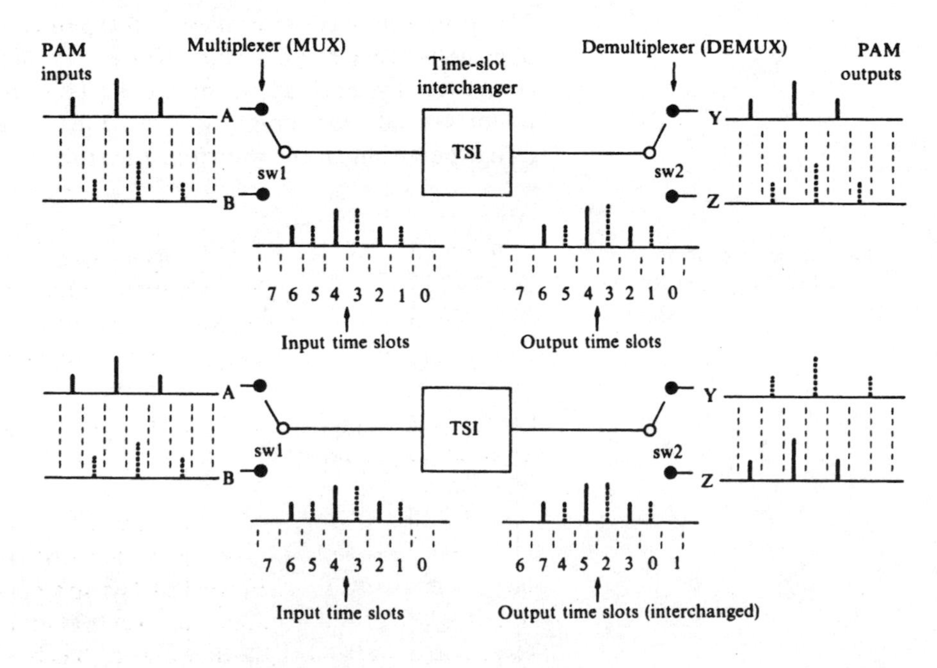

Figure 10.11. Time-slot interchanging of two PAM signals.

The two switches sw1 and sw2 in Fig. 10.11 alternate between their contacts in synchronism, so when sw1 is contacting input A, sw2 is contacting output Y, and when sw1 is contacting input B, sw2 is contacting output Z. Sw1 is a 2:1 multiplexer (MUX) and sw2 is a 1:2 demultiplexer (DEMUX). The period spent by each switch in one position is one time slot, and during this period a sample pulse is transmitted. This means that two PAM signals in TDM enter the time-slot interchanger (TSI). If the TSI leaves the order of time slots unaltered, pulses from A will be routed to Y and pulses from B routed to Z. If the TSI interchanges the time slots, pulses from A will be routed to Z and pulses from B to Y. The same scheme would work for PCM, with a binary coded train of pulses (serial word) taking the place of one sample pulse of the voice signal. Figure 10.12 shows how TSI can be extended to 32 PCM signals.

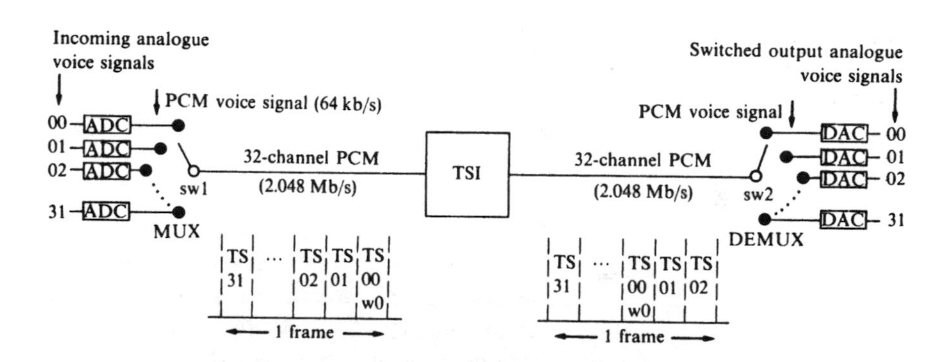

Figure 10.12. Time-slot interchanging of 32 PCM voice signals.

In the example shown in Fig. 10.12, an eight-bit sample of voice signal 00 is routed to output 02 by interchanging time slots 00 and 02. This interchange takes place on every sample of signal 00 until a complete message has been routed from input 00 to output 02.

The PCM *transmission* system used by BT determines the frame size in its TSI *switching* system. This means that a common TDM scheme is used in both signal transmission and switching. In Chapter 7 we saw that, in the CCITT 32-channel PCM transmission system used by BT, 32 time slots of 3.9 μs duration (each containing an eight-bit word representing the sample size of one of the 32 signals) constitute one frame of 125 μs. After each frame 30 voice channels will have been sampled (two frames are used for signalling and synchronization). The sampling rate is therefore $1/(125\ \mu s) = 8$ kb/s, which is well in excess of the Nyquist rate of 6.8 kb/s required for commercial speech. The bit rate from each ADC is 8×8 kb/s $= 64$ kb/s.

In Fig. 10.12 the MUX (sw1) stays on each of the 32 inputs for 3.9 μs, during which time it collects a sample word of eight bits. That means that the switch takes $32 \times 3.9\ \mu s = 124.8\ \mu s$ for one 'revolution', during which time it outputs one sample from each channel in TDM at a bit rate of 32×64 kb/s $= 2.048$ Mb/s. After a word has been time switched, a DEMUX (sw2) routes each eight-bit sample to the appropriate output. Each sample is converted back to analogue form by a DAC.

Action of the time-slot interchanger

The action of the time-slot switch in switching from input 00 to output 02 is shown in Fig. 10.13. The serial-in, parallel-out (SIPO) register converts

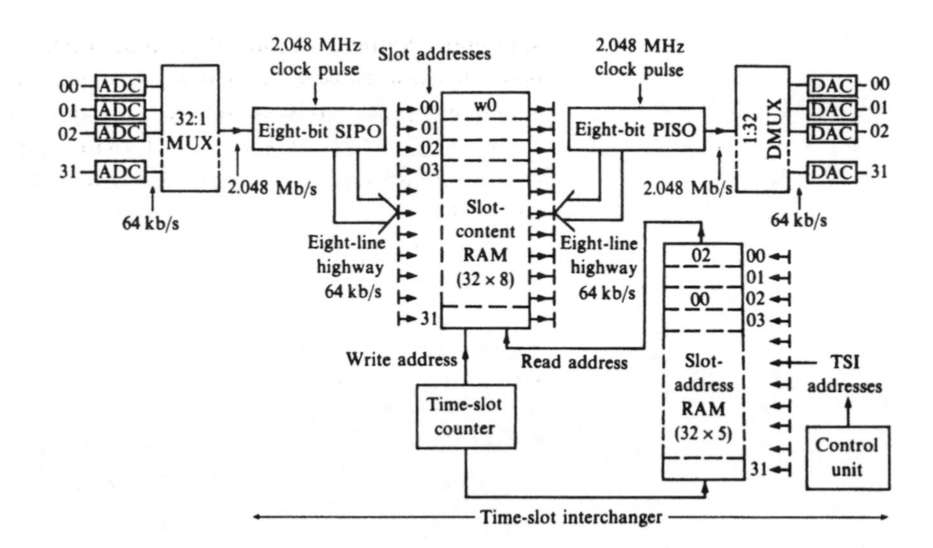

Figure 10.13. Action of a time-slot interchanger.

the eight-bit serial word in each time slot into parallel form for storage in a slot-content random access memory (RAM). Successive samples are read into addresses 00 to 31 at the rate of 64 kbyte/s, having been routed there via an eight-line highway. The samples are read out in whatever order is needed to switch the contents to the required output. Thus, to get a word from input 00 to output 02, the TSI instructions from the *control unit* must interchange slots 00 and 02 by storing 00 and 02 in addresses 02 and 00 respectively of the slot-address RAM. The effect of this is that, when the slot address RAM is read sequentially, the word w0 (which is a voice sample coded in eight bits) will be read out in time slot 02 and so will appear at output 02. The read-out order is determined by the contents of the slot-address RAM. Since only 32 addresses are required, an address word need be only five bits long ($32 = 2^5$). Thus 32 time slots could switch up to 32 inputs between 32 outputs in the time it takes to write and then read the slot-content RAM. The switch is non-blocking for up to 32 inputs.

Exercise 10.6
Check on page 250

Draw a figure, similar to that of Fig. 10.13, for a six-input system so that calls at inputs 00, 01, 03, 04 and 05 are routed respectively to outputs 04, 02, 05, 00 and 01.

10.5 System X

System X is the system used by BT in the switching centre hierarchy of the PSTN shown in Fig. 10.2. It consists of a number of subsystems, two of which will now be described: the *remote concentrator unit* (RCU) and the *digital switching subsystem* (DSS). An RCU pulse code modulates the subscriber analogue speech signals and then concentrates the local traffic before passing it on to a DCCE using 32-channel PCM. DCCEs and DMSUs use a number of DSSs to switch incoming PCM speech signals to unused output channels. A DSS takes in 32 lines, each carrying 32 TDM channels, giving a total of $32 \times 32 = 1024$ time slots, which may be interchanged. Since a DSS has 1024 time slots, the 32 time slots made available by time-interchanging 32-channel PCM are not sufficient. In System X, a switching capacity of 1024 is achieved by a combination of time interchanging and space switching known as *time–space–time* (TST) switching.

Time–space–time switching

A 32×32 TST switch is made up as shown in Fig. 10.14. To see how the TST switch works, we consider the 3×3 time–space–time switch example of Fig. 10.15.

Each time switch can switch 32 channels so, when combined with a 3×3 cross-point switch via highways, 3×32 channels will be switchable without

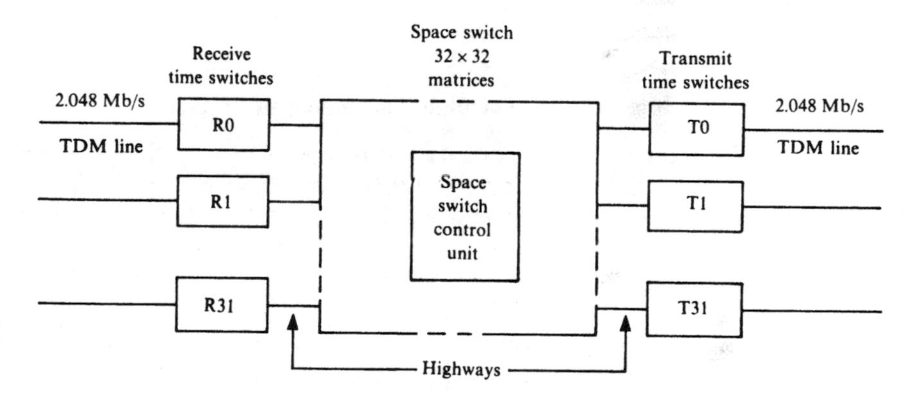

Figure 10.14. A 32×32 time–space–time switch.

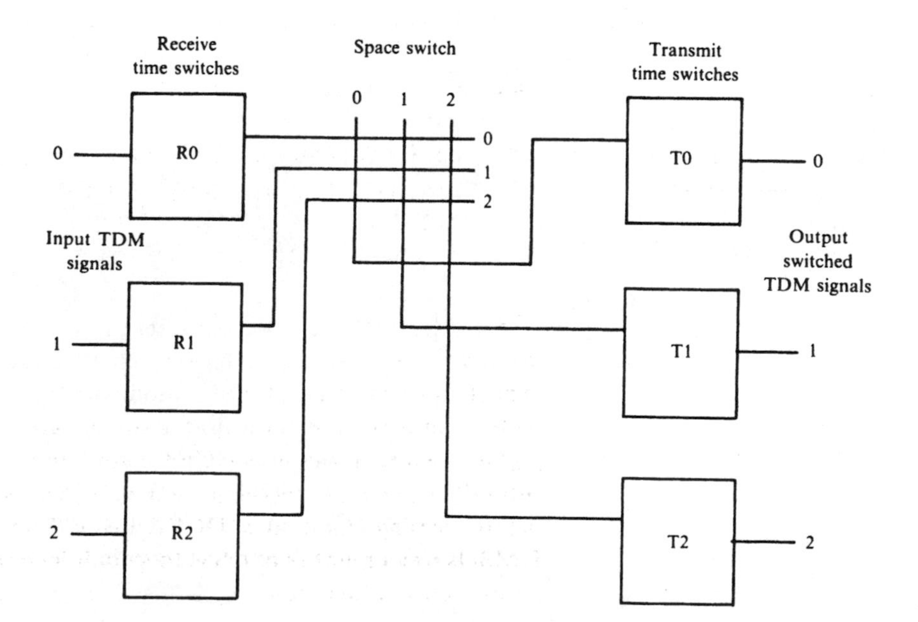

Figure 10.15. A 3×3 time–space–time switch.

blocking. Suppose we wish to switch the signal whose samples are in time slots 00 of input 0 to time slot 02 of output 1. This could be done by R0 inter- changing time slots 00 and 02 as was done in Fig. 10.11 and, at the same time, enabling the gate at the cross-point of row 0, column 1 of the space switch. T1 would simply pass the interchanged and space-switched signal without further interchanging. Once such an interchange has been made, row 0 and column 1 of the space switch are available for another interchanged signal.

Exercise 10.7
Check on page 230

Suggest another way to switch an input signal of Fig. 10.15, whose samples are at time slot 00 of input 0, to time slot 02 of output 1 using:

1. Interchanges in T1 only
2. Interchanges in both R0 and T1

Explain what advantage accrues when method 2 above is used.

The digital switching subsystem

The main function of the DSS is to connect any one of up to 1024 incoming time slots to any outgoing time slot in either direction via *digital line terminations* (DLTs) as shown in Fig. 10.16.

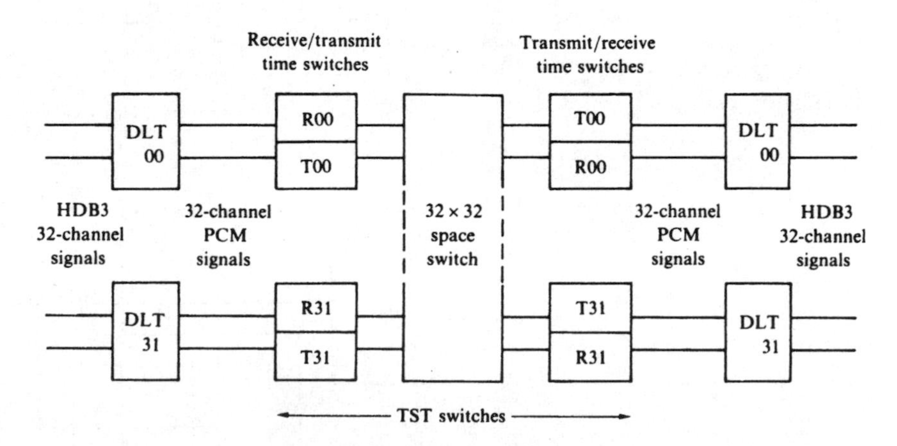

Figure 10.16. Duplex TST switch scheme used in a DSS.

In order to achieve duplex operation, there must be a TST switch for each direction. Hence, we have receive/transmit and transmit/receive time switch pairs for each of the 32 input/output lines. In the space switch two cross-points are used at a time, one for each direction.

Since a PCM line uses HDB3 code (refer to Chapter 7) and the switch uses binary code, the left-hand DLTs of Fig. 10.16 must convert HDB3 to binary and the right-hand DLTs must convert binary to HDB3 code.

All DLTs in a DSS are synchronized, so that all time slots 00 for every

PCM line in the exchange are sent out at the same time followed by time slots 01, 02 and so on to 31. Time slots on different incoming lines do not arrive at the same time because of the different distances between exchanges. DLTs must align all incoming signals by storing and simultaneous onward transmission to a receive time switch and detect errors by parity checking, as detailed in Fig. 10.17.

The time switch receives all the time slots from 32 DLTs and these 1024 slots are stored in a *receive-speech store* RAM. The DLTs are synchronized so that the receive time switch accepts the same time slot from all 32 DLTs simultaneously. Four groups of SIPOs convert the nine-bit serial samples

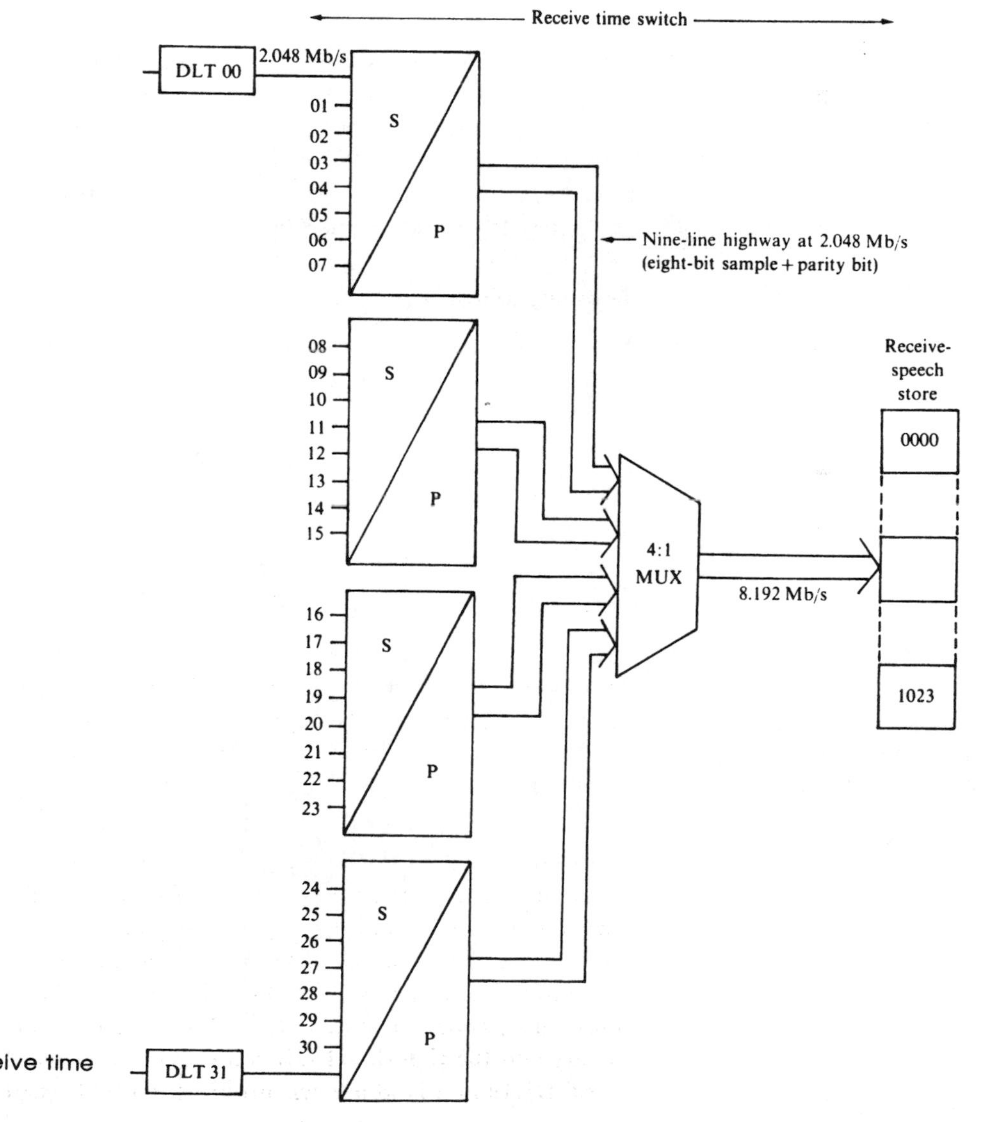

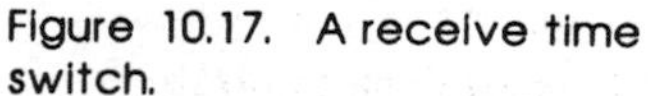
Figure 10.17. A receive time switch.

into parallel for storage in the 1024 receive speech store locations. It is an extension of the principle illustrated in Fig. 10.12, with the one 2.048 Mb/s input line replaced by 32 such lines. This means that the nine-bit highways into the MUX work at 2.048 Mb/s while the highway to the store operates at four times this bit rate. One frame from each of the 32 DLTs is read into the receive-speech RAM during a frame period of 125 μs.

Exercise 10.8
Check on page 251

> We saw, in the time-slot interchanger serving one 32-channel PCM line (Fig. 10.13), that 32×8 slot-content RAM and a 32×5 slot-address RAM were required. What changes in the size of these RAMs would be needed in a DSS?

For N inputs there has to be N write and N read operations going on continuously, which suggests that we need two RAMs, which are written to and read from alternately as shown in Fig. 10.18.

With the switches in the positions shown in Fig. 10.18, the upper RAM is being written to and the lower RAM read from. In the next frame the switches reverse. This process goes on continuously with each RAM being alternately filled and emptied in 125 μs.

The space switch

The space switch is a matrix of $32 \times 32 = 1024$ cross-points, one matrix for each of the eight sample bits plus a parity bit, making

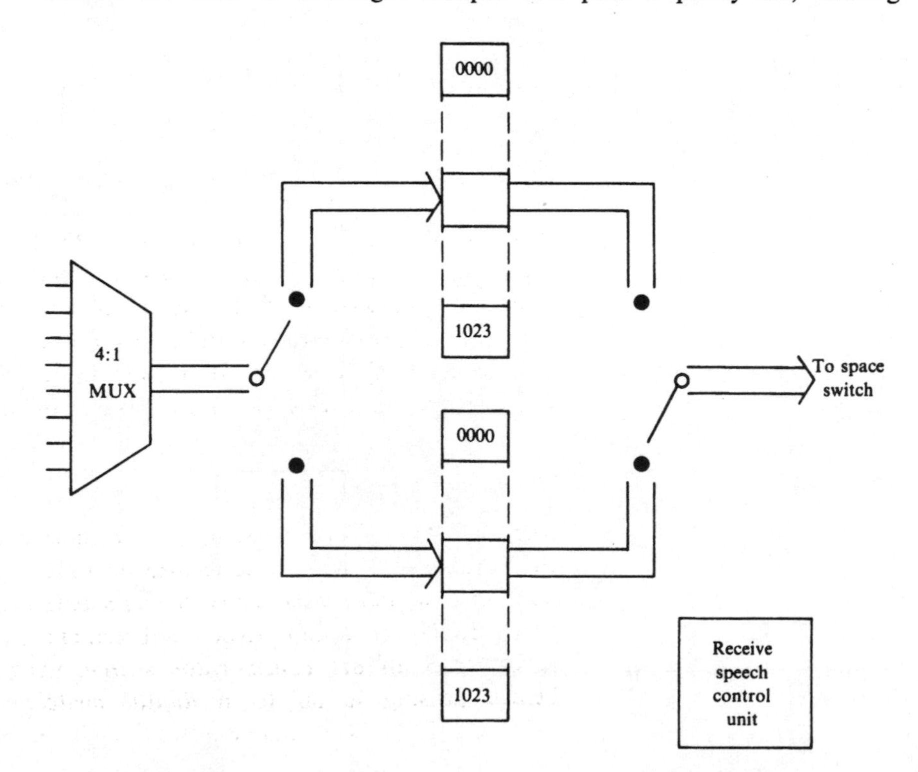

Figure 10.18. A receive-speech store.

nine matrices in all. Each gate operates for 125 μs/1024 = 122 ns, a period known as a *cross office slot* (XOS). Any vacant XOS may be used to switch a call sample.

For duplex operation, two paths (forward and reverse) must be set up across the space switch so two XOSs must be allocated to each speech connection. XOSs 0 to 511 are paired with XOSs 512 to 1023. That is, 0 with 512, 1 with 513 and so on up to 511 with 1023. The transmit speech store is filled from the space switch and works in the opposite way to the receive switch with a 1:4 DEMUX in place of the 4:1 MUX.

Exercise 10.9
Check on page 251

Calculate the number of interconnections that could be made with a 96 × 96 space switch in a 32-channel PCM system and the corresponding XOS rate.

Summary of TST switching in a DSS

1. A time slot sample is stored in one of the two receive-speech RAMs.
2. During the allocated XOS time, the receive output control unit connects the above RAM location to the space switch in which the required cross-point gate has been enabled by the cross-point control unit.
3. The contents of the above RAM location are read into the allocated XOS of one of the transmit RAMs via the enabled gate.
4. The transmit output control unit reads out the stored samples to a PCM output line of the DSS.

Other subsystems in a DCCE

Some of the other subsystems that make up a DCCE are shown in Fig. 10.19.

The *message transmission subsystem* (MTS) carries out common channel signalling (see Sec. 10.5) and error correction functions. The *signalling interworking subsystem* (SIS) facilitates interworking with analogue exchanges and provides tones and information services. The *processor utility subsystem* (PUS) is the data processor that executes the software controlling the traffic and the local and remote switching subsystems.

The remote concentrator

The lowest unit in the switching centre hierarchy of the PSTN, shown in Fig. 10.2, is the remote concentrator unit (RCU), which consists of a number of *digital subscriber switching subsystems* (DSSSs) of Fig. 10.20.

An RCU for about 10 000 subscribers would require five DSSSs. A DSSS is an 8:1 *concentrator switch*, which concentrates the traffic before passing it on to a *digital switching subsystem* (DSS) in a DCCE. The DSSS first digitizes each input analogue signal (each

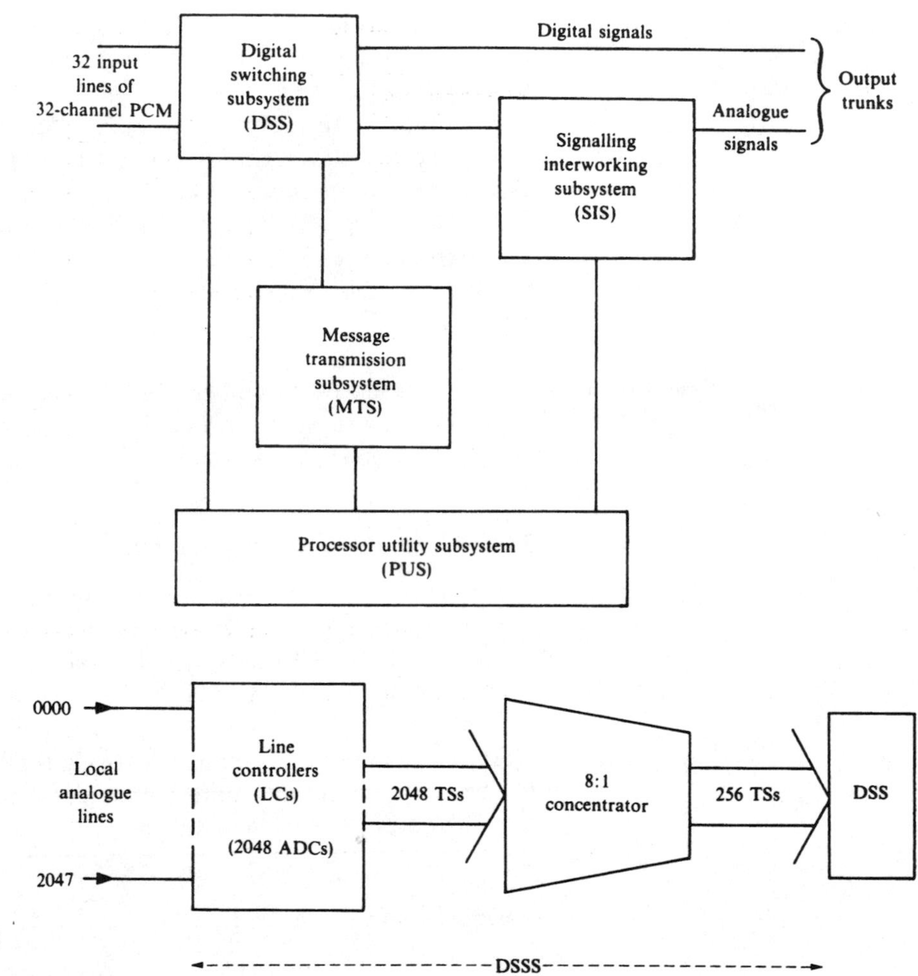

Figure 10.19. A DCCE arrangement.

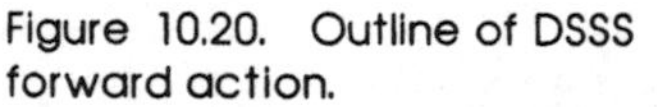

Figure 10.20. Outline of DSSS forward action.

subscriber line has its own ADC in the exchange) to produce a 32-channel PCM signal along each of 64 lines between the line controllers (LCs) and the concentrator, that is $32 \times 64 = 2048$ time slots. The 8:1 concentrator switch takes in this maximum 2048 time slots and concentrates the traffic from these into $2048/8 = 256$ time slots. That is, eight output lines each of 32 slots. Since two of the 32 slots are used for signalling, this means that 240 of the 2048 subscribers can call simultaneously. To establish two-way communication, the DSSS must also perform in the reverse manner to that shown in Fig. 10.21.

After multiplexing 32 of the 64 kb/s PCM voice signals we obtain $2048/32 = 64$ lines each carrying 32 time slots (that is 32-channel TDM). Each subscriber has a RAM location, which holds a speech sample while a call is in progress, so a 2 kbyte store is required. The connection control unit (CCU) holds the program with details

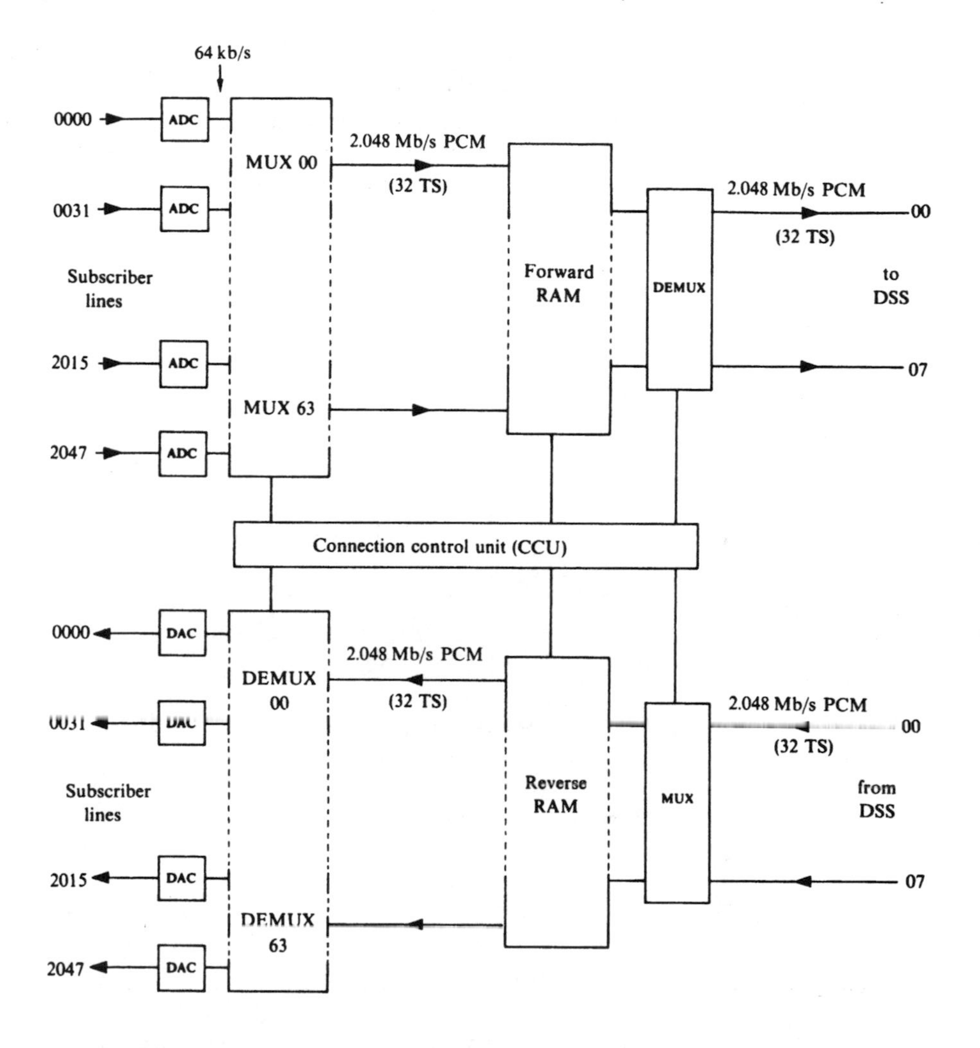

Figure 10.21. Digital subscriber switching subsystem.

of which store location is to be switched to a particular time slot. The contents of a store location (containing a sample byte) are read out on a nine-bit parallel highway to a MUX. The RAM locations are scanned cyclically. The reverse speech time slots are scanned in synchronization with the forward scanning. A reverse sample is written into the subscriber's reverse store location at the same time as a forward sample is being read from the forward store location.

Common channel signalling

Telephone signalling refers to those actions which are taken to initiate and route a call, meter it for billing purposes and, finally, clear the line. The state of a call such as indicating that the number is unavailable (NU) or engaged (ENG) is also part of signalling. These actions are shown in Fig. 10.22.

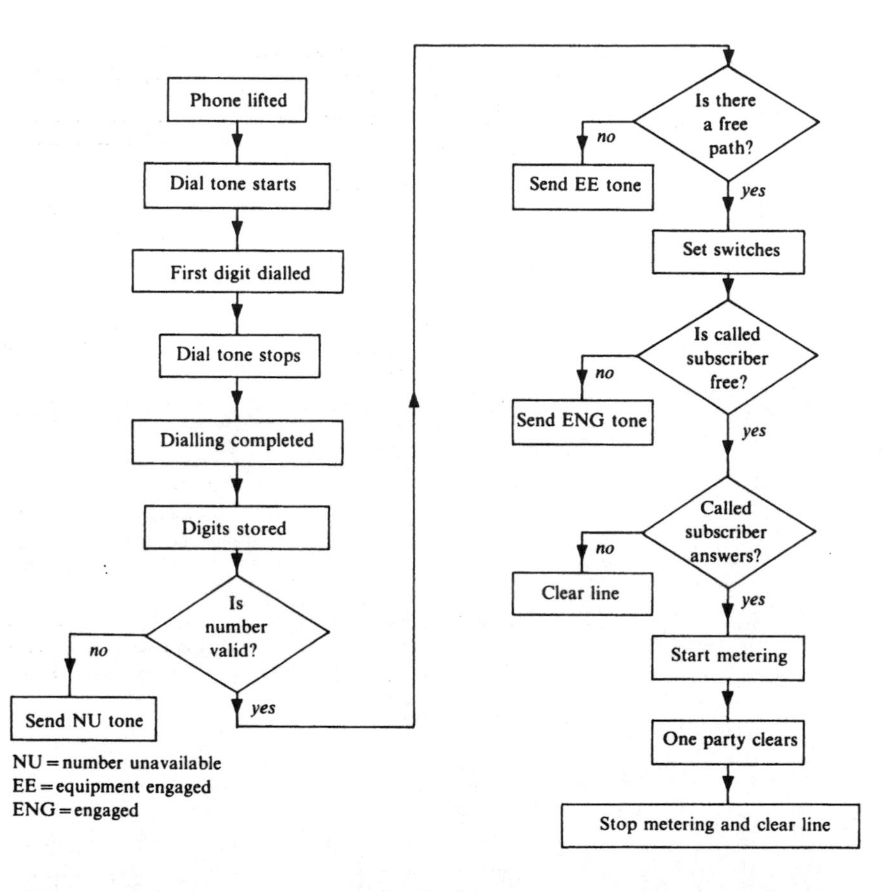

Figure 10.22. A signalling flowchart.

The actions outlined in Fig. 10.22 must be carried out with signals within the system. These signalling signals are kept separate from the speech signals by being confined to a common channel earmarked for signalling. Figure 7.26 indicates that in the CCITT 30-channel PCM system, an eight-bit signalling word is transmitted in time slot 16 of frames 01 to 16. This time slot provides the signalling channel for all voice signals. Each speech channel is allocated four signalling bits, as exemplified in Table 10.3.

Table 10.3 Examples of coded signals

Nibble	Forward	Backward
0001	trunk offer	manual hold
0011	circuit seized	number answered
0111		circuit free
1111	circuit idle	circuit busy

The first nibble (bits 0 to 3) of the signalling word is used for channels 01 to 15 and the second nibble (bits 4 to 7) for channels 16 to 30. Frame 01 serves channels 00 and 16, frame 02 serves channels 02 and 17 and so on. The principle of common channel signalling is explained in Fig. 10.23.

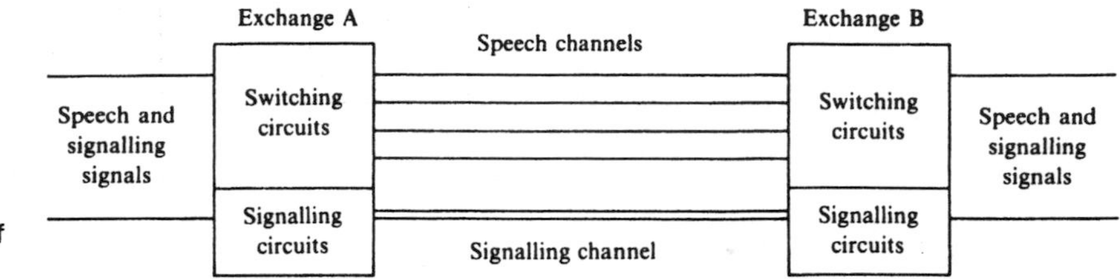

Figure 10.23. The principle of common channel signalling.

In System X common channel signalling is carried out as shown in Fig. 10.24. The MTS extracts the signalling information from time slot 16. This information is input to the system when a subscriber dials. The dialling process generates electrical signals in the form of either pulses or tones. The way signals are generated when a subscriber dials will now be described.

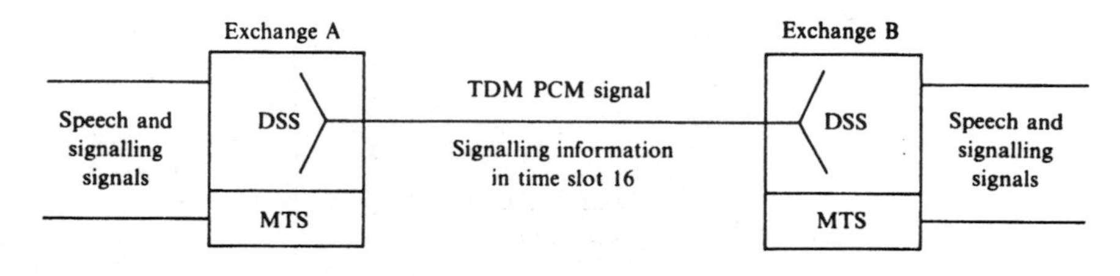

Figure 10.24. Common channel signalling in System X.

The operation of a telephone handset

Telephones with dials have been in existence for a long time, but push-button (keypad) telephones are now replacing them. The dial type works on the principle that when the dial is released from a selected number position it returns to its rest position under spring loading. In doing so it opens and closes a switch a number of times equal to the number dialled. The dial takes one second to do a full turn through the sequence of numbers 0, 9, 8 to 1, so dialling the number 071 would generate the signals shown in Fig. 10.25.

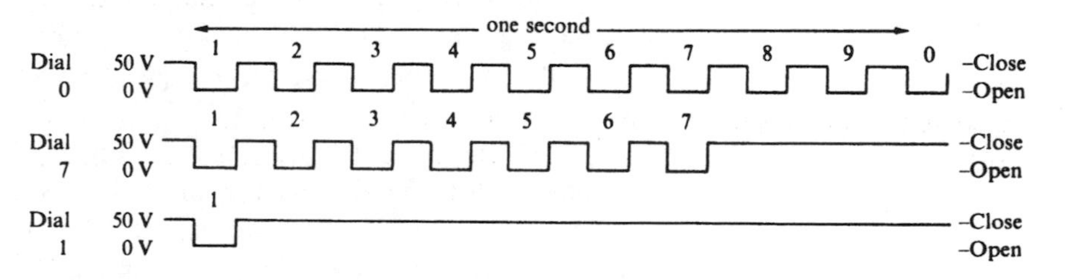

Figure 10.25. Dialling signals for number 071.

When the dial is released from position 0 (zero), ten disconnections are made in one second in the loop between the telephone and the line switch as the dial returns to its rest position. When the dial is released from position 7, seven disconnections are made but only one disconnection is made for position 1. This process of breaking the circuit between the handset and the exchange is called *loop-disconnect* (LD) signalling.

In push-button phones the numbers are selected from a keypad and either converted to LD signals or tones of various frequencies—a technique called *multi-frequency* (MF) signalling. Fig. 10.26 shows an MF keypad.

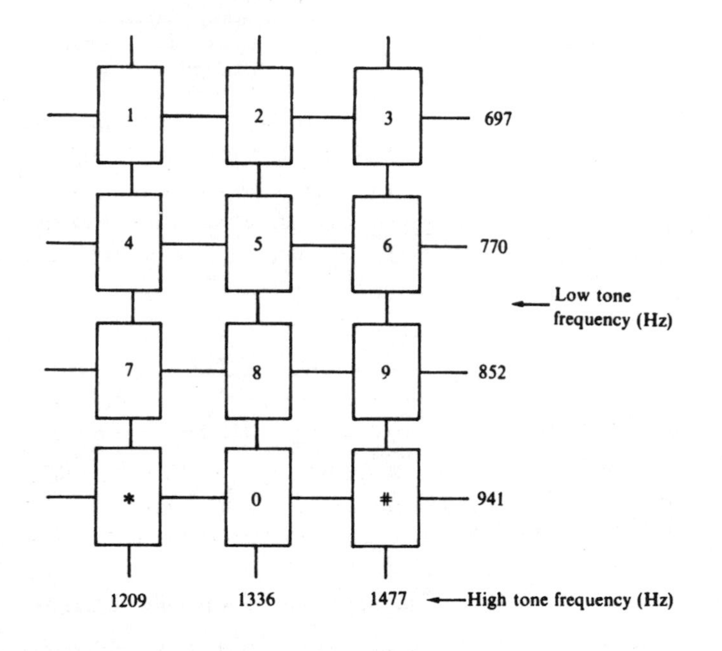

Figure 10.26. Keypad frequencies.

This two-tone selection process works on the cross-point principle. When any key is pressed, two tones of different frequency are connected to the call-signal circuit. For example, when key 0 is pressed, two tones of frequency 941 Hz and 1336 Hz are generated, and while key 1 is pressed two other tones of 697 Hz and 1209 Hz are generated. These pairs of tones perform a similar function in digital exchanges to the LD signals in analogue exchanges. They form part of the signalling signal in time slot 16 after they have been converted into binary code. Many modern handsets have both LD and MF capability, selected by a two-position switch, but there is no advantage in using MF unless the local exchange is digital.

The standard keypad layout of Fig. 10.26 has two more keys than are needed for the numbers 1 to 0. The * key is used either for muting (silencing the caller's voice so that the listener at the other end cannot hear what is being said) or for invoking the star services offered by System X. All push-button phones have at least a single-number electronic memory so the # key can be used for *last number recall*—engaged numbers can be recalled without the tedium of full redialling. In phones with 10-number memory storage this key is also used to recall any of 10 stored numbers that the subscriber has programmed in. Each key from 1 to 0 is allocated a telephone number by the subscriber. Pressing the #

key and then the abbreviated number key automatically calls the stored telephone number.

Exercise 10.10
Check on page 251

State whether LD and MF signalling are either analogue or digital techniques.

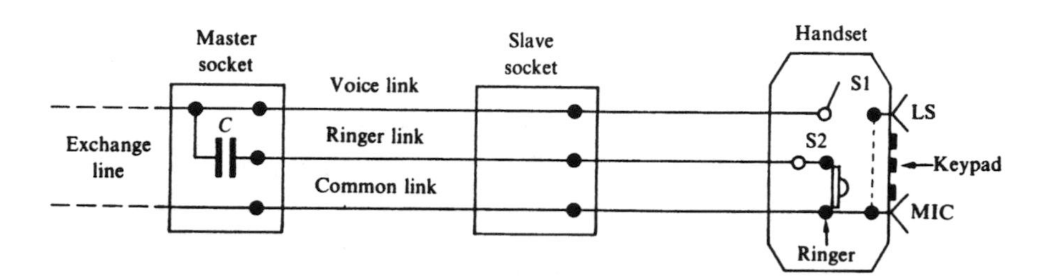

Figure 10.27. Master/slave socket/handset circuit.

A handset is linked to the local exchange as shown in Fig. 10.27. The direct exchange line (DEL) is terminated in the subscriber's premises at a master socket provided by the authorized public telephone operator (BT in the UK). This has a three-wire outlet for connection to the handset plug lead consisting of: a voice link which connects the LS and MIC via switch S1, a ringer link which connects the 'bell' via switch S2 and a common link. When the handset is resting in its cradle or 'on hook', S1 is open and S2 is closed. The call signal then operates a 'ringer' in the handset. Taking the handset 'off hook' releases the spring-loaded switches so that S1 closes and S2 opens, connecting the LS and MIC to the exchange and disconnecting the ringer. Recall from Fig. 2.1 that a battery in the exchange powers the MIC amplifier. A 1.8 μF capacitor (*C*) prevents battery current drain via the ringer while the handset is 'on hook'. *C* passes the alternating current, which activates the ringer.

Slave sockets may be wired in parallel with the master socket to provide for plug-in extensions anywhere in the premises. The slave sockets are similar to the master socket without the capacitor. Each ringer has a *ringer equivalence number* (REN) of 1 to 1.5. A DEL supplies enough current to ring four 1 REN ringers. If the REN capacity of 4 is exceeded, by having too many telephone extensions, the volume of a ringer will be reduced and some ringers may cease to work.

In the simplified arrangement shown in Fig. 10.27, the MIC and LS are connected in series so that the MIC sound is heard in the LS of the caller as well as in the LS of the called. This 'sidetone' reassures the caller, enabling him or her to monitor the transmitted sound level. The electronic hardware within a handset ensures that the right sidetone is heard, as well as providing MIC output amplification and keypad decoding.

10.6 Private branch networks

Many organizations require a telephone network for internal use but which can be connected to the PSTN. These private branch networks

(PBNs) are normally switched automatically for internal calls and either automatically or manually for external calls from the PSTN. Any caller on a network with a private *automatic* branch exchange (PABX) simply has to dial 9 for an outside line and follow this by dialling the number required as though the extension was a direct line to the local exchange. Each telephone extension may have its own number if the appropriate rental is paid to BT or Mercury; otherwise an operator must be employed by the organizations to route calls. A typical PABX extension makes about four 3 min calls during the busiest hour of the day—about 0.2 erlang. A PABX must be designed to cope with this level of traffic from each extension.

The term *call routing apparatus* (CRA) is used to describe equipment that is installed in a subscriber's premises to provide switching and call handling on the branch network. Such systems are PABXs and *key telephone systems* (KTS).

A key telephone is one with a number of control and switching keys. In the absence of a PABX, this system allows for any extension user to answer incoming calls and transfer them internally if necessary. The system also allows for free intercommunication between extensions. Programmed routing facilities may be added to such systems so that only certain extensions ring (perhaps in an ordered sequence) on an incoming call. The latest keyphones offer many facilities such as abbreviated dialling, paging (if additional radio equipment is added), loudspeech and hands-free operation.

A modern PABX takes the form of a small cabinet, containing racks of circuit boards, operated from a console of switches with indicator lights. It centralizes many of the distributed facilities offered by a keyphone system. Many of these facilities are alterable by the operator at the main console. Many systems nowadays offer keyphone functions combined with a PABX and are referred to as *hybrid systems*. This type of system is intended to occupy pride of place in a company reception area where it will enable a receptionist to double as a telephone operator. Typically, it will have a display that shows the status of each call and a console with function keys; a capacity of perhaps 24 exchange lines, eight internal private lines and 80 extensions, with the ability to monitor calls from all of these. The system operates under *stored program control* (SPC) with all the hardware enclosed in a small cabinet that can be placed anywhere within 200 m of the console. SPC is a system whereby switching or call routing is performed by electronic switches (hardware) under the control of a program (software) stored in a computer memory. Its great advantage is that the facilities offered may be changed fairly easily by amending the program.

Keyphones used with a modern PABX may also offer other useful features such as: call again mutual party (CAMP)—an automatic call-back facility—call queuing, distinctive ringing and extension hunting. Other features that a PABX can offer and are also offered to BT subscribers who are connected to System X local exchanges are: call diversion, call barring, call waiting advice, charge advice, conferencing, reminder call, abbreviated dialling and last number recall. These so-called 'star services' are invoked by pressing the * key on the handset.

Two variations of the PABX are automatic call distributors and dealer board systems. In the former, incoming calls are automatically made available to the first free extension. This is a particularly useful facility for customer service and sales departments where rapid response to outside calls is important and where intercommunication is not. Dealer boards are desk-top units linked to databases that give up-to-the-minute financial and commodity information. They often provide two handsets so that a dealer can buy and sell in one breath!

10.7 Radio telephones and pagers

The radio telephone has been used for many years as a means of establishing a 'wireless' link between two telephones. Wire links are not necessary—they came about because the radio spectrum that could be exploited was too small to accommodate all the subscribers. Modern electronics has changed this situation.

The simplest form of radiophone is the *cordless telephone* (CT) for use over a range of 100–200 m on a subscriber's premises. It consists of a base unit, connected to the PSTN like a normal telephone, and a handset, which is linked 'cordlessly' to the base unit by an FM VHF radio wave. Duplex operation is achieved by having transceivers in both units as shown in Fig. 10.28.

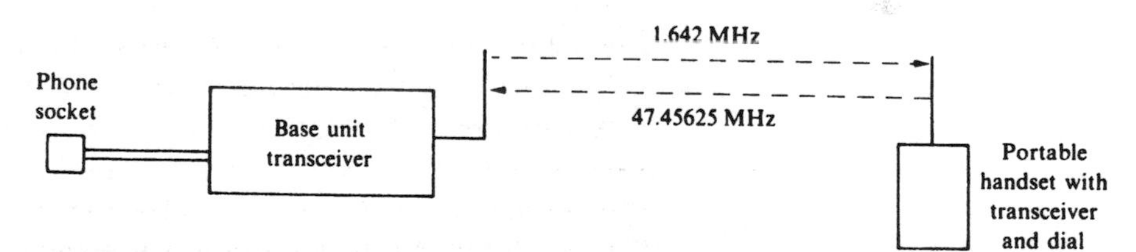

Figure 10.28. The principle of a cordless telephone.

The handset is able to receive from and transmit to the base unit at a maximum range of about 100 m. Thus, the cordless phone allows normal telephone communication via the PSTN while the handset is being moved about the subscriber's premises. It also allows the handset carrier to be 'bleeped' or *paged* by pressing a key on the base unit. Some cordless phones provide full intercommunication between the base unit and the handset. The handset is powered by rechargeable batteries so it must be returned to the base unit, which contains a mains charger, at regular intervals. Eight frequency channels have been allocated for cordless telephones as shown in Table 10.4.

With so few channels available and the comparatively long range of VHF waves, cross-talk can be a problem in urban areas. A second

Table 10.4. Operating frequencies for cordless telephones

Channel no.	Base transmit frequency (MHz)	Handset transmit frequency (MHz)
1	1.642	47.45625
2	1.662	47.46875
3	1.682	47.48125
4	1.702	47.49375
5	1.722	47.50625
6	1.742	47.51875
7	1.762	47.53125
8	1.782	47.54375

generation of cordless phones (CT2) is being introduced that will operate in the band 864–868 MHz with high-quality digitized speech. More channels will be available at UHF and cross-talk eliminated because each handset has a unique identity code, which is exchanged with the base unit every second or so to retain the link. CT2 technology, specified by BS6833, uses *time-division duplex* (TDD) to provide two-way digitized voice communication on one of 40 radio channels, allowing about 7000 CT2 handsets to be used per square kilometre without cross-talk. Speech is transmitted in 1 ms *data packets* (see Chapter 11 for explanation of data packets). Transmitter power is limited by law to 10 mW, giving a maximum range of about 200 m. CT2 handsets will be usable on the *telepoint* network.

Wide area mobile telephones

The natural extension of the cordless phone principle is the wide area mobile telephone—a system that enables a subscriber to wander anywhere in the country and remain connected to the PSTN. The necessary radio spectrum needed to make this possible in the UK has been made available. This extends from 890 to 915 MHz for base transmission and 935 to 960 MHz for mobile transmission. The standard adopted for making use of this allocated bandwidth is the *total access communications system* (TACS), which is an adaptation of the *advanced mobile phone system* (AMPS) used in North America. TACS uses FM with a peak frequency deviation of 9.5 kHz. FM gives high SNR at low transmitter power and its capture effect feature is most useful, for reasons to be mentioned later.

Exercise 10.11
Check on page 251

Calculate, using Carson's rule, the bandwidth required by the TACS standard to transmit commercial speech. Then work out the number of duplex speech channels available in the allocated bands: 890 to 915 MHz and 935 to 960 MHz.

In urban areas the number of subscribers far exceeds the number of channels available, so the same frequency channels must be re-used repeatedly in small regions or *cells*. Those cells using the same carrier frequencies must be out of radio range of one another. Ingenious electronic circuitry must ensure imperceptible hand-over as the carrier frequency is changed when the *mobile* (telephone) moves between contiguous cells. *Cellphones* operate by short-range UHF communication with *base radio stations* each serving a cell a few kilometres in diameter, the actual size depending mainly on population density. The base station and the mobile contain multi-channel transceivers. Most use of the cellular system is made by people in vehicles, but it can be used wherever mobile communication is required, such as on building site work. In order to keep those cells using the same carrier frequencies out of range of one another, the cells are grouped in *clusters* as shown in Fig. 10.29.

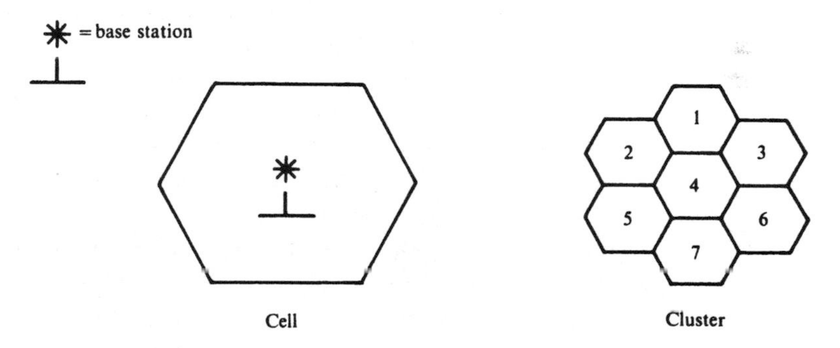

Figure 10.29. A single cell with base station and a seven-cell cluster.

Each cell's base station, which houses a low-range transceiver, is connected by land line to a *mobile exchange* (MX), which is, in turn, linked to the MXs of other clusters and a PSTN main switching centre as shown in Fig. 10.30. The term 'mobile exchange' is slightly misleading because this

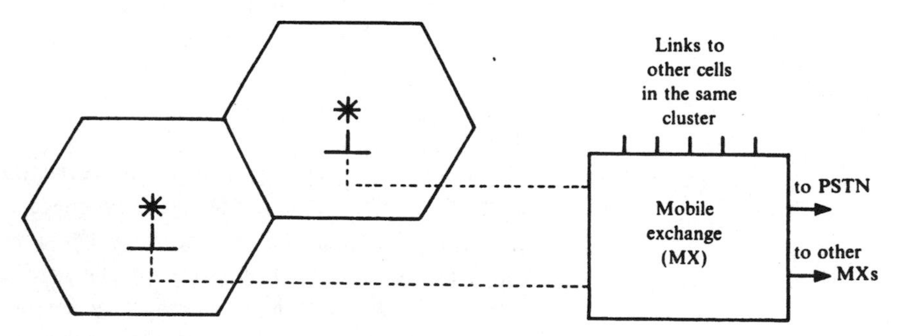

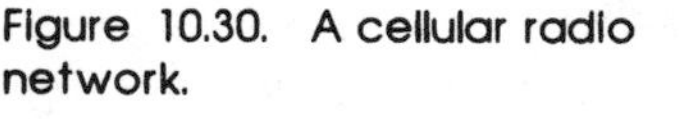
Figure 10.30. A cellular radio network.

exchange cannot move! It is called a mobile exchange because it serves the mobiles while they are in that MX's cluster. MXs should be thought of as cluster exchanges. Each base station has an omnidirectional aerial so, although the cells are conventionally represented by hexagons, the broadcast area covered by a cell will have an approximately circular boundary with a base station at its centre.

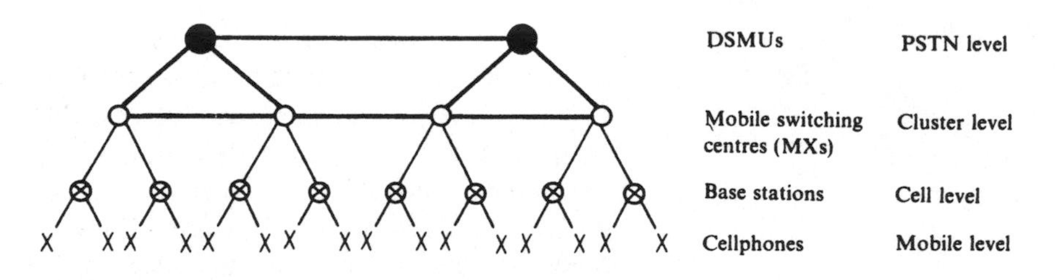

Figure 10.31. Routing and switching hierarchy in the cellphone system.

The cellular radio network has the hierarchical structure shown in Fig. 10.31. Each cell is given a number of base/handset frequencies in the same manner as a cordless phone. No two cells in the same cluster have the same frequencies. If these clusters are located around the country with the pattern shown in Fig. 10.32, the same frequency pairs can be re-used providing the clusters are large enough for the same-numbered cells to be out of range of one another.

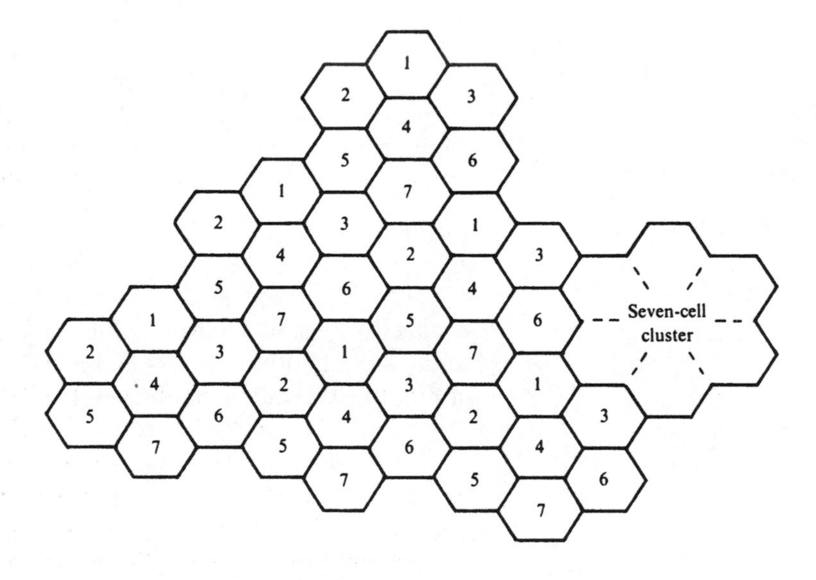

Figure 10.32. An area covered by seven-cell clusters.

The number of cells per cluster is restricted by the requirement that the clusters must fit together like jig-saw pieces. Some arrangements that do this are 4-, 7-, 12- and 21-cell clusters. Clusters with a small number of cells will tend to suffer co-channel interference unless the cells are large in area. Most of the UK is covered by seven-cell clusters, coverage being provided by two similar systems, under DTI licence, namely: Cellnet (owned jointly by BT and Securicor) and Vodaphone (a subsidiary of Racal). Each has been allocated 300 channels, leaving 400 in reserve for a future pan-European system. Initially these companies set up fairly large cells to economize on the number of base stations that had to be built. As the demand for cellphones increases, these large cells are split into smaller cells as shown in Fig. 10.33.

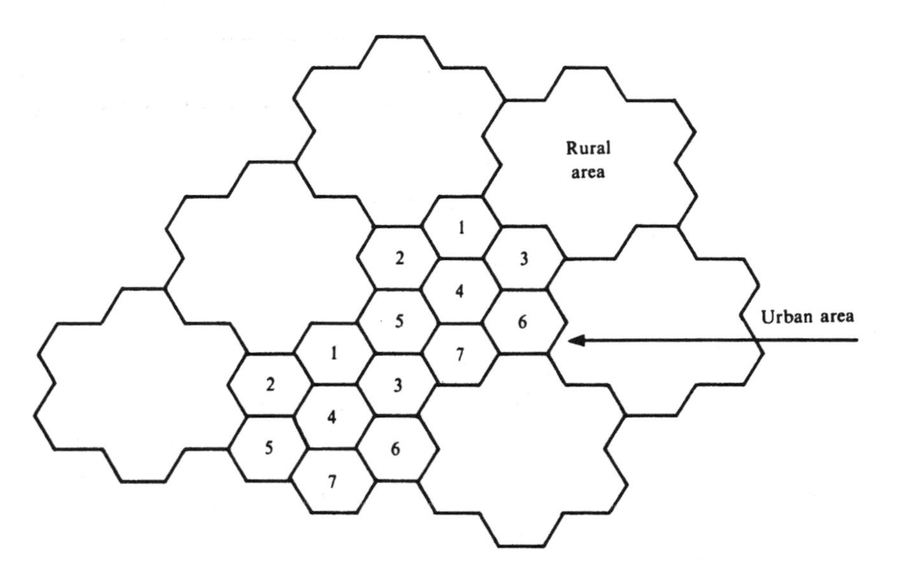

Figure 10.33. Cell splitting.

The area of a cell depends on the likely user density and is about 3 to 50 km^2—equivalent to a cell radius (r) of about 0.5 to 8 km from the base station—but in inner-city areas, street-sized microcells, serviced by transceivers mounted on lamp posts, are being used to reduce congestion.

Exercise 10.12
Check on page 251

Suggest how the area of a cell may be varied.

System operation

There are several distinct procedures involved in cellular operation, which will now be described briefly. When a mobile is turned on, it searches for a strong *dedicated control channel* (DCC) and a strong *paging channel.* The mobile then goes into its *idle* state in which it continues to monitor the paging signal. If this signal falls below a critical level, as it will do whenever the mobile enters a new cell, the mobile searches again for other strong channels. This procedure enables the mobile to 'listen in' for calls. The mobile must also *register* its whereabouts with its home base, updating its location each time it crosses a cell boundary.

In order to make a call from a mobile, the required number is keyed into the mobile and the call initiated by pressing a SEND key. This accesses the base station, which, if free, allows the mobile to transmit. If the base station has no free channels, the mobile backs off for a random time and re-attempts until a free channel becomes available. The mobile automatically tunes to this channel frequency and the call is set up by the MX. During the call this channel is held against other channels, so long as the mobile remains in the cell, by the capture effect of FM. When the handset is replaced after the call, the mobile transmits an 8 kHz signalling tone for 1.8 s to the base station and then returns to its idle state.

When a call for a mobile is received, the MX *pages* all the base stations near the mobile's current location (known from registration). This is done by transmitting a paging call on the paging channel of each base station. When the mobile receives a paging call, it automatically accesses the network. The mobile is then allocated a channel by the base station, to which it automatically tunes. The base then transmits a continuous 8 kHz signalling tone, which is turned off when the mobile user answers the call, after which the call is set up.

If, during a call, the mobile moves into a new cell, the signal level will fall and this initiates an *in-call hand-off*. The base station monitors the signal level from the mobile and informs the MX when the level falls to a critical value. The MX instructs all the surrounding base stations to measure the mobile's signal level and transfers the call to the strongest channel. All this occurs in about 400 ms, so the cellphone users are hardly aware of a break. Channel shifting is done automatically under the control of microprocessors built into the equipment.

The number of frequencies allocated to a cell

About 100 subscribers can be serviced by 10 channels. In a rural area there might be, on average, two subscribers per square kilometre, so in a 50 km^2 cell only 10 channels would be needed and a cluster could contain therefore 30 cells, although 7 is optimum. By contrast, in a densely populated urban area there might be 100 subscribers per square kilometre, so all the available voice channels would have to be distributed among seven cells—that is about 40 per cell, allowing for signalling channels. Since we need 10 channels per square kilometre, this means that the cell area (A) would be given by $40/A = 10$, so $A = 4\ km^2$.

Exercise 10.13
Check on page 251

What would be the maximum cell area in a location with an average of four subscribers per square kilometre if 20 channels were allocated to each cell?

The cellular telephone system provides the ultimate telephone service in which a subscriber has access to the global system with the freedom to move almost anywhere and remain contactable. It is, however, an expensive system. Equipment costs are high but, more significantly, it requires a large radiofrequency bandwidth—a scarce resource. This scarcity will ultimately limit the number of subscribers that can be handled in a given area. Fortunately, most people do not require such a mobile telephone service. They would be satisfied with a service in which they could be simply alerted while on the move and make calls without having to go to a fixed telephone. National *paging* and *telepoint* respectively provide these services at a much lower cost than the cellular system.

Paging systems

A pager is basically a radio receiver that can detect signals from a base paging transmitter. Unlike the cordless handset, it is not designed to transmit back to a base receiver—it is merely an alerter and collector of messages. On-site paging can be provided as part of a private branch network as shown in Fig. 10.34. This figure gives a simple illustration of the principle of paging.

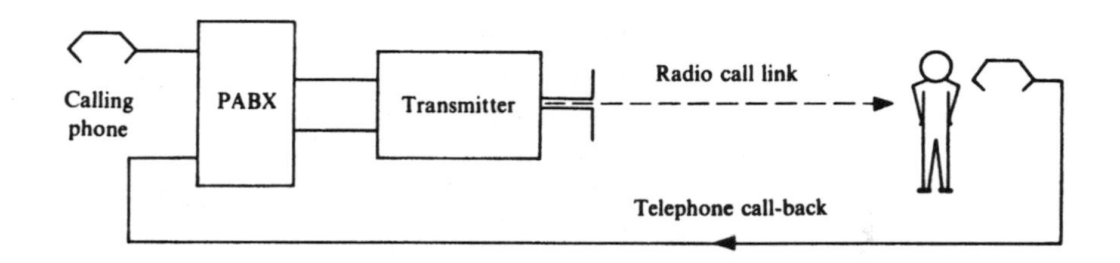

Figure 10.34. A pager as part of a PBN.

BT offer a radiopaging service in about 40 geographical zones, which cover most of the UK, each served by a transmitter operating in the 153.025 to 153.475 MHz band. The simplest pager gives 1 or 2 s 'bleeps', unless muted, when called. A *two-tone* pager gives the subscriber a choice of contact points, usually home or work, or the ability to distinguish between urgent and non-urgent calls. Up to four tones are possible. A *voice* pager bleeps and takes a recorded message. *Visual* pagers bleep and display simple messages on a liquid-crystal screen. Messages can be sent to the pager, which has its own unique number, by either phone or telex services. Pagers use FSK and a CCIR (see Chapter 11) radiopaging code. A desk-mounted, 1 W transmitter gives a range of about 250 m, which can be extended by mounting the aerial on a mast. A Home Office licence is required for an in-house paging system.

A calling system for a national radiopaging service must be able to provide a very large number of addresses. The present BT pulse dialling arrangement will give way to a European seven-tone paging signal whose format is shown in Fig. 10.35. In 1990, BT launched the transatlantic paging service called 'Metrocast'.

	First call number							Line clear	Second call		
Pager number	3	7	4	9	2	1	1		6	8	5
Digit frequencies	f_3	f_7	f_4	f_9	f_2	f_1	f_1	f_c	f_6	f_8	f_5
	300 ms							200 ms			

Figure 10.35. Examples of the European seven-tone paging signal.

Telepoint

The telepoint service provides the call-back link of Fig. 10.34 without the caller-back having to go to a fixed telephone. The service is a public version of the cordless telephone principle with base units or *telepoints*, linked to the PSTN, being provided in public places as telephone call-boxes are at present. The UK service is implemented by four licensed consortia trading under the names Phonepoint (BT, STC, *et al.*), Zone-phone (Ferranti), Callpoint (Shaye Communications, Motorola and Mercury) and BYPS (Barclays Bank, Phillips and Shell). Users of the service will simply move until they are in line-of-sight of a telepoint sign, switch on their handsets and key in their personal identification number (PIN). A central computer, belonging to the licensed operator, will verify the number, sanction transfer to the PSTN and monitor the call for billing purposes. When a *common air interface* (CAI) standard has been agreed and implemented, a subscriber will be able to use any of the four services.

Trials of CT2 with CAI are being made in other European countries but the *digital European cordless telephone* (DECT) standard, based on carriers around 1.6 GHz, has been formally adopted and CT2 will be interfaced with it when it becomes operational. The CT2 handsets are usable with private base units, in the same way as CT1 sets, with each base being able to interconnect up to six handsets and thus acting as a small, cordless PABX with six extensions. Indeed, the CT2 handset may be the telephone of the decade. Its relation to the expensive cellphone is analogous to that of the personal computer to the mainframe computer.

Personal communications

Personal communications networks (PCNs) provide another means of satisfying the demand for small and easily portable telephones. Although based on the cellular idea, PCNs differ in two respects. The handsets are much lighter, owing to greater digitation, and there is more scope for network expansion because 170 MHz of bandwidth is available (more than three times that given to the other UK cellular systems). The 1710–1880 MHz band has been allocated and a standard, digital cellular system (DCS 1800) has been agreed by the Groupe Spéciale Mobile (GSM) of the European Telecommunications Standards Institute.

A PCN comprises a grid of contiguous *macrocells* covering a country. These cells vary in size, depending on traffic, from about 8 km radius in rural areas to less than 1 km in towns. To cope with traffic peaks, the macrocells are split into *microcells* covering such areas as railway stations, airports and shopping centres. *Picocells* are used to provide a high-quality service to individual buildings. The hierarchy of cells relies on frequency re-use (as in any cellular system) plus dynamic allocation of channels (as in telepoint systems) to allow for time-varying traffic patterns. The PCN operators are allowed to use millimetre radio wave links (38 GHz band) to control handover and dynamic frequency allocation.

Summary

The railway companies, with their natural interest in signalling and their existing network structure, were probably the first to use a private telephone system. Such private systems were soon adopted by other businesses so that simple hard-wired intercommunicating networks were to be found within a factory site or between branch offices of the larger prosperous companies. Private telephone companies soon stepped in to provide a telephone service to other organizations and the general public, using manual switching instead of hard wiring to reduce the number of links. These private operators gradually connected their networks and merged together into larger business units, two developments that culminated in the PSTN and BT respectively. Finally, national systems have been linked so that nearly all 215 countries in the world can now be reached by international direct dialling, which establishes very high-quality links over vast distances. This global cover is now being facilitated by optic fibre submarine cables and microwave satellite links.

Many private networks are connected to local exchanges of the PSTN. These private branch networks (PBNs) are miniature versions of the public network, to which they are connected, and the two types of network have developed technologically together. When the public network went over to automatic switching, so did the larger private networks and the PABX came into being. In recent years PBN technology has led the way because it is economically easier to replace small private networks with modern equipment than it is to modernize the vast PSTN.

The major technological leap in telephone communications has been the replacement of the analogue transmission and switching systems by compatible digital systems based on CCITT 30-channel PCM standards. In the UK, BT has introduced System X, a TST system based on standard subsystems that can be used in all levels of exchange. This system uses HDB3 coding and common channel signalling.

The ideal telephone, which can be carried around without restriction, is gradually being realized by various forms of radiophone. The cellular system gets closest to the ideal, but is expensive. A compromise arrangement, which combines public paging with telepoint, is likely to become the choice of most mobile users. The same CT2 handsets will be usable in the cordless telephone systems at home and in the workplace.

The PSTN is not fully occupied by voice communication and this leaves time for it to be used for other services, which add value to the network. These VANS are mainly data services, which will be described in the next chapter.

▶ Checks on exercises

10.1 Immediately on recognition of 0 in the STD code number 091, the Fulham RCU would connect to a free trunk route to a London DCCE. The remainder of the STD code (91) would be used by that DCCE to establish, via a DMSU, a connection to a Tyneside DCCE. Meanwhile, the digits 514 0332 are sent along the route to enable that Tyneside DCCE to connect to Sunderland AFC's switchboard.

10.2 Glass is an electrical insulator, so there is no sense in which a fibre can be likened to a conducting wire. Light might be thought of as a 'radio wave' of exceptionally high frequency. Optic fibre communication uses the idea of light wave guidance, so the statement accords with this idea. It is really a modern version of the North American Indian's rail-tapping (digital) technique, mentioned in Chapter 1, in which sound waves are guided along the railway line with the consequent targeting of information and increase in range.

10.3 In an FDM analogue microwave link, all the layers would be involved except 5.

10.4 Peak 1 is the mid-morning burst of activity, which occurs when the workers have properly woken up and sorted out their daily tasks. Peak 2 is the rush to clear things up before leaving work. Peak 3 is mainly domestic traffic, which has been waiting for the cheap-rate period.

10.5 $$A = [4/(60 \times 60 \times 24)] \times 2 \times 60 \times 8000 = 44.44 \text{ erlangs.}$$

10.6 The figure required for a six-channel system that routes calls at inputs 0, 1, 3, 4 and 5 to outputs 4, 2, 5, 0 and 1 respectively is shown in Fig. 10.36. The words in the content RAM are written sequentially and read in the address order 4, 2, 5, 0 and 1. Note that it is the content of the time slot, that is interchanged not the time slot, as the term 'time-slot interchanging' might imply.

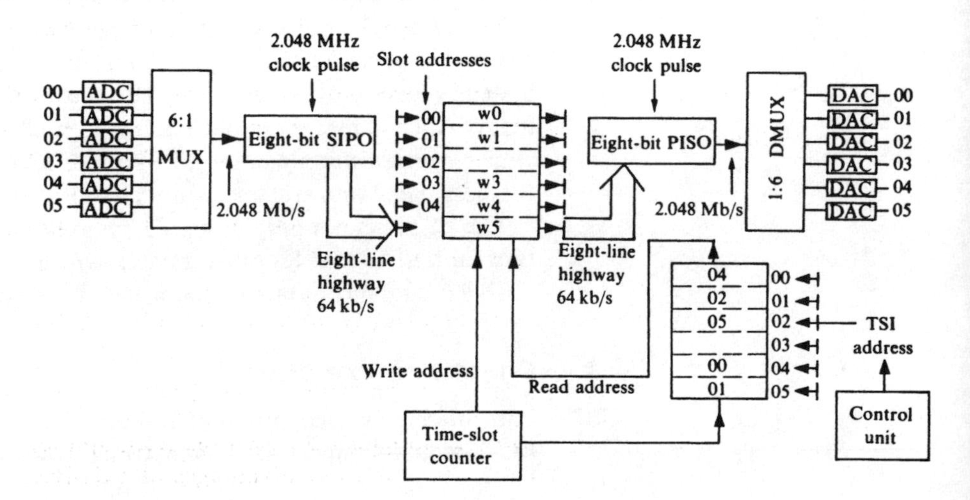

Figure 10.36. Action of a six-channel time-slot interchanger.

10.7 To switch the signal whose samples are in time slots 00 of input highway 0 to time slots 02 of highway 1 the TST switch must either:

1. Interchange time slots 00 and 02 in T1 only and enable the gate at the cross-point of row 0, column 1 during time slots 00, or
2. Interchange time slots 00 and N in R0, while enabling the gate (at the row 0, column 1 cross-point) during time slots N, and then interchange time slots N and 02 in T1.

The advantage of having many time-slot interchange possibilities such as are offered by process 2 is that the software can select the required cross-point whenever it is available (that is, when row 0 and column 1 are not being used to switch between other time-slot switches). This advantage is the reason for using both receive and transmit time-slot switches.

10.8 When the number of 32-channel input lines increases from 32 to 1024 and a parity bit is used, we must store 1024 nine-bit words, so the slot-content RAM increases in size from 32×8 to 1024×9 and is known as a *speech store*.

The slot-address RAM now has to store 1024 addresses instead of 32. Since $1024 = 2^{10}$, 10 bits will be needed to specify 1024 addresses. The slot-address RAM must therefore be increased from 32×5 to 1024×10.

10.9 With a space switch of 96×96, 2×96 time-slot switches are required, each of which can interchange 32 time slots. The number of interconnections that can be made using 30 voice channels will be $96 \times 32 \times 30 = 92\,160$. The XOS rate will be 4096 Mb/s $\times$ 3, which is what limits the size of the space switch.

10.10 LD is a digital technique and MF signalling is analogue. You may have realized the curious fact that digital signalling is used with analogue speech signals whereas digital transmission uses analogue signalling!

10.11 Carson's rule gives a bandwidth $B = 2(f_{dm} + f_t) = 2(9.5 + 3) = 25$ kHz. The number of duplex speech channels available is (915 − 890) MHz/25 kHz = 1000. In practice, some channels must be used for control and signalling.

10.12 The area of a cell may be varied by varying the transceiver transmitted power and sensitivity.

10.13 The cell area (A) would be given by $20/A = 0.4$, so $A = 50\ \text{km}^2$.

Self-assessment questions

10.1 Distinguish between the telegraph and the telephone.
10.2 What are the advantages of a tie-line?

10.3 Compare the hierarchical structures of the PSTN, System X and Cellnet.
10.4 Describe the international telephone number structure.
10.5 Describe a seven-layer telephone link protocol model.
10.6 Define the erlang.
10.7 Explain the functions of a concentrator, a distributor and an expander.
10.8 What feature of the CCITT 30-channel PCM transmission system has made time-slot switching possible?
10.9 Illustrate the action of a time-slot switch by showing how two RAMs may be used to route inputs 0, 1 and 2 to outputs 1, 2 and 0 respectively.
10.10 Why is time-slot switching supplemented by space switching?
10.11 What is the point of using receive and transmit time switches with a space switch?
10.12 What are the functions of the MTS and PUS in System X?
10.13 Describe briefly how common channel signalling operates in System X.
10.14 List the advantages of a PBN.
10.15 Give examples of call routing apparatus.
10.16 Illustrate how a pager and a cordless telephone may form a complete mobile telephone system.
10.17 Compare the system of question 10.16 with the cellular radiophone system.
10.18 Draw a cell network consisting of three contiguous four-cell clusters.
10.19 Sketch the routing and switching hierarchy of the cellular system.
10.20 How is a mobile contacted by an MX?
10.21 Describe the in-call hand-off procedure.
10.22 List the device technology improvements that have been mainly responsible for the recent developments in telephone communications.

11

Data communication systems

Exchanging digitized information

Objectives

When you have completed this chapter you will be able to:

- Outline the main features of data telecommunications
- Describe serial and parallel data formats and their interconversion
- Compare synchronous and asynchronous character frames
- Compare circuit switching, message switching and packet switching
- Describe a typical packet frame used in a wide area network
- State what is meant by a datagram and a virtual circuit
- Describe the various forms of access to a packet-switched exchange
- Describe a packet-switching network and a typical network routing algorithm
- Describe the packet-switching protocol X25
- Describe the Open Systems Interconnection seven-layer protocol model
- Describe the WAN services: viewdata, teletext, teletex and FAX
- Describe the functions and topologies of local area networks
- Compare the various forms of access to a local area network
- Describe how data networks are interconnected via gateways
- Outline an integrated services digital network
- Describe how satellite links are used in wide area networks
- Give some examples of various international standards

As the World's economy grows, more and more data is produced. This has created an ever increasing demand for faster data communications locally, nationally and globally. Data must be conveyed between adjacent pieces of equipment such as a computer and printer, or equipments in a local area must be networked. Often, however, data must be carried between remote equipments via telephone and telegraph links. These long-distance links limit the data transmission rate, so new high-speed links are being developed. In many countries, efforts are being made to achieve integration between audio, video and data networks so that in future all these forms of telecommunication can take place over a high-speed integrated services digital network.

11.1 A data link

Long-distance data communications has its roots in the telegraph and telephone systems: early computer networks adapted the data transfer methods used by the telegraph. Much data is still sent, over a wide area, via the PSTN but this method is limited by the inherently slow transmission rate allowed by the PSTN bandwidth. National and global data transfer is now accomplished by packet-switched wide area networks (WANs). Local area networks (LANs) also use data packets but are non-switched networks. LANs, like WANs, enable computers to 'converse' with one another. They also enable many terminals to access large common memories, and thus use the same software and data, as well as share expensive equipment, such as high-quality printers and plotters. LANs cover distances up to a few kilometres and, unlike WANs, have large bandwidth and low propagation delay. Access to a WAN is random but LAN access is either random or controlled by scheduling techniques such as polling, empty packet circulation or token passing. If random access is allowed, arrangements must be made to deal with data collisions in the network.

High-speed data networks, based on optical fibres, are now being used to interconnect LANs over regions where there is high data traffic such as the City of London, university campuses and large research establishments. These so-called metropolitan area networks (MANs) can be linked to WANs, as can the LANs, via protocol bridges or gateways. WANs, MANs and LANs usually transmit their data at baseband.

During the time when packet switching was being developed, computer manufacturers were developing different systems for data transfer, many of which would only work on their own equipment. Users who had acquired a variety of equipment were unable to transfer data easily because much of the equipment was incompatible. Standardization was needed. Various rules or protocols were adopted in different countries but it became clear that a more general approach was needed with the aim of allowing the exchange of data between as many different systems as possible. The International Standards Organization developed the Open Systems Interconnection reference model of a network. The model is based on several manufacturers' protocols. It represents a data network as a layered structure with up to seven layers being defined in order to give a comprehensive framework for developing future protocols which will facilitate the interconnection of equipment from any manufacturer.

Satellite links are being used increasingly for international data communications. These use packet radio techniques and operate in the microwave bands. Radio waves at certain of these frequencies penetrate clouds with little energy loss and, using FDM, provide sufficient channels for the many users. This poses special access problems and makes more urgent the need to achieve international agreement on telecommunications standards.

A basic telecommunications link is shown in Fig. 11.1. The transducer is often a computer, which might be regarded as a data-to-information converter. It processes input alphanumeric data, under the control of a stored program, and outputs binary-coded information to an adjacent or remote computer terminal, consisting of similar units such as a keyboard for input and a monitor screen for displaying the output information.

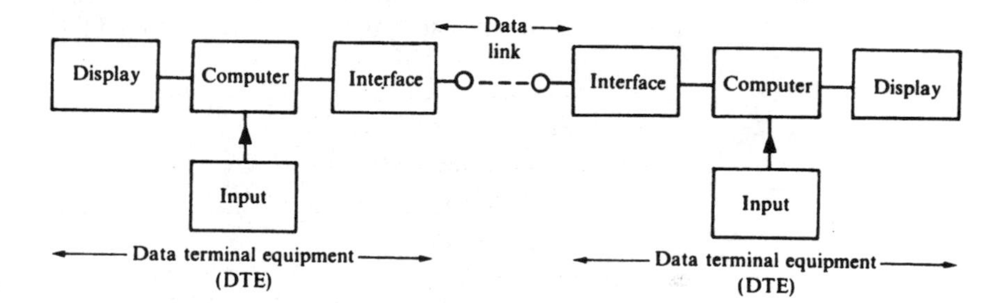

Figure 11.1. A basic data telecommunications system.

The electrical pulses that carry the information, by virtue of their pattern or order, constitute a data signal. Such pulses may also carry audio data using PCM, so data communications is not restricted to alphanumeric data.

There are two main differences between human and computer communications, which affect the techniques used. The signals between humans are analogue and are slowly processed by the humans: the signals between computers are digital and are processed by the computer at a rate that might be greater than the link can manage. An example of this is the use of the PSTN for data transfer. This network was designed to cope with low-frequency analogue voice signals and is, at the local level, too slow for efficient computer communications. In Sec. 4.6, various forms of modems (the interfaces shown in Fig. 11.1) are described. These devices enable data to be sent on the PSTN by low-frequency sinusoidal carrier signals. The fastest of these modems can manage a transfer rate of 9600 b/s; cheap modems are much slower.

The replacement of the analogue exchanges of the PSTN by System X digital exchanges and the use of glass fibres in place of copper wires will increase the channel bandwidth of the PSTN from a few kilohertz to a few megahertz, enabling computer data to be exchanged directly over the network at rates of some megabits per seconds (Mb/s).

Exercise 11.1
Check on page 300

Calculate the maximum transmission bit rate that can be used if the channel bandwidth is 3 kHz and its SNR is 10 dB. What bandwidth is required to transmit data at 64 kb/s assuming the same SNR?

11.2 Data formats

The central processing unit (CPU) of a computer transfers data internally using parallel links, called data buses, in which each bit of each word is transmitted simultaneously along its own link. In a long-distance communications link, data is transmitted as a serial stream of bits because it becomes uneconomic to use parallel links as the distance increases. A common feature of all computer/line interfaces is, therefore, a parallel-to-serial and serial-to-parallel converter known either as a *universal asynchronous transmitter and receiver* (UART) or *universal synchronous transmitter and receiver* (USRT). These are produced in IC form and have the functional structure shown in Fig. 11.2.

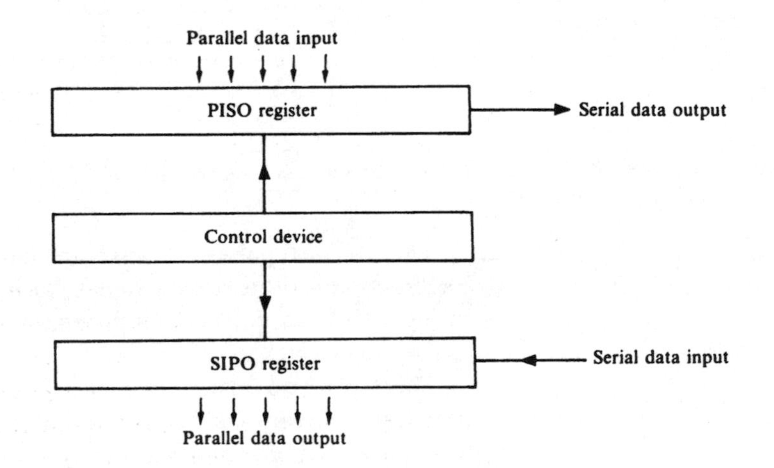

Figure 11.2. A UART/USRT system.

The transmitter is basically a parallel-in, serial-out (PISO) register. The data is shifted out by pulses from the controller at the rate of one bit per shift pulse. To receive data, the controller transfers its shift pulses to a serial-in, parallel-out (SIPO) register, which will contain a byte of data after eight pulses. The controllers of the USRTs at each end of a synchronous link would have to be synchronized by a common clock pulse generator.

In short-distance data communications, such as that which occurs between a computer and a printer, both serial and parallel links are offered. The plugs and sockets for the serial and parallel connections have been standardized and many terminals have both RS 232 serial interfaces and Centronix parallel interfaces. Details of these are given in Appendix E.

The cheapest connecting arrangement would be a one-way, serial, asynchronous type because only one signal line would be needed between, say, a computer output point and printer input. The computer, in this case, would be referred to as the *data terminal equipment* (DTE) and the printer as the *data communications equipment* (DCE). In practice, these two pieces of equipment intercommunicate. When the printer 'talks' to the computer, in order to acknowledge receipt of data for example, it becomes the DTE and the computer becomes the DCE. This implies that another data link is necessary between the printer output point and the computer input, thus making duplex communication possible. One of the main duplex operations is *handshaking*, which is the sort of process that goes on when the slow printer tells the fast computer when it can take more data and the computer responds until told to stop.

Character framing

The basic unit of information is the bit, but in data communications it is more appropriate to deal in *characters* since we are concerned with driving terminals such as VDUs and printers for much of the time. There are two *character framing* arrangements used for recognizing the separate characters sent as part of a bit stream. These are asynchronous and synchronous framing, the former being used for slow data transfer (300 to 9600 b/s) between printers and modems, the latter for the high data rates used between data processors.

The international character code, used for short-distance communication between VDUs, keyboards and processor, is that recommended by the American Standards Committee for Information Interchange (ASCII). This uses seven code bits plus one (optional) parity bit and so has 128 different key options consisting of 93 alphanumeric characters and symbols and 35 control operations such as 'carriage return' and 'end of text' (see Appendix D). An asynchronous frame for ASCII symbol 'M' would be put together by the UART transmitter as shown, using the bipolar NRZ form, in Fig. 11.3.

The negative voltage start bit triggers the shift pulse generator in the controller of the UART at the receiving end. The controller then shifts the

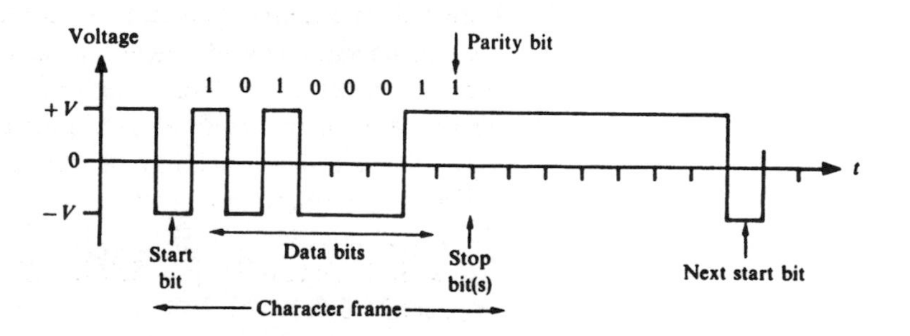

Figure 11.3. An asynchronous character frame.

serial input into the SIPO, stopping the shift pulse generator when the seven data bits have been received, on the negative voltage transition of the stop bit. In slow devices like printers, two stop bits are used to allow more time for the mechanical parts to operate: VDUs need only one stop bit.

Asynchronous framing is used for intermittent transmission. When large amounts of data, such as computer files, are transferred, characters are sent in long sections at a high bit rate, so the channel appears to be filled with a continuous stream of pulses as shown in the data format of Fig. 11.4. The receiver must be able to break up this bit stream into characters, so synchronous framing is used in which bursts of the ASCII SYNC code (00010110) are transmitted at regular intervals between data blocks (large frames) of continuous data.

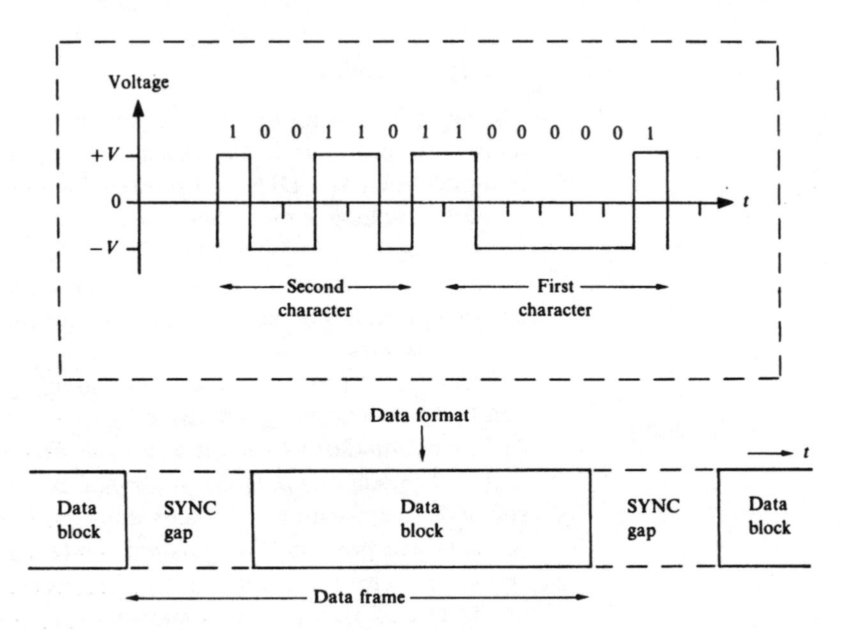

Figure 11.4. Synchronous character transmission.

The operations that occur between the start and stop bits in asynchronous data transfer have to be synchronized! In synchronous data transfer, the whole operation must be synchronized. Initially, the USRT controller detects the SYNC pulse pattern after the first few patterns have been received. During subsequent transmission, the character frames are recognized as a successive group of seven or eight bits. The transmitter and receiver clocks are initially synchronized during this start period and are periodically resynchronized by sending SYNC bursts between the frames. Serial and parallel links may be either asynchronous or synchronous.

11.3 Multi-terminal systems

In order to link more than two terminals, we must consider how they can be interconnected and the data routed between the appropriate terminals.

This same problem is considered for telephone communication in Chapter 10. The PSTN is, effectively, a circuit-switched network, designed for human conversation and the time taken to establish the physical link between subscribers is short compared with the average conversation period. This is not the case for computer 'conversation' over the PSTN—a computer can often 'say' all it needs to in a few milliseconds! To avoid the time wasted in circuit switching, dedicated computer communication networks are *message switched*. The two types of switching are compared in Fig. 11.5.

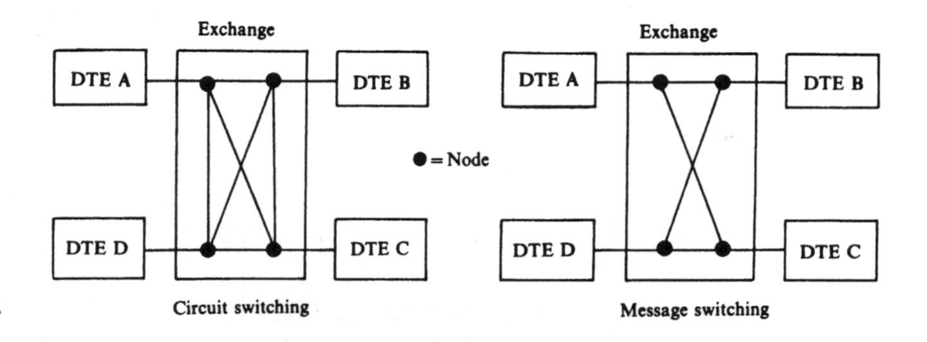

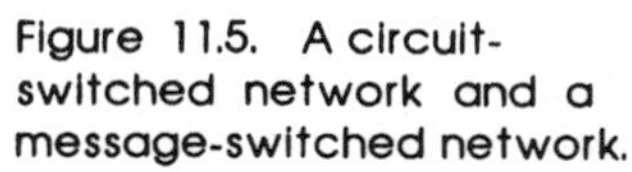
Figure 11.5. A circuit-switched network and a message-switched network.

A network user's *data terminal equipment* (DTE) is connected to *nodes*, which are electronic switches in a circuit-switched exchange and computers in a message-switched exchange. In the latter case, nodes are referred to as *data circuit equipment* (DCE). The circuit-switched exchange provides a system in which a calling DTE is directly connected to the called DTE. In the partially connected network on the right of Fig. 11.5, DTE A can 'talk' directly with DTE B and DTE C but can communicate with DTE D only by going via DTE B or DTE C or the node computer. DTE A must address its *message* to D and then DTE B or C or the node will store the message, inspect the *address* and forward the message to this address when a route is available. A DTE that is capable of storing and forwarding messages is called a *host*—this may be a mainframe or personal computer (PC). As well as being able to communicate with non-directly connected DTEs, a computer in a message-switched network can address several other computers and thus 'talk' to several at once—a most useful feature of message switching. A switch-connected network has real circuits: a message-switched network consists of *virtual circuits* carrying messages of the kind shown in Fig. 11.6.

At each node the destination address in the *header* is inspected. The

SOH	Address header	STX	Message text	ETX	BCC

SOH Start of header
STX Start of text
ETX End of text
BCC Block check characters

Figure 11.6. A typical message format.

node will search for a free route through the network to that destination, the message being placed in a queue in the node's buffer memory until a route is found. The messages keep circulating in the buffer and the node inspects the buffer each time the message comes round, so making maximum use of the network. A priority code in the header would allow urgent messages to jump the buffer queue. The delays involved in the nodes when messages are circulating are generally much less than the delays that occur in network switching.

Exercise 11.2
Check on page 300

Using the message format of Fig. 11.6, which is that of IBM's binary synchronous communication (BSC) protocol, extend the idea by drawing the frame format for a message whose length requires the text to be sent in two blocks, using the control codes listed in Fig. 11.6.

Suggest how the receiver might reply to the sender on:

1. Successful transmission
2. Receiving an erroneous block

11.4 Packet switching

The disadvantages of message switching appear when long messages are sent. These may exceed the memory of the node buffer or they may monopolize the network and prevent other messages getting through in reasonable time. If a network handled only small messages these disadvantages would not arise. Messages are therefore split into standard *packets* (Fig. 11.7), typically of about 1 kbyte (1024 bytes).

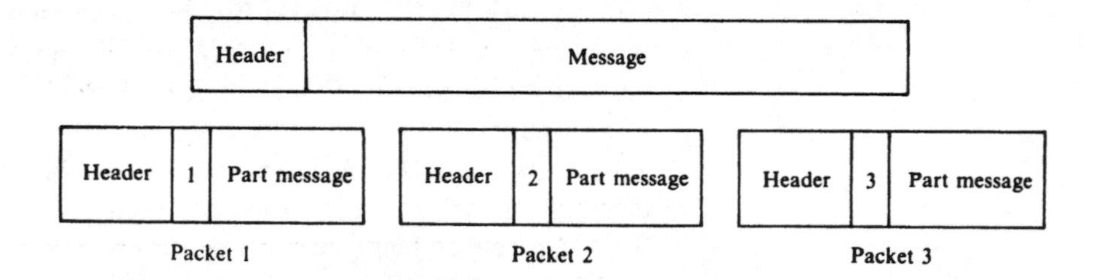

Figure 11.7. Long message split into three packets.

Packets are switched by partially connected exchanges so each must have a destination address in its header. Each packet is sent independently through the network and reassembled at the destined node to give a complete and unaltered message. This process, called *packet switching*, has become a standard way of sending data. It is quite possible for consecutive packets from one terminal to take different routes and to arrive in a different order to that in which they were sent. For this reason, each packet must contain not only a part message and address, but a number giving the order in which it was despatched. Several packets constitute a *datagram*. A typical packet would have the format shown in Fig. 11.8.

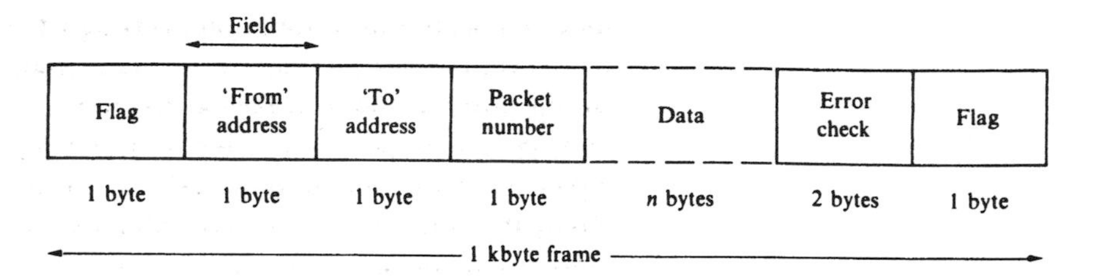

Figure 11.8. Typical packet frame.

Each *field* of a packet *frame* has a particular function. Those at either end of the frame are the *flag fields* and contain the code 01111110 to 'flag' or indicate the ends of the packet. For identification purposes, the flag fields are the only places in the packet where a sequence of six 1s is permitted. After five consecutive 1s anywhere else in a packet, the hardware inserts a 0 after them, at the packaging stage. This is known as *bit stuffing*. As the receiving hardware scans an incoming frame, a 0 following five 1s is removed. If the sequence of five 1s is followed by another 1 the flag may have been found. The seventh bit is then inspected—if it is a 0, the flag has been found and the data is left unchanged; if the seventh bit is another 1, then the frame is in error and is aborted.

Exercise 11.3
Check on page 300

The idea of a transmitted packet includes the following sequence, after bit stuffing; 1111101100100111110000001. What was the original data?

Access to a packet-switched exchange

Methods of accessing a packet-switched exchange (PSE) are shown in Fig. 11.9. A packet terminal, which might be a computer or a dedicated packet processing device, can be directly connected to a private line providing the line has been designed to handle bit rates above 9.6 kb/s. If the PSTN is

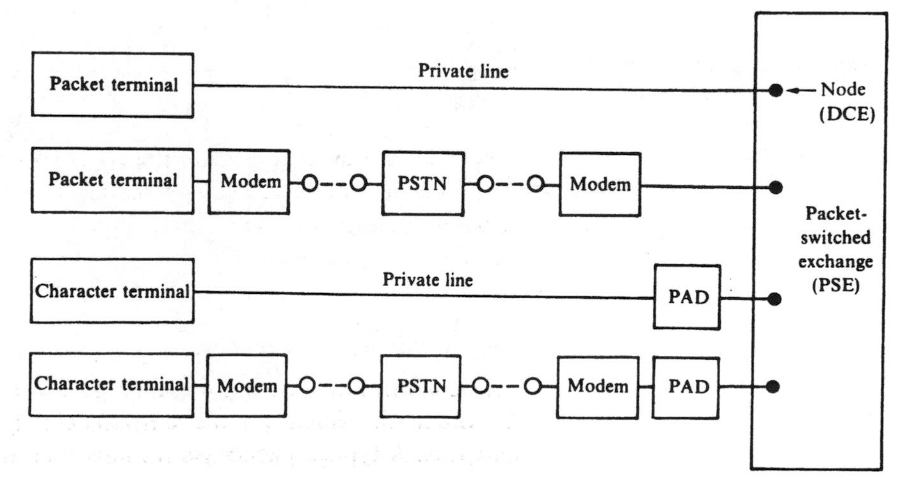

Figure 11.9. Forms of access to a PSE.

used, modems will be required and this will bring the bit rate down to one of the standard rates: 300, 600, 1200, 2400, 4800 and 9600 b/s depending on the sophistication of the modem.

A character terminal, such as a teletype or telex machine, must be connected to a node in the PSN via a *packet assembler/dissembler* (PAD). These terminals send messages one character at a time to the PAD, which converts them into packets. A packet terminal may be a mainframe or personal computer (PC), programmed to form packets, or a specially designed packet terminal. These do the work of a PAD, so they may be connected directly to the PSN via a private line or the PSTN.

Packet routing

Instead of allowing packets to take different routes as they become available, an alternative procedure is to set up a *virtual circuit* in which a connection is maintained during an exchange of all the packets making up a message, with the result that all packets are delivered in the order sent. This alternative would have to be used if voice packets rather than data packets were involved because some delays, inherent in datagrams, might be unacceptable in human conversation.

In the examples shown in Fig. 11.10, packets 1, 2 and 3 are to be sent from terminal T1 to terminal T4. In the left-hand network a virtual circuit has been established via T5, so a correctly ordered message arrives at T4. In the right-hand network the packets have taken different routes that happen to have been available when they were transmitted by T1. They arrive at T4 to give a datagram made up of packets in the sending order 132.

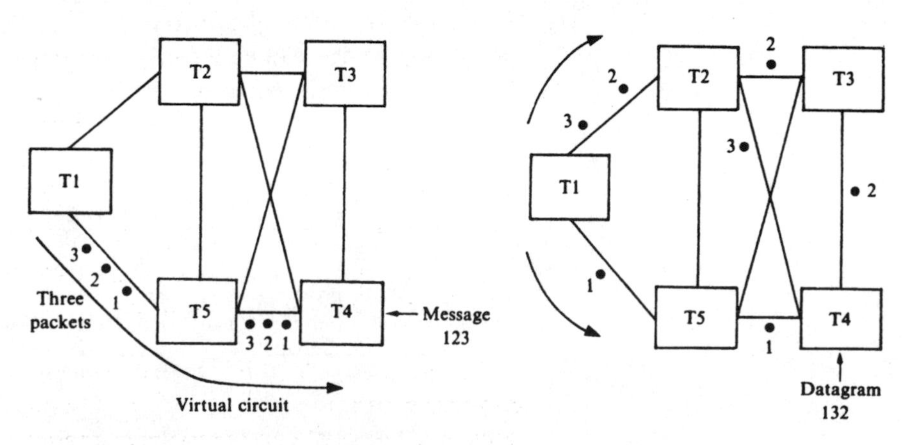

Figure 11.10. Packet routing in a PSN.

Packet-switched networks

Any organization may have its own *packet-switched network* (PSN), just as it may have a private branch network (PBN), or it can use the public *packet-switching service* (PSS) provided by BT. This public PSN consists of several partially connected *packet-switching exchanges* (PSEs) with their nodes linked to host computers or terminals in their area. The node computers give the local exchanges the address of the calling host or terminal and store and forward packets to their destination. The procedure is similar to that used in circuit-switched centres, and the network layout is typically that shown in Fig. 11.11.

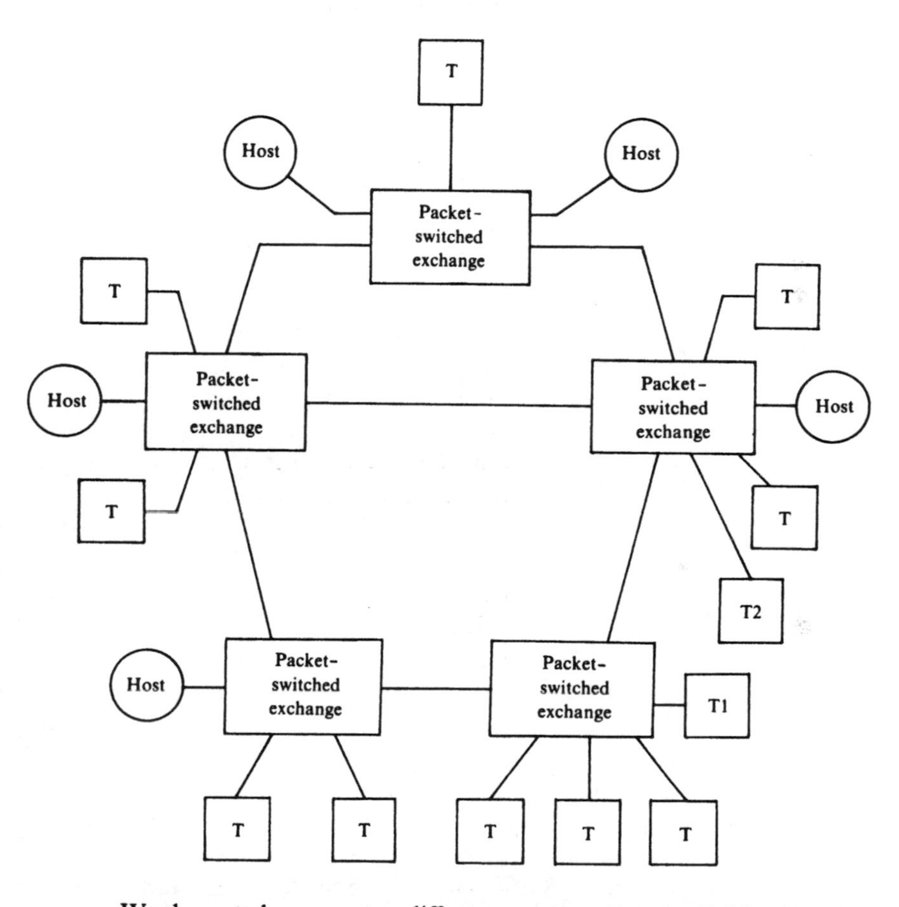

Figure 11.11. A packet-switched network.

Exercise 11.4
Check on page 300

Work out how many different routes are available between terminals T1 and T2 in Fig. 11.11 and compare this with number available in a fully connected PSN.

The terminals (T) of Fig. 11.11 may be either the packet or character terminals shown in Fig. 11.9—these rely on the node computers to do the switching. The host computers have enough processing power to do their own switching and are not reliant on the exchange nodes. This independence makes for faster data exchange especially when the network is congested. PCs started out as little more than terminals but are becoming ever more powerful hosts, using network routing algorithms of the kind exemplified in Fig. 11.12.

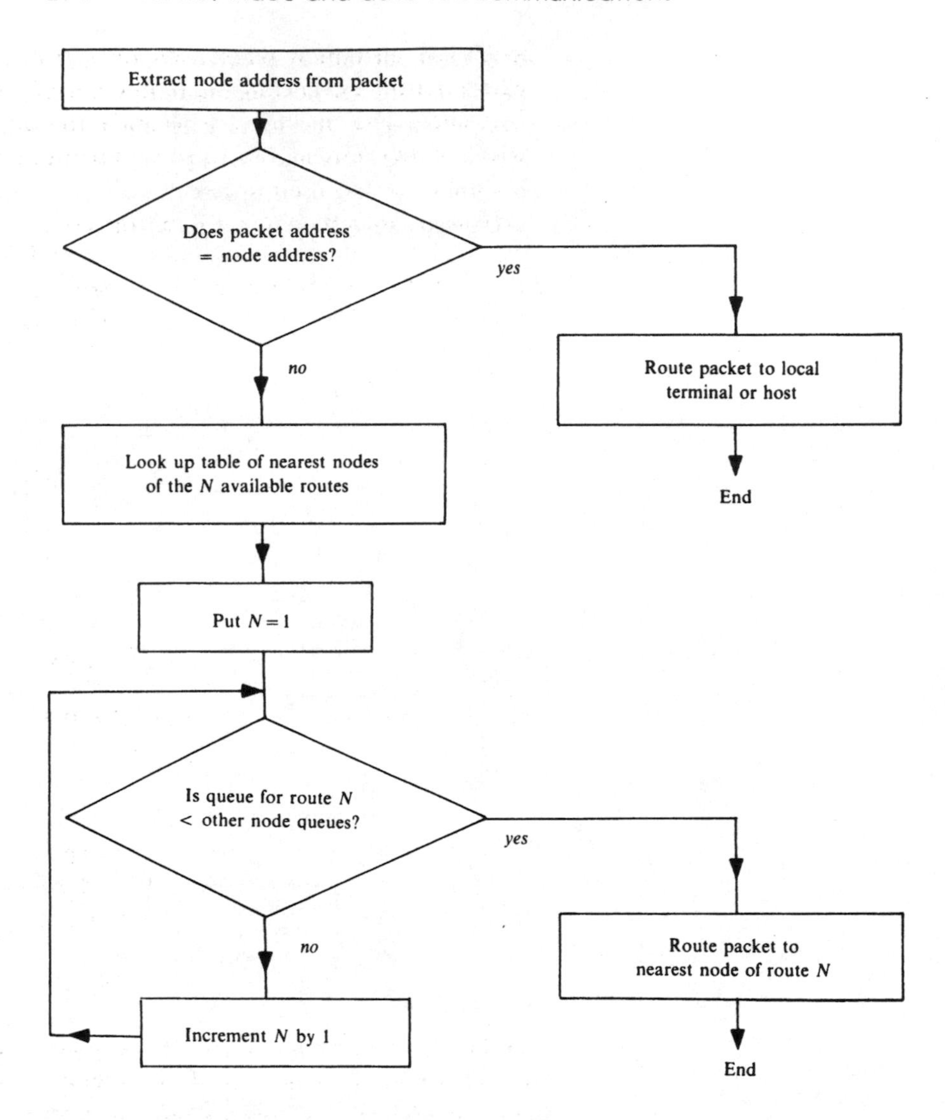

Figure 11.12. A network routing algorithm.

The links from terminals and hosts to the exchange are short and the traffic is such that they need not transmit data at a high rate. By contrast, the trunk links between exchanges must be 'high-speed'. These are usually leased from BT and are referred to as *value-added networks* (VANs) since the packet service adds value to the links, which are already providing other telecommunication services. Such data networks are also known generally as *wide area networks* (WANs). BT's PSS is a WAN, as is TYMNET, which is operated by a US company of that name. The first international WAN to be operated was the US Defense Department's Advanced Research Projects Agency network ARPANET. This network carries voice and data packets. Unlike TYMNET, it uses only datagrams with adaptive routing worked out by routing algorithms in the hosts. The algorithms incorporate estimates of the shortest time to the destination node. The maximum packet length is restricted to 2 bytes in order to reduce the time delays in the reordering of datagram packets. Batches of eight successive packets are allowed to take the same route, giving a pseudo-virtual circuit. These speed-up procedures make the system suitable for voice communications.

Packet switching, like all other aspects of telecommunications, requires coordination between the users at either end. There must be agreement on the packet format and the terminal equipment so that the user can design, or buy, a terminal that can handle the packets. This brings us to the topic of protocols.

11.5 Protocols

In a data communications system, such things as the number of bits in each character, the bit rate and the voltages used are matters of standardization. Such features are specified by *protocols*, which are rules or recommendations governing the way links are established and data is transferred. A protocol for setting up a voice telephone link is illustrated in Fig. 10.6. If two directly connected pieces of equipment have to work together, they must be designed to an agreed specification, that is, they must use the same protocols, otherwise they have to be interfaced with a *protocol converter*. A simple example of such a converter would be an adaptor that interfaces a round-pin plug and a square-pin socket. If the two pieces of equipment followed the same protocol one would have matching plugs and sockets and so not need a converter. An example of a non-physical protocol would be the data frame specifications for the USRT shown in Fig. 11.4 or the message format of Fig. 11.6.

Layered protocols

The simplified data communications system of Fig. 11.1 shows two directly connected data terminal equipments (DTEs). To allow a number of DTEs to exchange data, this arrangement needs to be extended to that shown in Fig. 11.9. Direct and packet-switched links are contrasted in Fig. 11.13.

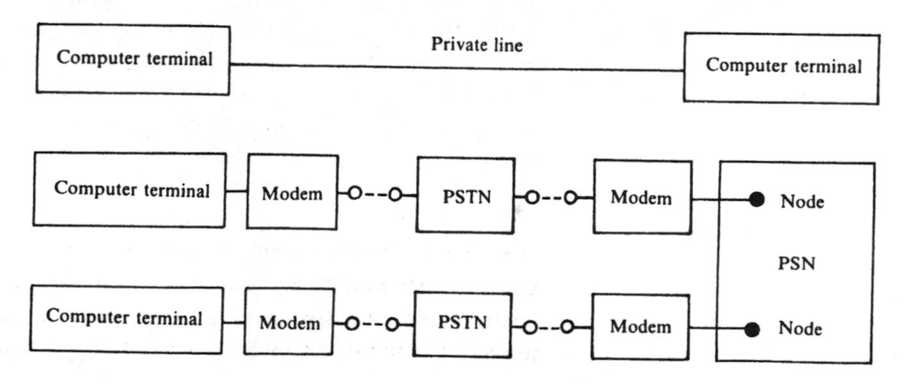

Figure 11.13. Direct and packet-switched links.

In the upper part of Fig. 11.13 we have two computers linked directly via a two-wire line. When the data is entered and processed by one computer, all that needs to be done to transfer this data to the other terminal is to enter a transmit instruction. When two computers are connected via public networks, as shown in the lower part of Fig. 11.13, there are several layers of control needed to effect data transfer. These are:

1. The *physical links*, consisting of the cable and communications hardware
2. The *data format and link controls* used to ensure error-free flow of data through these links and the system
3. The overall or end-to-end control

Each of these three layers is specified in a protocol. The basic layer is the physical link specification. Riding on this feature we have data in a specified format being moved around the PSN in a specific, coordinated way. This view has led to protocols being structured in *layers* or *levels.*

11.6 Packet-switching protocols

Data networking protocols have historically been dictated by one or two major American computer manufacturers whose equipment initially dominated the market. Other manufacturers either followed suit or developed their own protocols. As a result there were various protocols in use, lacking compatibility and thus hindering technological advance. The solution to this problem lay in the standardization of network features: a difficult task, requiring an impartial body, consulting with the companies and user organizations who had eventually to agree to use the resulting standards. Harmonization involves many stages of trial and improvement before a standard is finally published. Two bodies that have taken on this task were already active in other fields. These bodies are: the *Comité Consultatif International de Téléphonie et de Télégraphie* (CCITT) and the *International Standards Organization* (ISO).

The CCITT standardizes public telephone networks world-wide. One of their most important recommendations for users of data communications is the X25 packet switching protocol, which covers the interface between the user's equipment and the network. The ISO have adopted a more general approach to standardization in their seven-layer protocol, which makes provision for all possible levels of present and future standardization.

The CCITT recommendation X25

Recommendation X25 is a data network access protocol that specifies how a virtual circuit is set up between a network user's DTE and the network provider's DCE. It defines the hardware connections and software procedures for setting up calls, controlling packet flow and clearing calls. Host computers implement this protocol. Terminals, which are not intelligent enough to do this, use PADs, which interface with a DTE according to CCITT recommendation X29 (another protocol). X25 consists of three levels as follow:

- *Level 1* is the hardware specification linking the terminal to the PSN. These specifications are contained in recommendations X21 and X21 bis ('bis' means 'second part').
- *Level 2* is the software specification linking the terminal to the PSN. This is a version of an existing link access protocol (LAP) and it provides for reliable data transmission between adjacent nodes.
- *Level 3* specifies the packet formats and procedures for packet flow between nodes in the PSN.

The procedures of level 3 are interesting because they show how computer telecommunication compares with human telephone conversation. A virtual circuit may be established by the PSN, just as a real circuit is set up in a telephone exchange, by calling (dialling) and waiting for the called party to answer. Information is then exchanged (talking) and the call finally cleared (handset replaced).

In the case of computer 'conversation', the above procedure must be broken down further, with each step precisely defined as follows. The calling terminal sends a *call-request packet* (CRP) containing the called address and channel number, which identifies the virtual circuit to be used. The PSN or host computer converts this packet to an *incoming-call packet* (ICP), which is sent to the called terminal. This packet contains the calling address and channel number. The called terminal can accept or refuse a call. Refusal involves sending a *clear-request packet* (CQP). Acceptance is indicated by returning a *call-accepted packet* (CAP), which contains a call-accepted code and the channel number. The CAP appears at the calling terminal as a *call-connected packet* (CCP). The virtual circuit is now set up, so data exchange can take place by exchanging data packets (DPs) of the kind typified in Fig. 11.8. The call can be cleared at any time by either terminal sending to the other a CQP containing a call-clear code and the channel number. Each CQP is immediately acknowledged by a *clear-confirmation packet* (CFP). This clear-request procedure is repeated as the packet goes from one exchange to the next until the CQP reaches the called terminal, where it becomes a *clear-indication packet* (CIP). All these steps are summarized in Fig. 11.14.

Exercise 11.5
Check on page 300

Draw a diagram, similar to that of Fig. 11.14, to show what happens when a called terminal refuses a call.

Recommendation X25 provides a set of standards for three layers of interaction between a host computer and its local switching network node.

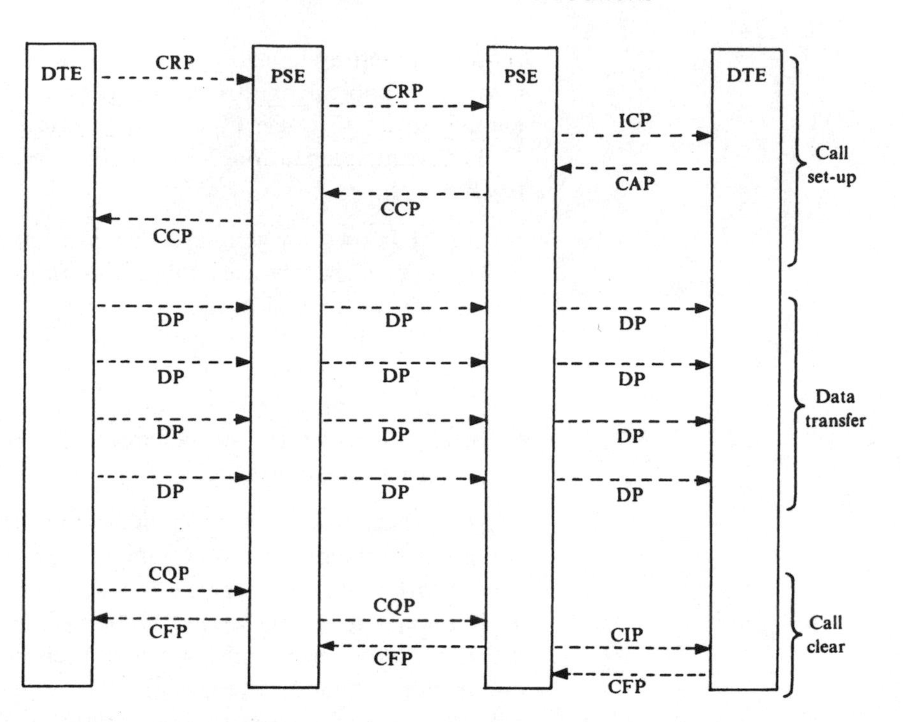

Figure 11.14. Virtual circuit switching steps.

X25 is a low-level access protocol. In order to standardize the higher levels of data communication systems, an extended protocol is needed. A model for this, on which actual protocols can be based, has been provided by the ISO.

Open Systems Interconnection

Open Systems Interconnection (OSI) is a set of seven protocol layers, analogous to that of Fig. 10.6, which has been recommended by the ISO to coordinate the design of future protocols so that systems based on these protocols will be compatible. Open systems are 'open to all' in the sense that they can exchange data with systems produced by other manufacturers. Figure 11.15 shows the order, names and functions of the layers of the OSI model.

The OSI model follows the same principle as used in the telephone link seven-layer protocol model of Fig. 10.6. It incorporates X25 as its lowest three layers. To explain the function of the seven layers, we take a typical example of data transfer in which two bankers are exchanging data, and show what happens to data as it is controlled in the various layers.

Layer 7 ensures that all the bankers will see at their terminals are the prompts and enquiries of this layer and the messages received from the other end. The tasks performed by this layer are *applications*. Both terminals must be identified to set up the connection, and the form of dialogue agreed. For the bankers, this involves keying-in their user numbers and giving details of the type of transaction to be made. These

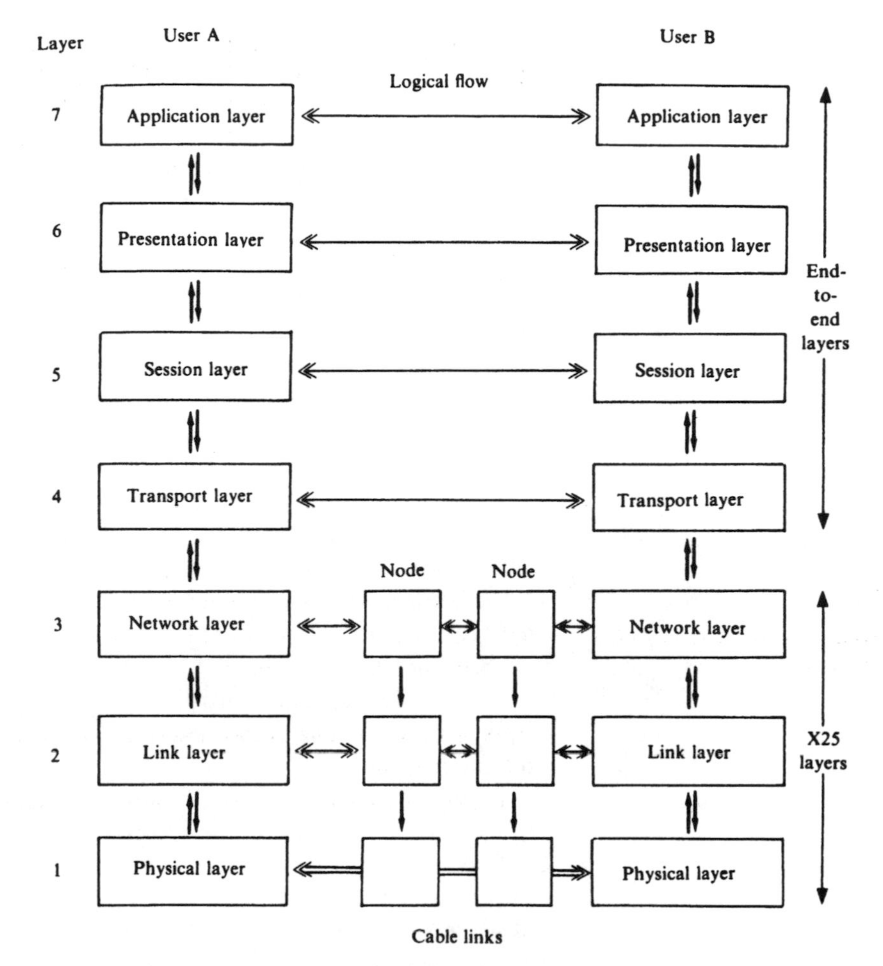

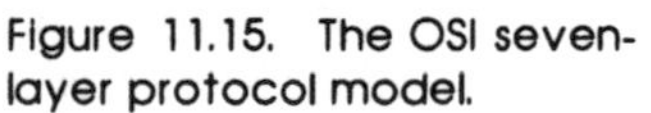
Figure 11.15. The OSI seven-layer protocol model.

are *common application-service elements* (CASEs). In addition to these, the layer takes care of *specific application-service elements* (SASEs) such as the transfer of files and database access. In general, SASEs ensure the use of correct procedures for particular applications (which might be non-banking) such as purchasing or stock control. Once both bankers have identified themselves, the next step is the presentation of the data in a form suitable for its exchange—this is the task of the next layer.

Layer 6 is concerned with *presentation*. The two banking terminals of our example may not use the same data format, so this presentation layer decides on a form that preserves meaning while resolving syntax or language differences. It then performs the appropriate transformation on the data so that a 'common language' is found for 'dialogue' between the terminals. In practice, this service will often be provided as a menu containing a selection to suit the application. When one banker keys-in the type of transaction required, the appropriate format would be selected. The data transformation chosen at this level could also provide special facilities like data compression for transmission efficiency and data encryption for security.

Layer 5 covers various tasks that are best performed on virtual circuits and can be set up when a *session* is requested. The bankers would use this facility for interactive communications. Once a session has been requested, the session layer establishes connections and terminations, synchronizes the dialogue, controls the data direction and closes down the transaction.

Layer 4 allows the network nodes to *transport* data reliably, sequentially and independently of any intervening nodes. It is the layer that is involved in the packeting of messages and their reassembly, and the setting up of datagrams and virtual circuits. It effectively hides the network layers below layer 4 from the layers above and provides what appears to be a direct end-to-end connection for the data passed down from the presentation layer.

Layer 3, as we have seen, divides the data into packets for transmission on the *network*. The transaction data is split into packets by this layer, which then decides the route and sets up a connection between end-points. If the bankers were in different countries, for example, the packets woul be delivered by the PSS to an international *gateway* connecting the national PSN of the other banker.

Layer 2 sets up and controls the *links* between adjacent nodes in the PSN. The inevitable transmission errors are corrected by detection and retransmission.

Layer 1 specifies the network hardware features such as voltages, pulse rates and formats, cables, connectors and interfaces.

Thus the VDU data, which a banker wishes to telecommunicate, goes from top to bottom through all these layers until it is finally in the form of a digital signal, which is then conveyed over the physical link to the open system at the other end. This electrical data then climbs up through the layers of the destination system so that the original VDU data appears, in the correct form, at the receiving banker's terminal. What takes place between layers is called a *service*. A service, in this context, is a statement of the tasks that can be performed by a layer N for the various parts of layer $N \pm 1$.

Exercise 11.6
Check on page 300

State which layer of the OSI protocol model would be associated with:

1. The choice of ASCII code
2. The management of an airline reservation system
3. Deciding which user is to talk next
4. The use of datagrams or virtual circuits between nodes
5. The routing of packets between networks

Protocols may be regarded as analogous to a common language used at the interface of different subsystems. It is rather like German and Italian speakers using English for communication. If neither of these speakers shares a common language, however, they will need a 'presentation layer', which might take the form of two interpreters one of whom understands German and English and another proficient in Italian and English. These two interpreters form an English language 'presentation layer' supporting the German and Italian speakers. These speakers form the 'applications layer', which is a 'virtual' communications layer in that the German and Italian speakers are speaking directly to one another but actually using the interpreters as the 'real' link layer. This link would need other supporting layers if the interpreters conversed via microphones and earphones—this would be the 'physical' link layer.

11.7 Wide area network services

Having looked at the principles of WANs, based on either the PSTN or specially designed data links, we now consider the various data- and message-carrying services that these networks make available.

Datel is the modem rental service that BT provides for data transfer on the PSTN. Until the advent of packet-switching networks, this was the only public network available for transferring computer data. Dial-up connections are charged at normal call rates and are still probably the best option for occasional data transfer. *Leased lines* are also available, giving a permanently connected circuit on the PSTN. Grades of leased lines, offering various data rates, with either half- or full-duplex communication, are offered at rentals based on their quality. Leased lines only become economical with high-volume use. The main categories of leased-line digital services, known collectively as *X-Stream* (based on System X), are as follows.

1. *KiloStream*—a single-channel service with data rates up to 64 kb/s offering such services as links between VDUs and teleprinters at 2.4 kb/s, credit verifications and slow-scan TV at 11.8 kb/s, high-speed facsimile (FAX) at 9.6 kb/s and computer/speech data transmission at 64 kb/s.
2. *MegaStream*—a 32-channel service, with data rates at multiples of 2.048 Mb/s (nominally 2 Mb/s) and offering direct connections between digital PABXs. *SatStream* is a global version of this service using satellite links.

In addition to these services offered by BT, Mercury Communications Ltd (MCL) offer data services, primarily for the business user, which compete with most of BT's X-Stream services. Its trunk network is based on two optical fibre loops operating at 140 Mb/s. The packet-switched MCL 5000 service was recently introduced to rival BT's PSS, and MCL provide a satellite data service and a *metropolitan area network* (MAN) to service the London financial institutions.

The *Packet-Switched Service* (PSS) was introduced by BT several years ago, initially for international data communications. It has since been developed as a national network as well. All PSS connections are full-duplex and can handle data at speeds of up to 48 kb/s compared with a voice-grade telephone line which only manages 2400 b/s at best. A packet terminal can be linked directly to the PSS, on what BT call *Datalines*. Most manufacturers offer software that converts their microcomputers into packet terminals.

Apart from peak periods, the telephone system is seldom fully loaded, so other services can be provided with the idle capacity. The PSTN carries a range of non-voice traffic. Most of this is data. The non-voice services provided by the telephone networks have proliferated to an extent that they have earned themselves a special collective title—*value-added network services* (VANS). You may come across these extra services referred to by the more restrictive name of *VADS* (the D, of course, is data). Here are a few of the VANS that are available.

Videotex

Videotex covers *viewdata* and *teletext* services. Viewdata is the generic name given to a service in which information may be called up via the PSTN and displayed on a monitor or TV screen. BT's version is called *Prestel* and was among the earliest of the VANS to be implemented in the UK. Other operators have since followed with alternative private services. In its simplest form, viewdata provides access to many thousands of pages of information, a page usually being one or more VDU screen-loads, from which a subscriber may select the material required for display. The vast volumes of data to be stored and handled at the provider's end requires the use of a computer system that maintains the data in one or more databases, depending on the number and variety of topics treated.

Teletext, like viewdata, also provides pages of information from a database, but it is broadcast over the air for reception on a domestic TV set. Unlike viewdata, however, information can only be received—there is no return link to the database. The database is quite small (some hundreds of pages) and interaction with the user is limited to keying a page number, selected from a menu. Because teletext data is sent in between the data for each separate TV field, there can be a frustrating wait for the demanded page to appear. Viewdata's response times are rather quicker.

The teletext specifications are common to the BBC and ITV. Information is transmitted and displayed on the TV screen in 'pages', each page being comprised of: a header, giving the name of the service—Oracle or Ceefax—along with the page number, date and time and 23 lines of information, each line being up to 40 characters in length. The data is carried in transmission by scan lines 17, 18, 330 and 331 (see Sec. 6.3). The teletext data is always present in the video signal. However, only a TV receiver that has been modified by the addition of a decoding unit can display them (Fig. 11.16). Moreover, since the information required has to be selected, a selector unit of some kind is needed. The latter is usually a

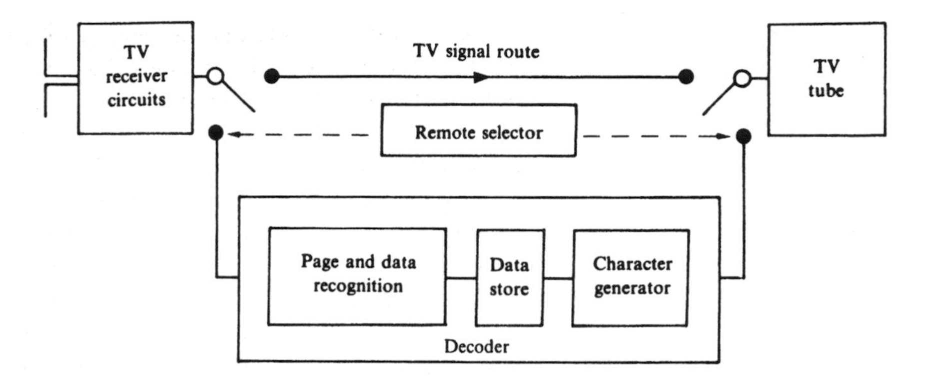

Figure 11.16. Simplified teletext decoder.

more elaborate version of the TV remote control unit. It has extra keys enabling the user to switch between TV programmes and teletext and provides other teletext selection functions.

Exercise 11.7
Check on page 301

How many pages of teletext are transmitted per second?

The videotex family tree is shown in Fig. 11.17.

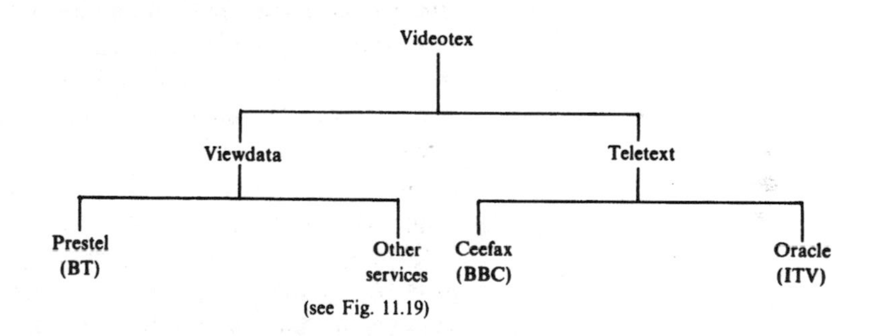

Figure 11.17. Some members of the videotex family.

Teletext is a one-way information service. Viewdata is also an information service, but it is fully interactive, that is, it has two-way communication. Further, it can provide much more data than teletext, having a databank of several hundred thousand pages. Data and requests are transmitted along the telephone lines—not over the airways. Like teletext, however, the data may be displayed on a TV receiver screen.

Since Prestel was introduced, other operators have followed with similar private services. Unlike teletext, Prestel's services are charged to the user; they are not included in the TV licence fee nor the phone line rental cost. A relatively modest subscription charge provides access to Prestel's many thousands of pages. Figure 11.18 shows a block diagram of the Prestel system.

Note that subscribers are connected to a local computer serving their district. The local computer does not hold all the Prestel data that might be required by a subscriber on any given enquiry and may have to retrieve

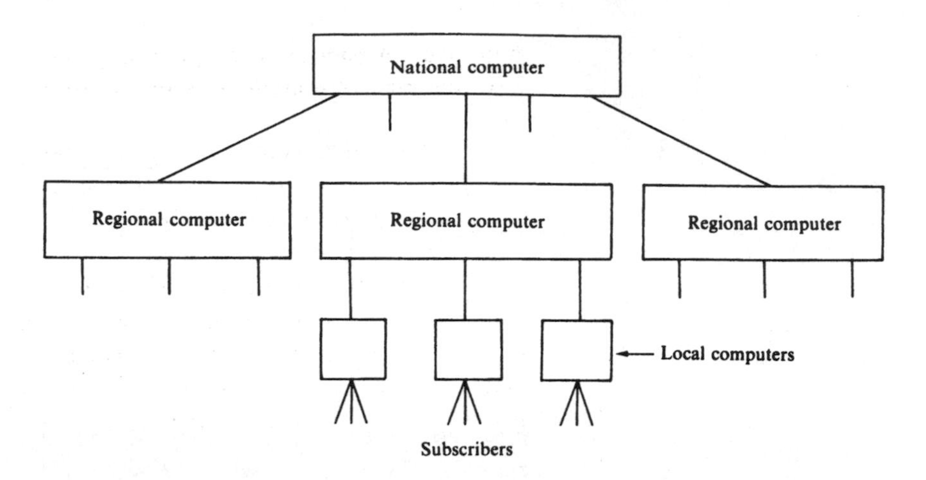

Figure 11.18. The Prestel system.

items from a regional or even the national computer. Although the data may have had to come from the other end of the country, the subscriber pays at the local telephone charge rate.

Receiving Prestel data

The simplest Prestel terminal consists of a domestic TV receiver plugged into a small keypad unit that is connected to the telephone line via a modem. To gain the full benefit of Prestel's features, however, a PC is needed as a terminal. The Prestel service is activated by dialling the telephone number of the local service, which then responds by asking for the caller's identity. The caller's number is keyed in and the system checks that it is valid (fully paid-up).

Once the subscriber is connected to the service, information is selected, as in the case of teletext, by choosing from a menu. Because there are vast volumes of data on Prestel, menus have several levels—selecting an item may bring up a sub-menu and choosing from that may present a sub-sub-menu and so on. Users who know the page number of the wanted data may go straight to it by specifying that number. Since only the requested data is transmitted, information appears on the screen faster than is typically the case with teletext. Information in Prestel is provided by various organizations, usually on a commercial basis. Therefore, some pages carry a viewing charge, which is indicated in the top right-hand corner of the page. The variety of topics presented on Prestel is very much greater than can be squeezed into the teletext service. Since the user–system communication is two-way, it is possible for users, for example, to place orders for goods and make banking transactions at home.

Both teletext and viewdata can transmit colour graphics as well as the more usual textual information. The teletext picture displays tend to be rather primitive, being constructed of rectangular mosaics. Much better pictures can be obtained from Prestel, but the line transmission costs are high. An A4 page-sized graphic may need one million bits to describe it. For this kind of graphic display, X-stream services are required.

Exercise 11.8
Check on page 301

To appreciate what high-resolution graphics via Prestel implies, work out the time to transmit a 1 Mbit picture using the 1200 b/s transmission rate.

The viewdata subscriber who does not want to do anything but look at standard information such as train times, theatre programmes, stock exchange prices and the like, can make do with a simple numeric keypad and a domestic television receiver rather than a PC. However, a computer is required if you want to realize the full potential value of viewdata. Any size of computer will do, from the large mainframe down to the smallest PC.

Other Viewdata services

Viewdata services span a wide range of interests—commercial, professional and leisure—and may be publicly available (like Prestel) or privately available. The private systems include such items as supplier–distributor networks, stock listing and ordering, and sales promotion information. Access to those systems operated commercially may be confined to restricted groups of users called *closed user groups* (CUGs). These are effectively in-house services, which may, however, be offered to other interested parties on a fee-paying basis.

The other two categories also have CUGs and cater largely for *special interest groups* (SIGs); the professional systems tending to be particularly orientated to computing and allied matters such as programming and communications. Training and education material is also provided. Leisure systems cover a multitude of hobbies, pastimes and interests from *multi-user (computer) games* (MUGs), through theatre information, to wine and food SIGs (Fig. 11.19).

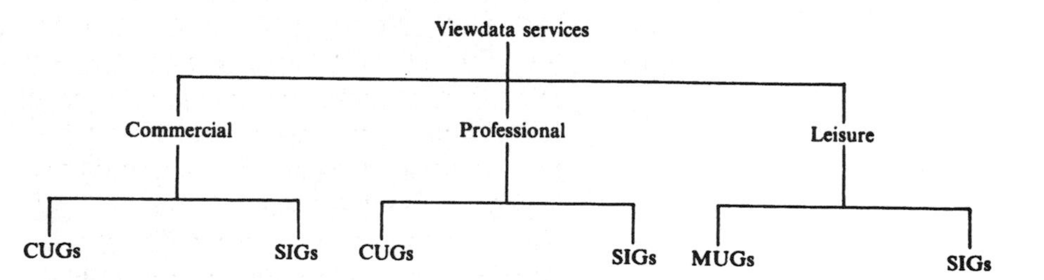

Figure 11.19. Viewdata groups.

Getting access to group systems involves dialling a specified phone number, then, if appropriate, selecting the page that marks the entry point for the required group and, finally, logging-in with an assigned password or personal identification number (PIN). Access fees are payable.

Some very powerful computer processing lies behind the operation of a large viewdata system and it calls for mainframe capability at source. The system has two major activities to carry out: communication with thousands of users while concurrently storing and accessing data that may be subject to frequent updating. Some several hundred *information providers* (IPs) supply the pages on Prestel and are charged for the space they occupy. The IPs use the opportunities offered by the system in ways suited to their business activities. They may offer public access to their pages for a price (the page viewing fee noted above), or restrict access to a 'club' of subscribers (CUGs), who are given the code to reach the relevant pages. They can use the pages to sell their goods and services. The interaction between system and subscriber, which has just been discussed, could be used to exchange information between viewers—such a viewdata electronic mail (VEM) service exists.

Electronic mail

An electronic mail (EM) service is a communication system that will accept messages from one user and pass them on to another specified user, usually by storing them in the recipient's 'mailbox'. Like viewdata, EM systems are operated by a computer, or even a network of computers, and users 'log-in' by dialling a particular telephone number. However, EM does not need the vast databases called for by viewdata—it is not an information service but rather a kind of local postman moving at a rather faster pace!

Unlike viewdata, which delivers words and pictures, EM deals in words only and could be regarded as a word-processor communications system. Many large organizations have widely scattered locations, which are connected over leased lines to a central mainframe computer. The connection of their word-processing equipment to this network provides a private electronic mail service within the company, each word processor having its own identity code and mailbox.

The sender of a document via *email* (an alternative contraction to EM) addresses it with the recipient's mailbox code. If the addressee happens to be connected to the EM system when the document arrives, a message is displayed at the bottom of the adressee's screen to indicate its advent. Otherwise it will be stored either in a local or system memory, depending on how the mailbox has been set up. To preserve confidentiality, especially in cases where there may be more than one mailbox address on a terminal, access to the box may be granted only on the entry of the correct password.

Some of the many other features offered by EM include: recording the times of despatch and delivery of messages, automatic redirection of messages, multiple addressing, the filing and retrieval of messages, connection to radiopaging services and access to the Telex system. This last facility is important since Telex is still the most used world-wide data network service and reaches into the under-developed countries where EM is not likely to appear for some years.

Some email systems additionally offer non-electronic services such as posting, via normal mail, letters to addressees not subscribing to the email service or even delivering by courier if a document is very urgent. Translation services for many languages are also available.

The best known in the UK of the public 'scrolling systems', as EM is sometimes called, is *Telecom Gold*, operated by BT. Gold services, however, cover rather more than just plain text delivery. They are designed to provide an electronic office for the user, including word-processing, diary, phone message reminders, travel and hotel information and even a database management system.

Exercise 11.9
Check on page 301

Viewdata systems use the 1200/75 b/s standard. Assuming that an A4 page has about 3250 eight-bit characters, explain whether or not you regard this standard as suitable for electronic mail functions.

Bulletin boards

Bulletin boards (BBs) also exchange messages between users via mailboxes. The name derives from their primary purpose, which is to act as the equivalent of office notice board for users of the service. Hence the messages, even if directed to an individual, are open for all to read. BBs are many and varied, often run by amateurs for fun, but they are also operated commercially. Telecom Gold, for example, has a BB, and many BBs serve CUGs and SIGs, so, when an interesting topic arises on the board, a version of teleconferencing can develop. Anyone can join in and their contribution will be added to the queue of messages awaiting display. Many BBs also furnish what might be described as a 'chat line'. This is about the nearest thing to a conversation on the phone that a data system can manage—it is interactive messaging between two users, with a keyboard and display replacing speech and hearing respectively.

Teletex

The name causes much confusion—the *Teletex* system has nothing whatever to do with the television services' *teletext*. It is a text transmission system, like Telex, but a lot faster and cheaper, with a far wider range of characters. It operates over PSTN but can be connected by gateways into the Telex and packet networks. As well as providing higher transmission rates, Teletex furnishes a much greater flexibility in text construction through its eight-bit character coding.

Exercise 11.10
Check on page 301

Telex uses a five-bit character code and transmits at 50 b/s. Teletex employs an eight-bit code and a 2400 b/s transmission rate. What is the ratio of character transmission speeds in the two text transmission systems?

For data transmission systems to provide an efficient service, they must incorporate error-checking processes and Teletex does this. No procedure or equipment that humans can devise will guarantee the delivery of completely error-free documents, but Teletex should avoid some of the worst afflictions of Telex, where messages can turn up 'YOMZTHIMG RIKE OHPS'!

It is worth noting that where Teletex connection goes directly to the PSS, transmission rates can rise to 48 kb/s, providing the receiving terminal can cope with this speed. As with Telex, there are dedicated equipments for connecting into the Teletex service, which also provide the word-processing facilities mandatory for cost-effective fast transmission of data. Again, as for Telex, a PC can do the job equally well if it is fitted with one of the various (BT-approved) adaptors that are on the market. Teletex is an established service in a number of countries; standards have been agreed through the CCITT and an international Teletex service is operational. The extent to which Teletex will expand is still an open question. The modernized Telex system is likely to be a major competitor while the hi-tech email facilities can offer things that Teletex cannot match.

Facsimile

Facsimile (FAX) is a transmission system for documents that are already printed or in manuscript. The principle of the FAX is described in Chapter 1. FAX transceivers are grouped according to CCITT specifications as follows. The slowest, group 1 machines, use FM and take about 6 min to send an A4 page. Compatible with these machines, and twice as fast, are the group 2 models, which use AM. Both types are still in service but are obsolete. Group 3 is the current generation, producing digital signals, which are sent over the PSTN via modems to produce good-quality A4 copy in about 1 min. The most modern types fall into group 4 and use a completely digital mode of transmission over 64 kb/s leased lines and can produce an A4 page in about 3 s.

Voice messaging

This is the concluding item of the value-added services unit. It is sometimes called *voice mail* or *voice store and forward*. It is system very much like EM in that a user dials into the service to record a message to be placed in a specified mailbox or to retrieve anything recorded in the user's own box. A sophisticated version of such a system is *Voicebank* offered by BT. A small keypad is used to select facilities.

11.8 Local area networks

A local area network (LAN) is a short-distance, high-speed, packet communications network designed to link computers and their peripherals within the same building or site. A LAN is to data communications what a PBN is to voice communications. It consists basically of a wiring system, which interconnects data handling equipment like VDUs, keyboards, PCs, printers, plotters and disk or tape storage facilities via suitable interface equipment. A LAN not only allows many users to share expensive or intermittently used terminal equipment, it also allows the sharing of software and data banks and facilitates the free exchange of data between terminals in the network. The rate of data transfer is about 100 kb/s to 100 Mb/s. The size of a LAN falls between that of a computer bus and a WAN (Fig. 11.20).

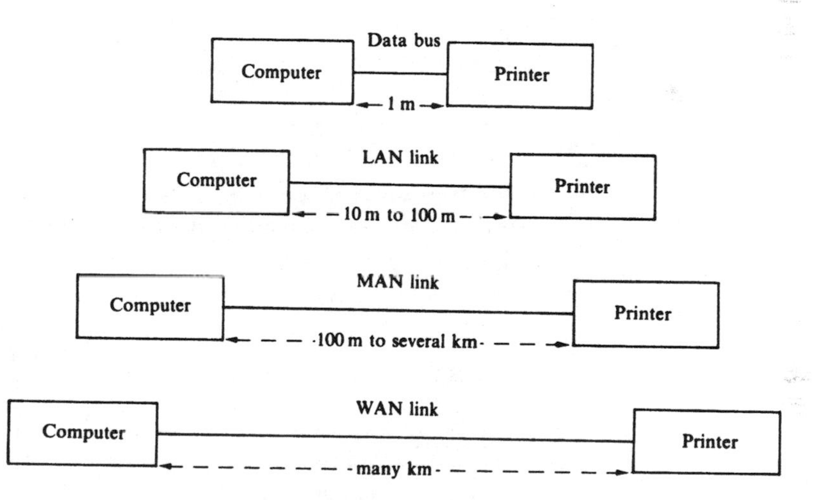

Figure 11.20. Comparison of LANs with other data networks.

The main three LAN configurations, the star, the ring and the bus, are shown in Fig. 11.21. In the star network, packets flow via a central control unit or host computer. The ring network moves data, always in the same direction, through each terminal in turn. Packets travel in both directions on star and bus networks. Various access protocols are used to ensure that each terminal gets a fair share of the network. Each cable used in a LAN provides a single communications channel working at baseband. We now consider the action of these networks individually.

Star LANs

In this configuration a central control computer (host) contains storage, processing and switching equipment to enable a number of terminals to intercommunicate. Access to the terminals is by the host asking each terminal in turn to se if it wants to 'talk' or 'listen' (these are jargon terms

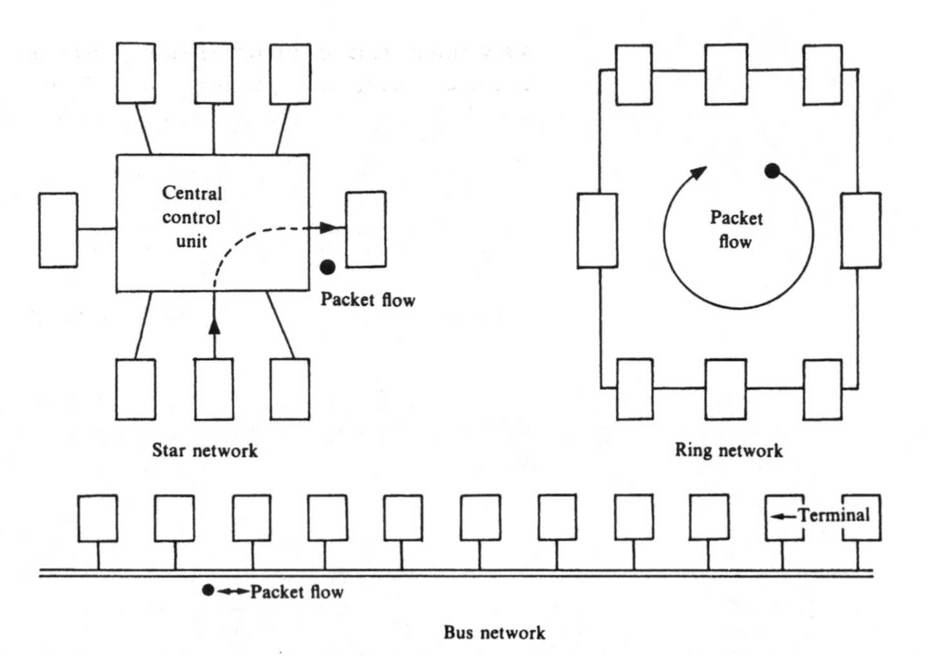

Figure 11.21. LAN configurations.

for transmit and receive). This 'asking' arrangement for time-sharing the LAN is known as *polling*. It can have terminal priority features built into it so that certain specified terminals get polled more often or in preference to others. If one terminal wants to 'talk' to another it must do so via the host, which acts as a switching node, routing the packets between the appropriate terminals. The network requires a large amount of cabling and tends to be slow.

The packet format for a star LAN can be fairly simple, consisting of a destination address, a sending address (so that the host can acknowledge receipt of the packet), data and check bits (Fig. 11.22).

Destination address	Source address	Data	Check digits

Figure 11.22. A simple packet format.

In addition to packet routing, a host, after polling a particular terminal, deals with requests from that terminal and will handle all the data processing and files. In particular, it will ensure that each user is allocated areas of disk storage with varying levels of protection from other users. In fact, much of a host's time is spent in disk management. This system leads to inherently variable response times, implying that the star LAN is not suitable for linking together word processors for example. The central role of the host in this topology suggests this type of LAN for connecting together a number of keyboard/VDU combinations to a central processor so that cheap terminals, which do not want to interact very often, can share processing power, software, data storage and the relatively expensive plotter/printer facilities of the system. It would be ideal therefore for connecting up a computer-aided design (CAD) system.

A digital PBN has a star configuration with a PABX at its centre. A PBN may function as a LAN, using data rates of 64 kb/s—the same as that used for PCM voice transmission. Or it may operate at a higher rate but use the same wiring as the PBN. StarLAN, developed by the American Telegraph & Telephone Company (AT&T), operates at 1 Mb/s and uses the same wire-pair cabling as that of the installed PBN.

Bus-type LANs

In this configuration continuous cables, or buses, are joined via repeaters and link equipment via transmitter/receiver circuits as shown in Fig. 11.23.

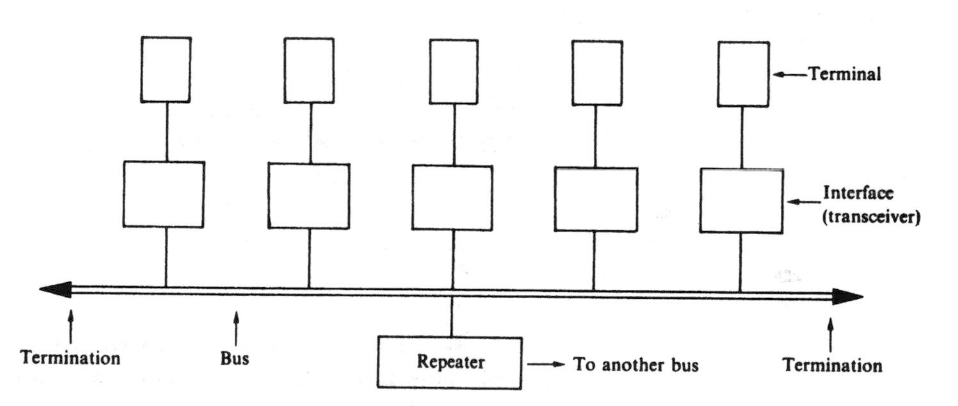

Figure 11.23. LAN bus configuration.

Ethernet is the most common example of a bus-type LAN. It was developed in the 1970s by Xerox and later taken up by the Digital Equipment Company (DEC) and the microprocessor manufacturer Intel in an attempt to establish a standard method of controlling the flow of data between terminals. It can use twisted-pair wiring but normally uses coaxial cable, terminated at each end to avoid reflections. It can handle high data rates (10 Mb/s maximum) over distances up to about 12 km in 500 m segments, each supporting up to 100 terminals.

Multiple access control

Unlike the star configuration, all terminals on a ring or bus share a common link—the ring or the bus. This means that not only can all terminals 'listen' to the network, but also that all are potentially competing for use of the ring or bus. This facilitates rapid data exchange, because a packet can be transmitted as soon as it is formed, but it implies *contention.* To try to avoid more than one terminal at a time transmitting to the LAN, a terminal must be able to detect that another is 'talking', back off, and then transmit after a network-agreed interval (9.6 μs in the case of Ethernet). This procedure is known as *carrier sense multiple access* (CSMA)—it is a 'listen before talk' operation. However, as terminal activity increases, the chances of two terminals transmitting within the data propagation time also increases and when this happens *collision* may occur, in spite of CSMA, as shown in Fig. 11.24.

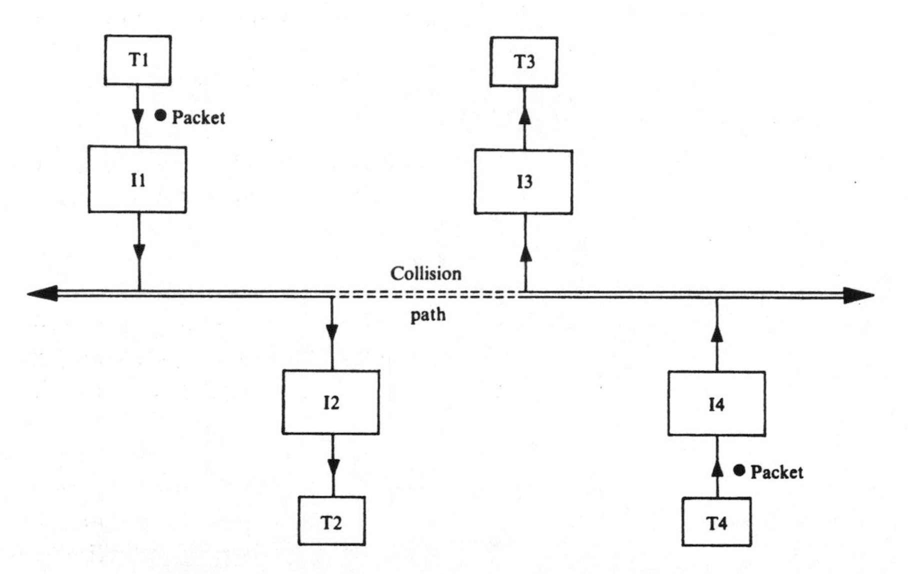

Figure 11.24. Bus collision.

In the example of Fig. 11.24, terminals T1 and T4 both sense the bus and, finding it is free, each transmit a packet destined for T3 and T2 respectively. Packets meeting in the collision path shown will be corrupted. This corrupted packet can be sensed, as it traverses the bus, by the two sending terminals which, having thus performed *collision detection* (CD), stop transmitting and back-off for a random interval before retransmitting the original packets. If collision occurs again the process is repeated with a longer random time delay before retransmission. All this happens so quickly that the network appears to give immediate access to all terminals. CD might be described as 'talk and listen'. The whole procedure of CSMA/CD is specified by the *Institute of Electrical and Electronic Engineers* (IEEE) standard 802.3. Random retransmission works satisfactorily at low usage but, as activity increases, a 'log-jam' occurs.

The Ethernet packet format

Since the network is asynchronous, it is necessary to synchronize the sending and receiving terminals each time a transmission path is established. Each packet therefore begins with a synchronizing byte or *preamble*, followed by destination and source addresses, layer information, data and cyclic redundancy check (CRC) digits as shown in Fig. 11.25.

The preamble consists of seven bytes of alternate 1s and 0s for synchronization. The start delimiter frame separates the preamble from the address

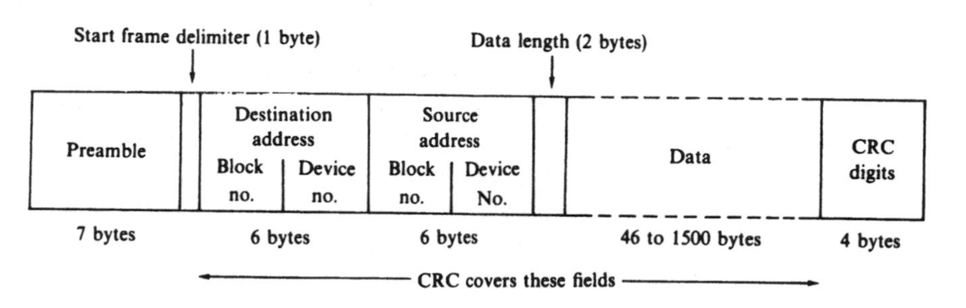

Figure 11.25. Ethernet packet format.

field by using alternate 1s and 0s and ending in two successive 1s. A block number in the address field indicates that several users in a group are to 'listen' or 'talk', otherwise a single device number is entered to specify the source and destination terminals. The two bytes after the source address specify the number of data bytes being supplied by the logic link layer (layer 2 of X25) to the next higher network layer. The minimum data field recommended by IEEE 802 standards is 46 bytes, to give time for the operation of the CRC procedure. Messages shorter than 46 bytes are stuffed with dummy bits to bring the message up to minimum length. A checksum field of 32 bits enables errors in the address and data fields to be detected. Each transceiver constantly listens to the bus and passes a packet to its terminal when it contains that terminal's address.

Scheduled access

The disadvantages of CSMA/CD is the long delay (log-jam) that builds up at high usage. When heavy traffic is the norm, scheduled access rather than random access to the network leads to greater efficiency.

There are two scheduled access techniques used to solve the contention problem; *token transmission* and *time-slot transmission*. In the former, a small packet or *token* circulates in the network, which, when accepted by a terminal, allows it to talk. When the terminal has transmitted, it releases the token back to the network where it may be 'grabbed' by another terminal, which can then take its turn to talk. The network may be accessed only when a 'free' token is in circulation and this means that only one token can be used in each network. In *time-slot transmission*, empty packets circulate around the network. These packets contain address and data fields and are equivalent to very long time slots. When a terminal wants to transmit it must wait for an empty packet to arrive at its interface. It then puts data into the empty packet data field, addresses it and passes it on to the network. As each full packet circulates its destination address is examined by each interface, in turn, until the destination terminal identifies its address and accepts the data in the packet.

Time-slot procedures

When a device is ready to send its data, it places the data in its buffer (a mini store). As the packet passes through the buffer interface, the device data is entered into the vacant data slot in the packet. Should there be no vacancy (indicated by a '1' in a full/empty bit position), the terminal buffer will wait until the next pass of a packet and try again. When data is entered into a packet, the packet is routed to its destination terminal where the data is extracted and the full/empty bit changed to a '0' before the packet is passed back to the network. At any particular time, there will be many packets in circulation, some empty, some full. A master station is required to call in any wrongly addressed packets that would otherwise go round the network indefinitely because no terminal would claim them.

In both the scheduled access techniques describe above, the idea of packets *circulating* in the network has been mentioned. This implies that such networks might appropriately have a *ring* configuration.

Ring LANs

In this configuration a continuous cable, in the form of a ring, links all the terminals via suitable interface circuitry. Two popular forms of ring network are shown in Fig. 11.26.

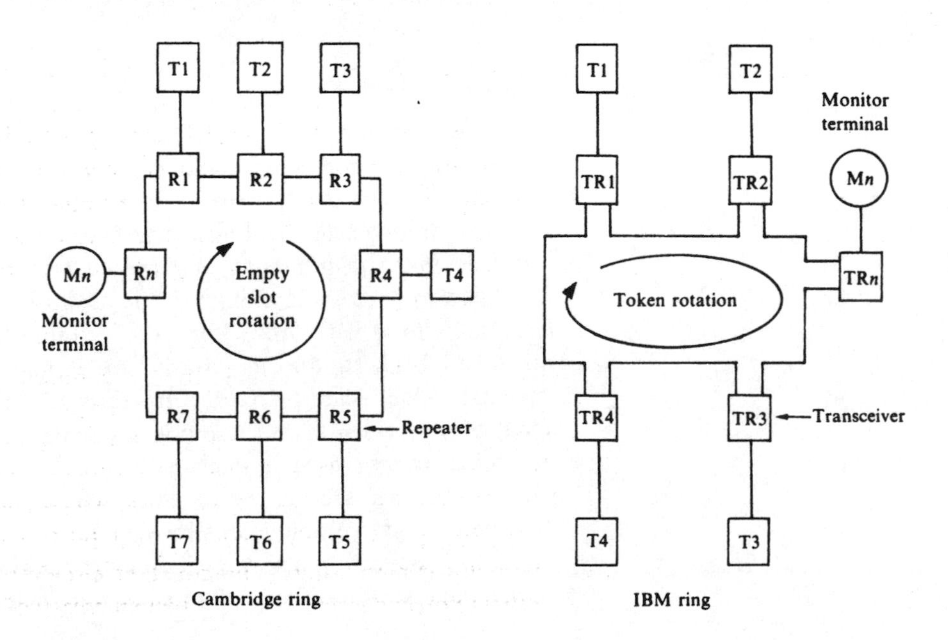

Figure 11.26. Two forms of ring LAN.

The *Cambridge ring*, developed at Cambridge University, uses either copper cable pairs, coaxial cable or optical fibre according to the distances and bit rates required. At present it works at 10 Mb/s with further development aimed at increasing this to 100 Mb/s. A succession of empty packets, of the form shown in Fig. 11.27, is circulated round the ring via repeaters (R), which are interfaced with terminals (T).

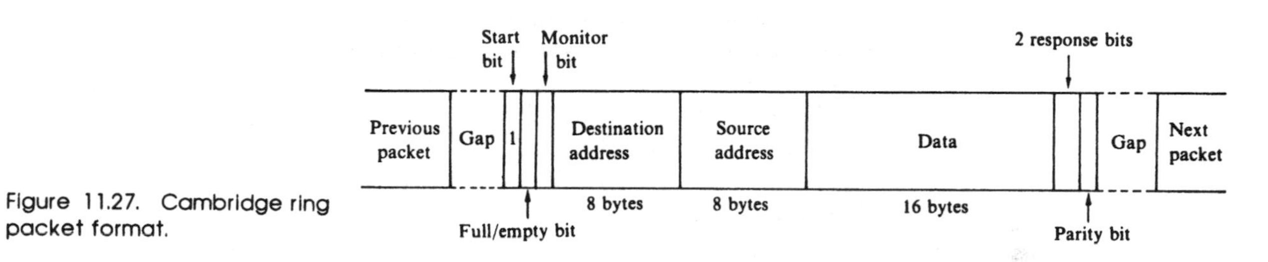

Figure 11.27. Cambridge ring packet format.

A start bit 1 marks the beginning of the packet. On receipt of an empty packet a terminal may seize it, if it has data to transmit, by substituting a 1 for the 0 in the full/empty bit position. It then fills in the address and data fields, setting the two response bits to zero and adding an appropriate parity bit.

On receipt of a full packet, a terminal will take one of three actions:

1. If it detects its own address in the source address position, it will set the full/empty bit position to zero and pass on this now effectively empty packet. The next terminal overwrites anything in the packet.
2. If it detects a source and destination address that are not its own, it will simply pass on the packet.
3. If it detects its own destination address, then, before passing the packet back to the network, it will read the data and set both response bits to 1, indicating to the sender that data has been accepted. If the terminal is busy with another packet, it will put 01 in the response field. If it is waiting for a further packet to make up a complete message, it will respond with 10. The sending terminal can then delay retransmission until the addressed terminal is free. An error in the received data is indicated by a 00 and this prompts the sending terminal to retransmit without delay.

The monitor bit is used by the monitor terminal to check that an unclaimed packet is not circulating in the network. If the monitor terminal detects 1 in the full/empty slot and 0 in the monitor slot, it will exchange the latter for a 1. This will be set to 0 when the empty packet is returned to its originator unless there is a fault in the system, in which case the monitor will detect 1 in both slots and will initialize the packet to prevent its permanent circulation. The constant circulation of packets, some empty, others not, suggests that a complete message of several packets will be built up sporadically at each terminal. The possible delay in completing a message makes the empty slot technique unsuitable for real-time communication.

The *IBM ring* operates at about the same bit rate as the Cambridge ring because it uses similar cable. Its method of access control, however, is quite different: it uses *token passing* to avoid contention. In this technique, a three-byte mini-packet or token (see Fig. 11.28) is generated by a designated node when the system is switched on. This circulates round the ring, entering each transceiver in turn (see Fig. 11.26).

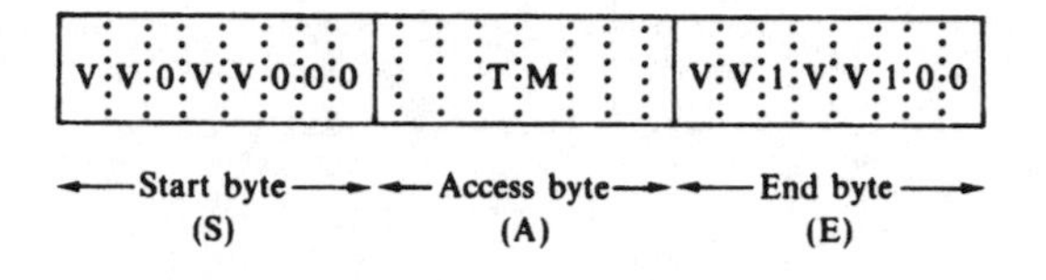

Figure 11.28. IBM ring token format.

Token passing procedures

The start byte identifies the start of a token and the end byte separates the token from the following packet. Violation bits (V) are used to limit the number of consecutive 0s and thus prevent loss of timing (see Sec. 7.9). The second byte gives access control. The first three bits of this byte give eight levels of priority in accessing the ring and the last three bits give eight levels of priority reservation for requesting the next packet. The M bit is used by the monitor terminal in a similar way to that of the Cambridge ring. When a terminal that has data to send examines a token having greater priority than itself, it passes on the token. Otherwise it changes the token bit from T to 0 to signify 'network busy' and passes on the token, together with all the packets it wants to transmit, to the next terminal.

When a terminal receives this 'network busy' token it examines the destination address on the following packets and passes on the token and packets if they're not addressed to that terminal. Otherwise, it copies the data in the packets, and passes on the token and the packets. When the sending terminal eventually receives back its copied packets, it absorbs them, changes the token bit from 1 to 0 and passes on the token, thus freeing the network from other packets. These token ring protocols are covered by IEEE 802.3 which, among other things, specifies the packet format shown in Fig. 11.29.

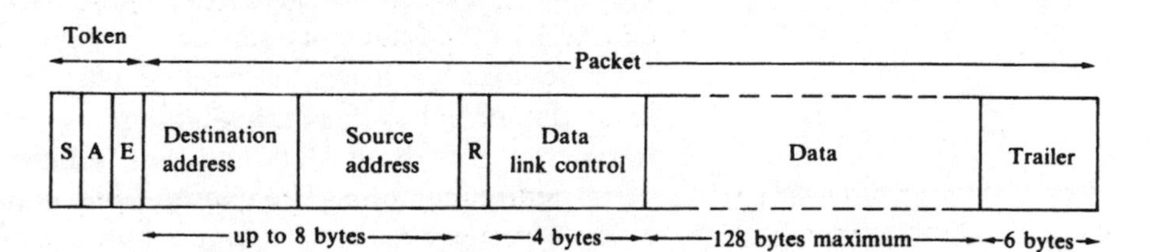

Figure 11.29. IBM ring token and packet format.

Token passing ensures that each terminal gets a fair share of the network, and is therefore ideal when activity is high. However, the time spent waiting for the token to arrive makes the system progressively more inefficient as the packet traffic reduces. The great advantage of the token passing technique is that there is no time lapse between successive packets in a message, so the IBM ring is suitable for real-time operation. Ethernet is, however, ideal for low traffic, so the optimum network might be one that used CSMA/CD until the traffic reached a certain level and then converted to token passing! The *deterministic* nature of the token passing technique is in marked contrast to Ethernet's random or *non-deterministic* nature. The unpredictability of packet arrival puts the Cambridge ring somewhere in between these two LAN types, making it *semi-deterministic.*

Exercise 11.11
Check on page 301

1. Suggest which of the two LANs, the Cambridge ring or the IBM ring, is the quicker at routing packets between terminals.
2. Could empty slot and token passing techniques be used in a bus LAN?

In a ring-type LAN each terminal must take part in the control of the data in the network. This implies that the ring topology is used when there are many intelligent terminals that need to 'talk' with one another. This type of network would therefore be ideal for computer graphics systems, for example, which require intelligent terminals with a fair amount of storage capacity and software support to enable graphics to be generated quickly on each VDU.

Glass fibres have much greater bandwidth and immunity to electromagnetic interference (EMI) than copper cables. The availability of cheaper fibres and transceivers is encouraging the development of empty slot and token ring fibre networks. One promising development is the *fibre distributed-data interface* (FDDI), which uses the token passing technique and can operate at 100 Mb/s over a single-loop range of 100 km and support up to 500 terminals. A range of this extent makes FDDI a metropolitan area network (MAN) rather than a LAN.

IEEE standards for LANs

The IEEE 802 family of standards (Fig. 11.30) coincide with the first three layers of the OSI model. They deal with the CSMA/CD and token access techniques in MANs and LANs and specify the use, for example, of Manchester coding (see Chapter 1) in 802.3 and a differential Manchester coding in 802.5. The 802.1 standards deal with systems management as well as internet working and therefore cover parts of all the X25 layers.

Exercise 11.12
Check on page 301

With what two main aspects of LAN operation would you expect IEEE 802.2 to deal?

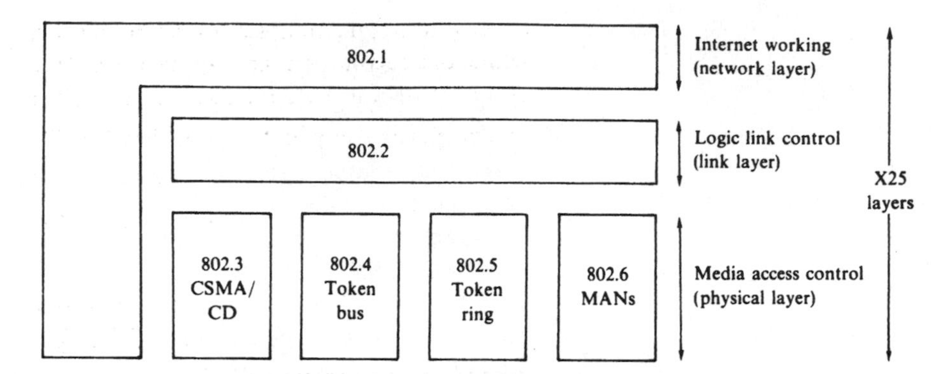

Figure 11.30. IEEE 802 protocols.

11.9 Data network interconnections

LANs and MANs may be interconnected over large distances using WANs as shown in Fig. 11.31. Having physically connected networks together, the remaining task is to transport the packets between differing protocol systems. As we have seen, the protocols used within a network to transfer packets between terminals are layers 1, 2 and 3 of OSI, and the remaining layers are used to convey data between the end users according to the

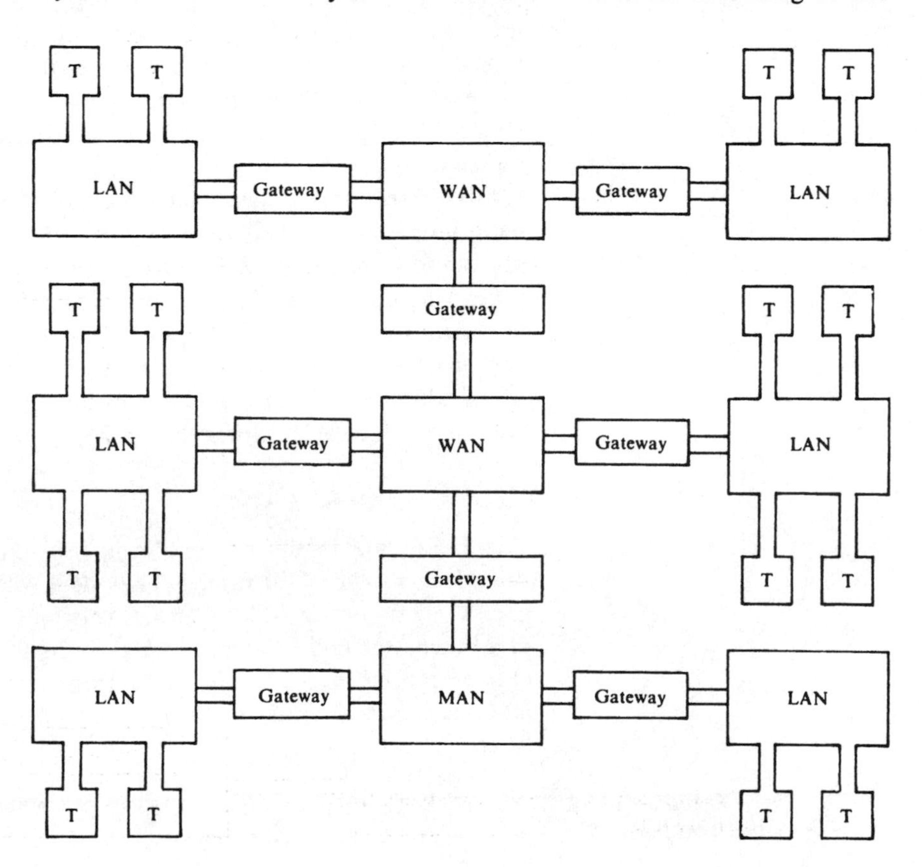

Figure 11.31. Data network interconnections.

application. The first set of protocols are therefore network-dependent while the second set are application-dependent. A *gateway* or *protocol bridge*, shown in Fig. 11.32, is used to effect the exchange of packets from networks using different protocols at layers 1, 2 and 3.

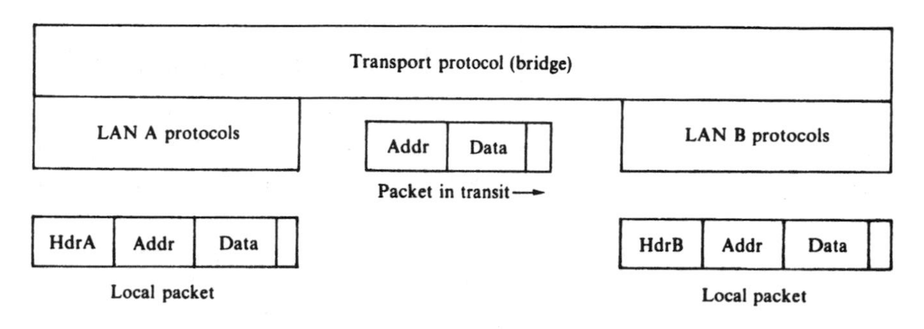

Figure 11.32. Action of protocol bridge.

11.10 An integrated services digital network

An integrated services digital network (ISDN) is a digital telecommunications network that links together all the audio, video and data networks and coordinates all the services they provide into one system as shown in Fig. 11.33.

An ISDN allows the telephone network to carry data with the same ease

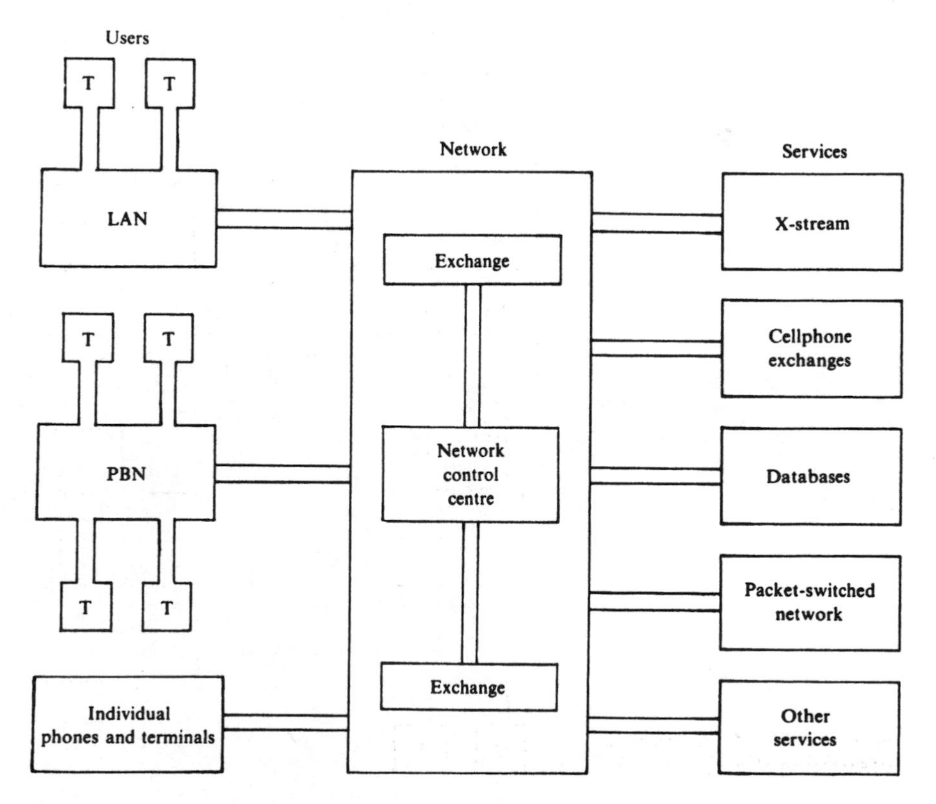

Figure 11.33. An integrated services digital network.

with which it now carries voice. It will also provide easy access to many services such as teleconferencing, teletex, videotex and high-speed FAX. In addition, telephone subscribers will have available, over a wide area, all the facilities offered by a modern PABX.

11.11 Satellite communications

Direct TV broadcasting by satellite, using microwaves, is described in Chapter 6. To justify the high cost of producing and launching a satellite, many other revenue-earning uses must be found for it. Wide area voice and data communications offer great scope in this respect, providing the problems of making the system available to as many users as possible are overcome. In this respect, the use of frequency bands in the microwave range, which can penetrate clouds, means that many channels are available for allocation when multiplexing is employed. Packeting PCM voice messages and data helps to make full use of a system. A satellite telecommunications link is the space equivalent of the terrestrial microwave line-of-sight link introduced in Chapter 1, as can be seen in the multi-purpose satellite system shown in Fig. 11.34.

The International Satellite Telecommunications Organization (Intelsat), formed in 1964, provides global coverage for telecommunications service

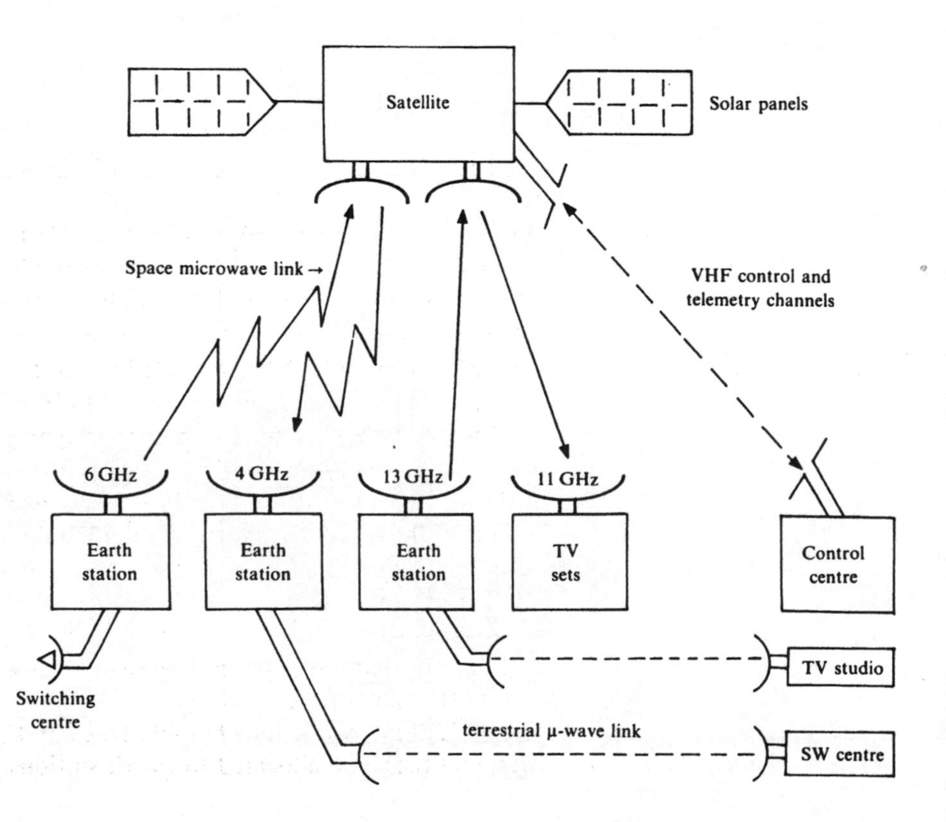

Figure 11.34. A satellite telecommunications system.

providers. Since then Intelsat has launched 34 satellites covering five generations of development. Intelsat VI was launched by an Ariane 4 rocket launcher in Autumn 1989. Satellite networks cover an enormously wide area. By putting satellites into geostationary equatorial orbits above the three main oceans, a world-wide voice and data service has been established. In each of the three main regions there is one primary satellite that provides the major linking role between the main international switching centres. Other satellites give back-up to this and provide services on high-traffic routes, special business services such as facsimile, video-conferencing and TV. Other consortia have launched satellites such as the International Maritime Satellite Organization (Inmarsat series) and the European Space Agency (Meteosat for weather forecasting and others for diverse functions such as facsimile, video-conferencing and TV).

A transponder carried by a satellite receives a signal at a given carrier frequency from the ground station, amplifies it and transmits it back to Earth using a lower carrier frequency. Typically the up/down frequencies are about 6/4 GHz (C band) for voice and data and 13/11 GHz (J and X bands) for TV. With about 36 000 km between Earth and a satellite, the communication path loss is high (about 200 dB). The ground stations must therefore use transmitter powers up to about 3 kW and very sensitive, low-noise receivers (SNR about 50 dB). Weight restraints limit the satellite transmitted power to about 7 W per transponder.

Exercise 11.13
Check on page 301

Suggest why the higher of the up/down frequencies is used for the up link.

A single satellite is able to give satisfactory communications cover over one-third of the world's surface using less total power than would be needed using terrestrial transmitters. Further, even though the signal attenuation is high, it is fairly constant, varying by only about 5 dB at most due to changes in atmospheric conditions. This is a very attractive feature. By contrast, the equivalent Earth-bound system may have an attenuation variation of about 30 dB. The lifetimes of the solar cells and rechargeable batteries limit the expected operational lifetime to between 10 and 15 years.

The transponders are locked to a particular carrier frequency and perform frequency recognition and down-translation only. They are insensitive to the form of modulation used, so all modulation and detection schemes are controlled by the ground equipment. In TV broadcasting the communication is one-way, but in voice and data exchange it is two-way. This is not implied in Fig. 11.34. As far as the latter is concerned, Fig. 11.35 is clearer.

Suppose that the satellite is designed to receive over the range 5800–6200 MHz and transmit over the range 3800–4200 MHz. The 400 MHz bandwidth is divided into ten 40 MHz bands, each served by a transponder, and these are allocated to Earth stations by Intelsat or the European

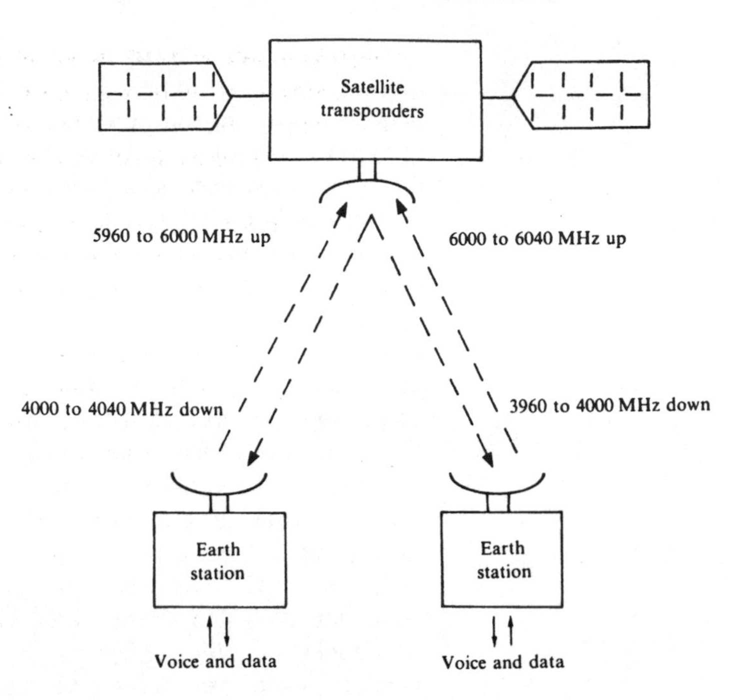

Figure 11.35. An example of a two-way satellite link.

Space Agency. Each station has its own 40 MHz up-band and 40 MHz down-band separated by 2 GHz as exemplified in Fig. 11.35. The one transponder is servicing 10 Earth stations. An increase in the number of ground stations would require extra satellites with different up/down frequency ranges.

Most forms of modulation have been tried in the various satellite links used to date. In the earlier systems the data is sent using FSK and the voice signals use SSB-AM with a single carrier per channel (SCPC). A number of these carriers make up an FDM signal, which frequency modulates an *intermediate carrier frequency* (ICF), which is then up-converted to the actual carrier frequency of around 6 GHz. Later systems take only pulses (TDM PCM voice signals and data) and modulate the ICF using QPSK. An Earth station system is shown in Fig. 11.36.

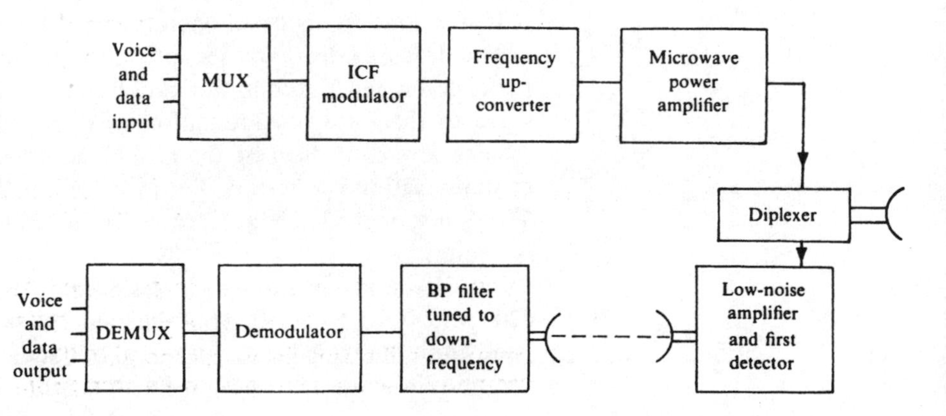

Figure 11.36. An Earth station system.

Each station's 40 MHz band gives a usable 36 MHz. If FDM is employed, this band is divided into channels of 38 kHz with 7 kHz guard bands giving a total of 36 MHz/45 kHz = 800 channels. If TDM is used, the number of 64 kb/s channels will be 36 MHz/32 kHz = 1125. The reasons for modulating an ICF and then up-converting to the EHF carrier, rather than modulating the EHF carrier directly, are given in Sec. 4.5. The up-signal is fed from a travelling wave tube (TWT) to a dish aerial via a multiplexer so that the same dish can be used for transmission and reception.

The recovery of voice and data signals from the down-signal requires a very low-noise amplifier, situated near the dish, and a band-pass filter tuned to the down-frequency of the station from which signals are being received. The system is similar to that, outlined in Chapter 6, for satellite TV reception.

The satellite transponder system consists of an amplifier/frequency translater/amplifier cascade for each 40 MHz band as shown in Fig. 11.37. The up- and down-frequencies are different to prevent any positive feedback between a transponder output and input causing instability.

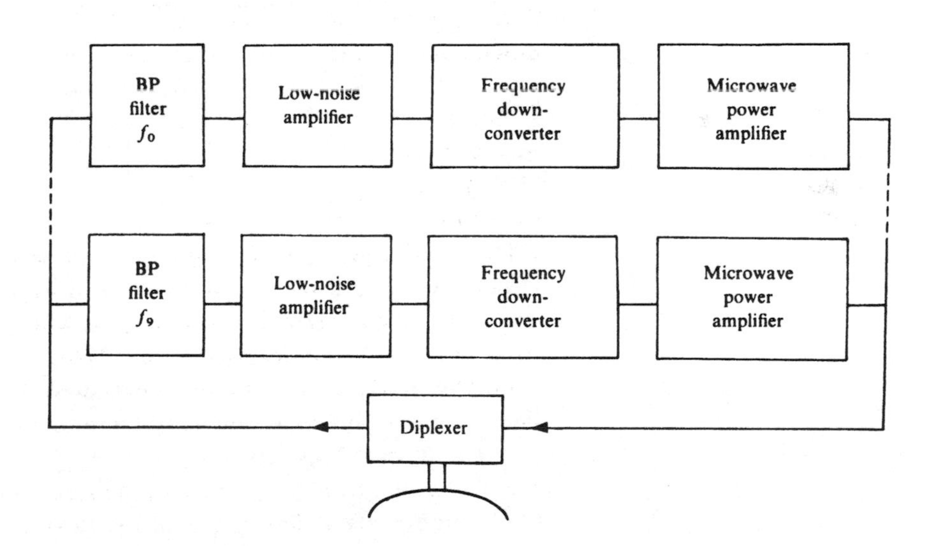

Figure 11.37. A satellite transponder system.

Exercise 11.14
Check on page 302

Calculate the frequencies f_0 and f_9 in Fig. 11.37 for a satellite designed to receive over the range 5.8 to 6.2 GHz.

The main disadvantage of FDM is the intermodulation distortion that arises due to the non-linearity of the TWT, which provides the microwave amplification in a transponder. This is controlled by reducing the TWT power output at high usage. TDM does not have this disadvantage.

Multi-access to a satellite

The satellite we have considered as an example provided ten 40 MHz bands, which could be allocated to 10 ground stations. This, in effect, gives a wide area network (WAN) interlinking 10 stations. Each ground station is allocated a channel centred on a different carrier frequency as shown in Fig. 11.38.

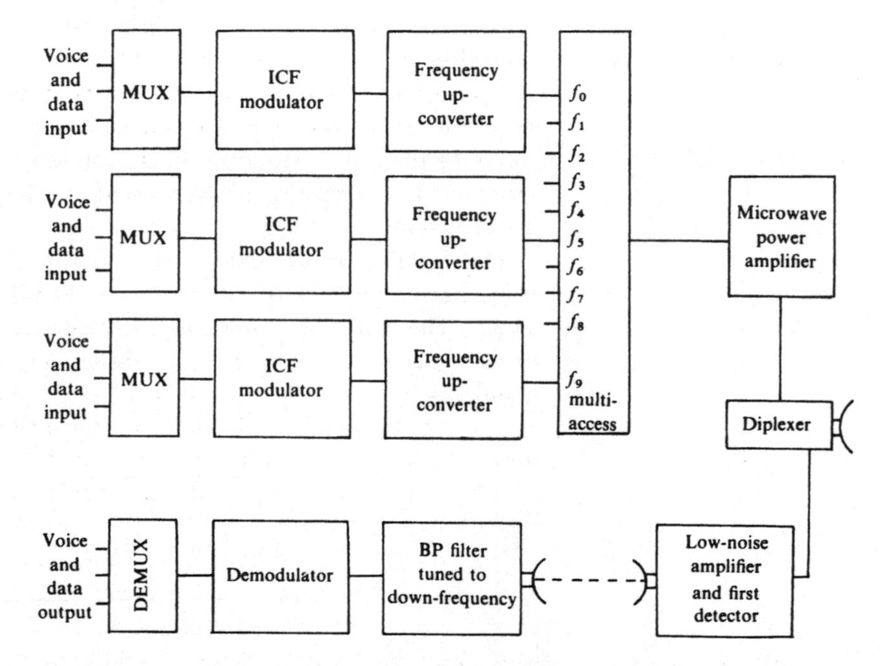

Figure 11.38. Frequency-division multi-access from 10 Earth stations.

The type of access shown in Fig. 11.38 is appropriate where FDM is used for voice and data signal. It is known as frequency-division multi-access (FDMA). FDM exacerbates the intermodulation distortion that arises due to the non-linearity of the TWT.

An alternative form of multi-access, used when TDM of the voice and data signals is used, is time-division multi-access (TDMA). Here, time slots are allocated instead of the frequency slots of FDMA. Each station uses the same carrier frequency, so intermodulation does not occur. Each TWT can operate at full power all the time, thus maximizing the SNR of the return signal at ground level.

Fully scheduled access to a satellite network, such as is provided by FDMA and TDMA, avoids the sort of collisions that are possible in random access networks. These networks function most efficiently when the traffic is high but are wasteful if a frequency band or time slot is idle. A random access network gives greater efficiency at low usage but the efficiency decreases as collisions become more likely with high demand. Random access is really an untimed or asynchronous form of TDMA. A protocol for random access to a packet radio WAN, called ALOHA, was developed at the University of Hawaii. This has been modified, to increase traffic, by allocating time intervals for the data packets by using a similar principle to the Cambridge ring. Indeed, some of the principles used in WAN access are the same as those described earlier in this chapter for LANs. For example, CSMA is also used in WANs but token passing is not. When random access is used, the data packets must be addressed and must use formats and protocols similar to those used in any packet-switched network. The basic idea of FDMA and TDMA is that signals between different stations are separated in either the frequency or time domains. Addressing is inherent in these systems because receiving stations can select the data intended for them by tuning or timing.

Another form of multiple access that is particularly suitable for data is *code division multiple access* (CDMA). All stations operate over the same carrier frequency range simultaneously. Each message is uniquely coded and the destination receiver is programmed to decode the code by *correlation detection.* The receiver extracts the wanted data signal from the noise and other signals by comparing the signal patterns.

The capacity of any radio communications system can be increased by *frequency re-use.* This can be achieved by applying the cellular radio principles described in Chapter 10 and by the use of *polarization* of the radio waves. This same technique is used in terrestrial TV broadcasting, where contiguous service areas might use different planes of polarization, either vertical or horizontal.

Intelsat satellites now provide about two-thirds of the world's trans-oceanic telecommunications. Over the past 25 years, 34 satellites have been launched covering five generations of development. An Intelsat IV satellite carries 20 transponders linked to 300 stations in 125 different countries and has a capacity of 25 000 voice circuits. Thirteen Intelsat V satellites are in orbit also with about 20 transponders but able to give increased service by more use of digital techniques. The first of the new generation, Intelsat VI was launched in Autumn 1989. It is able to provide the equivalent of 270 000 telephone channels. As more are launched they will gradually take over the work of the fifth generation, which are now coming to the end of their life. Intelsat VII satellites are now being planned for phasing in from about 1995.

The huge amount of global telecommunication of audio, video and data signals has lead to the setting up of a vast array of international standards. This chapter concludes with a review of some of the more important of these.

11.12 International standards

It is just as easy to make a phone call to the other side of the Earth as it is to call someone just across the street. The ease of making international telephonic connection conceals the great cooperative effort between many countries over many years that has made it possible. The routing of a call from London to Canberra, described in Sec. 10.2, can be used to show a few of the factors involved.

The call starts from London, specifying that it is to an overseas destination by using code 61 to identify Australia. Goonhilly has to access a microwave channel to the satellite using frequencies allocated to the UK. The satellite signals the Canadian station of the call to be relayed. The Canadian Earth station must detect and extract the call to Canberra and pass it on to the Pacific submarine cable from Vancouver. At Sydney the call passes into the Australian national network to be linked to the subscriber.

There are political, operational, technical and administrative issues that have had to be agreed to achieve the successful completion of such a call. For historical reasons, there are a variety of signalling systems in use throughout the world. That used by the transmitting nation may or may not correspond to the one employed in the receiving or the relaying country. The international network cannot be expected to cope with all possible systems directly. A standard international system and the appropriate interfaces between national networks have to be defined and agreed. The technologies differ sharply between segments of an international route—land line, radio, satellite and optic cable; the interfaces linking them also require agreed definition. Finally, there is the question of how payment for a call apportioned between the telephone administrations involved.

The field of telecommunications has a long history of world-wide cooperation. It began in 1868 with the formation of the *Comité Consultatif International Télégraphie* (CCIT). This later became the CCITT with the advent of the telephone. The *Comité Consultatif International de Radiocommunications* (CCIR) was formed, some time later, to deal with radio. The CCITT and CCIR discuss and settle most of the issues to do with global telecommunications. After the formation of the United Nations both of these bodies became parts of the *International Telecommunications Union* (ITU). Other bodies associated with global telecommunications are the *International Standards Organization* (ISO) and the *International Electrotechnical Commission* (IEC). In Europe, there's the *Comité Européen Postes et Télécommunications* (CEPT), the *Comité Européen de Normalisation* (CEN) and the *European Computer Manufacturers Association* (ECMA).

Telecommunication affairs have generally been conducted on a friendly and successful basis. Much of the cooperative effort occurs in the CCITT, which is the standards-setting organization of the ITU. Members of the committee are the representatives of governments and telecommunications administrations. The ITU promulgates its CCITT standards as recommendations in deference to national sensitivities. Full sessions of the CCITT are held every four years to ratify new or revised recommendations.

Some important CCITT standards

The CCITT recommendations are very comprehensive. They are divided into 24 groups, each designated by a letter of the alphabet—only 'W' and 'Y' remain unused to date. 'A' starts things off with the 'Organization of the Work of the CCITT' and 'Z' ends the collection with 'Programming Languages'. Browsing through the alphabet reveals, for example, that 'C' covers 'General Tariff Principles', 'G' covers 'Transmission' and 'I' deals with the 'ISDN'.

The 'V' series provides standards for data communication over the telephone network, and therefore includes modem speeds and operation modes. All value-added data services (VADS) quote the modem characteristics that are supported for connection to their service by referring to one or more of V*n*, where *n* comes from the list given below. For example, Prestel makes use of V21, V22, V22 bis, V23 and V24. All of these relate to transmission over the PSTN. There are other modem recommendations for leased lines.

The modem standards of the CCITT 'V' series are one of the cases where complete international accord on a single set of standards has not been reached. They are widely used throughout Europe and the rest of the world but not in the USA, where Bell standards (originated by the Bell Telephone Labs) are employed. The relevant Bell standards are largely but not completely compatible with the equivalent 'V' recommendations. Special requirements are laid down by BT for modems used in communication with the USA.

Recommendation V3 describes the 'international alphabet no. 5'. This is an example of CCITT taking over well established standards. Alphabet no. 5 is the same as the ASCII code used almost universally for communication between computers and defined originally by the *American National Standards Institute.* V24 defines the characteristics of circuits to interface DTE and DCE. It is better known as the *Electronic Industries Association (EIA) recommended standard RS232.* V25 defines the procedures for automatic calling or answering equipment operating over the PSTN. Appendix C gives a full list of the 'V' and 'X' recommendations.

'X'-series recommendations

This group deals with data communication networks. X3 details the PAD facility in a public data network, covering such things as packet assembly, call set-up and clearing. X25 specifies the interface between the DTE and the DCE when the DTE is operating over the public data network in packet mode. X28 gives the requirements for character terminals dialling up the PAD facility in a public data network: it is intended to cover such cases as Telex-packet interworking. X32 covers the interface between the DTE and DCE for terminals operating in packet mode, but accessing the packet network via the PSTN. X75 specifies procedures for the transfer of data on international circuits between packet-switched networks. There are many more, including a whole string, X2*xx*, that cover the ISO seven-layer protocols as recommendations, and X400, which defines ways of exchanging information between computer-based systems on a store-and-forward basis.

'I'-series recommendations

The ISDN is still in a development phase, but the CCITT has already produced a wide-ranging set of standards relating to it. The 'I'-series recommendations are presently structured as a set of subseries: I100 series deals with general ISDN concepts; I200 with service aspects; I300 with networks; I400 covers the user–network interface; I500 internetwork interfaces; and 1600 covers maintenance. No doubt, more will emerge in the future. It would appear that the CCITT is pursuing the laudable objective of ensuring that when the ISDN is established it will have the universality of the system it is meant to replace: the PSTN.

Other bodies involved in standardization

In addition to those organizations already mentioned in this section, several others are also involved with setting standards. Among these are the *Institute of Electrical and Electronic Engineers* (IEEE), a US learned society of professional engineers who include standardization among their many activities, and the US Department of Defense, who issue exacting *military* (MIL) standards. There are also national standards such as *British Standards* (BS) published by the British Standards Institution and the German industrial standards, *Deutsche Industrie-Normen* (DIN). Certain major manufacturers have also developed standards. Probably the most important of these, in the data communications field, are IBM, with its *System Network Architecture* (SNA), Xerox, who were responsible for *Xerox Network Systems* (XNS), and the Digital Equipment Corporation (DEC), who use *Digital Network Architecture* (DNA). The General Motors Corporation and the Boeing Computer Services Company developed, respectively, the *Manufacturing Automation Protocol* (MAP) and the *Technical Office Protocol* (TOP). All these groups will probably, in the fullness of time, bring their standards into line with the OSI model.

Summary

As the world's economy grows, more and more data is produced. This has created an ever-increasing demand for faster data communications locally, nationally and globally.

Data communications has its roots in the telegraph and telephone systems: early computer networks adapted the data transfer methods used by the telegraph. Much data is still sent, over a wide area, via the PSTN, but this method is limited by the inherently slow transmission rate allowed by the PSTN. National and global transfer is now accomplished by packet-switched wide area networks (WANs). Local area networks (LANs) also use data packets but are non-switched networks, which have either star, bus or ring topologies. LANs enable computers to 'converse' with one another and share expensive equipment such as high-quality printers and plotters. LANs cover distances up to a few kilometres and, unlike WANs, have large bandwidth and low propagation delay. Access to a WAN is random but to a LAN access is either random or controlled by scheduling techniques such as polling, empty packet circulation or token passing. If random access is allowed, arrangements must be made to deal with data collisions in the network.

High-speed data networks, based on optical fibres, are now being used to interconnect LANs over regions where there is high data traffic such as the City of London, university campuses and large research establishments. These metropolitan area networks (MANs) can be linked to WANs, as can the LANs, via protocol bridges or gateways. WANs, MANs and LANs usually transmit their data at baseband.

During the time when packet switching was being developed, computer manufacturers were developing different systems for data transfer, many of which would only work on their own equipment. Users who had acquired a variety of equipment were unable to transfer data easily because much of the equipment was incompatible. Standardization was needed. Various rules or protocols were adopted in different countries, but it became clear that a more general approach was needed with the aim of allowing the exchange of data between as many different systems as possible. The International Standards Organization developed the Open Systems Interconnection reference model of a network. The model represents a data network as a layered structure, up to seven layers being defined in order to give a comprehensive framework for developing protocols that will facilitate equipment connection.

Satellite links are being used increasingly for international data communications. These use packet radio techniques and work over broad frequency bands around 6 and 4 GHz for the up- and down-links respectively. Radio waves at these frequencies penetrate clouds with little energy loss and, using FDM and TDM, provide many channels. Multiple access of many ground stations to a satellite is provided by the assigned-access techniques of FDMA, TDMA and CDMA or by random-access techniques similar to those used in LANs. WANs pose special problems and make more urgent the need to achieve international agreement on telecommunications standards.

In many countries, efforts are being made to achieve integration between audio, video and data networks so that in future all these forms of telecommunication can take place over a high-speed integrated services digital network (ISDN).

▶ Checks on exercises

11.1 Channel capacity (C) is given by Shannon's equation $C = B\log_2(1+\mathrm{SNR})$, so the maximum transmission rate will be $3\log_2 11 = 3(\log_{10} 11)/0.3 = 10.4$ kb/s.

The required bandwidth (B) to transmit data at 64 kb/s will be $64\,000/(\log_2 11) = 64\,000/[(\log_{10} 11)/0.3] = 18.4$ kHz.

11.2 The frame format would be as shown in Fig. 11.39.

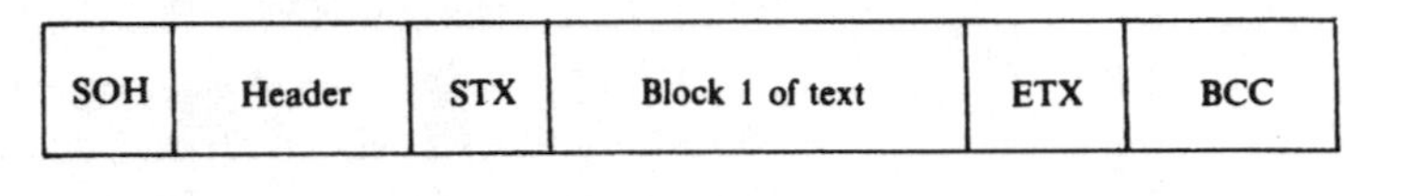

Figure 11.39. A frame format.

1. After each block of data the receiver must acknowledge successful transmission so the sender knows it can send the next block.
2. If a block is corrupted in some way, there must be a way to get the sender to retransmit it—some sort of negative response.

11.3 Here is the data with the 'stuffed' zeros pulled out:

$$\begin{matrix} & 0 & & 0 & \\ 11111 & \uparrow & 110010011111 & \uparrow & 00001 \end{matrix}$$

11.4 There are three different routes available between terminals T1 and T2. Seven routes would be available if all the PSEs were fully connected.

11.5 When a called terminal refuses a call, the procedure is to send a CQP back through the system as shown in Fig. 11.40.

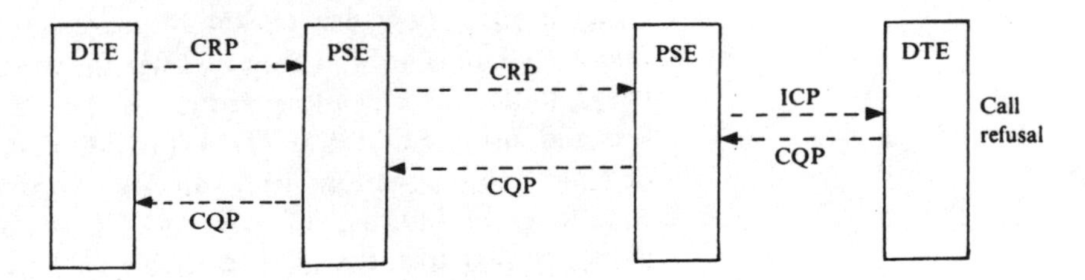

Figure 11.40. Call refusal procedure.

11.6 The tasks and their associated layers are:

1. The choice of ASCII code--presentation layer
2. The management of an airline reservation system—application layer
3. Deciding which user is to talk next—session layer

4. The use of datagrams or virtual circuits between nodes—network layer
5. The routing of packets between networks—transport layer

11.7 Each picture carries four teletext information lines, so a complete page is transmitted in six frames. A frame is transmitted in $1/25\,s = 40\,ms$. One Teletext page takes 240 ms to transmit—that is about four pages per second.

11.8 The time to transmit a 1 Mb picture is 1 Mb/(1200 b/s) = 13 min 53 s! Even with the top rate of 9600 b/s it will take 1 min 44 s to get the picture 'downloaded' (the term used to describe the transfer of data from a remote computer to a terminal).

11.9 An A4 page has about 20 kb, which would require over 4 min to send at 75 b/s, so this standard is not suitable for EM. The 1200/1200 standard is really needed for VEM and is available on some viewdata systems.

11.10 The arithmetic is trivial but the result is significant. The transmission rates are: 2400 b/s = 300 characters/s for Teletex and 50 b/s = 10 characters/s for Telex. So Teletex is 30 times faster than Telex—with that speed advantage it cannot help being cheaper to use than Telex.

11.11 The Cambridge ring has many empty packets in constant circulation, which gives each terminal quick access to the ring. In the IBM ring a terminal must wait for the single token to arrive before it can gain access. This means that the average access time is longer in the IBM ring than in the Cambridge ring, making the latter quicker at routing packets between terminals.

Empty slot and token passing techniques could not be used in a bus LAN because the access technique is dependent on the network configuration.

11.12 The two main aspects of LAN operation dealt with by IEEE 802.2 are:

1. Packet frame formats—this is logic link control
2. The type of access, CSMA/CD or token—this is medium access control

11.13 There are two main factors involved:

1. Transmission path attenuation increases with frequency so it is better to make up for this with a higher-powered transmitter on the ground rather than on the satellite.
2. The beamwidth of an antenna decreases with frequency, which means that it is easier to maintain the satellite antenna alignment if it transmits on the lower of the up/down frequencies.

11.14 The range 5800–6200 MHz is divided into ten 40 MHz bands, so $f_0 = 5820$ MHz and $f_9 = 6180$ MHz.

Self-assessment questions

11.1 State what is meant by data terminal equipment.
11.2 What two main factors limit the speed of data exchange over the PSTN?
11.3 What is the main advantage of message switching over circuit switching?
11.4 Why are messages packeted?
11.5 Outline the operation of a UART/USRT device.
11.6 In what type of application would it be more appropriate to use asynchronous rather that synchronous character framing?
11.7 Define the terms: 'host', 'virtual circuit' and 'datagram'.
11.8 Describe two forms of access of a PSE used by packet and character terminals.
11.9 What is the function of a PSE?
11.10 What is a VAN?
11.11 Describe the X25 protocol.
11.12 What is OSI and what does it have in common with X25?
11.13 List the WAN services available in the UK.
11.14 Define what is meant by 'videotex'.
11.15 Distinguish between 'viewdata' and 'teletext', giving an example of each.
11.16 Distinguish between 'teletext' and 'Teletex'.
11.17 What kind of network would be used to interconnect: a group of computer-controlled milling machines on the same site; a national cash dispenser system; the shared computer facilities in a science park; and the shared CAD/CAM facilities of two factories several hundred miles apart?
11.18 Explain, briefly, the two forms of multiple-access control used in LANs.
11.19 State how this multiple-access control is achieved in Ethernet, the Cambridge ring and the IBM ring.
11.20 What is an ISDN?
11.21 Compare and contrast the satellite links used in data and TV communications.
11.22 Why is FM preferred for the satellite/ground communications link?
11.23 Describe the ground station and satellite radio systems.
11.24 Compare the multiple-access techniques used in satellite communications with those used in LANs.
11.25 What are the CCITT recommendations and how are they designated?

12

Summary and future trends

Taking stock and speculating

In this short, concluding chapter, the principles outlined in the book are summarized and some suggestions are made as to the way in which telecommunications might develop in the near future.

Communication concerns the transfer of information from a source, the transmitter, to a sink, the receiver. The sink is made aware of thought or information at the source when a signal is transmitted through a communications channel. Signal energy takes the form of electromagnetic waves, either beamed from point to point or guided by a conductor or a glass fibre. Broadcasting radio waves into free space is an alternative to point to point transmission when simultaneous transmission from a single source to many receivers is desired. It is convenient to think of information energy flow in terms of an electric current or voltage where a wire is involved, otherwise the electromagnetic wave concept is more appropriate. The use of electrical transmission has made possible near-instantaneous communication over vast distances because radio waves travel at about 3×10^8 m/s. The power dissipated over such distances may be considerable, but highly sensitive receivers can recover information from weak signals with varying levels of fidelity depending on the application.

A transmitter must convert a signal into an electrical form suitable for the transmission medium and transmit sufficient signal power. Variations in the rate of flow of electrical energy can be sensed by a receiver. These power variations constitute a signal when they are controlled by an information source. A signal voltage (v) may be classified as periodic if it satisfies the equation $v(t) = v(t+T)$. The smallest value of T that satisfies this relationship is the period of $v(t)$. A signal for which no value of T can be found is non-periodic. The reciprocal $1/T$ is the signal frequency (f)—a feature of great significance in telecommunications.

Signals may also be classified as random or non-random (deterministic). The former cannot be specified as functions of time or frequency—it is possible only to ascribe each signal component a certain probability of occurrence. In gaussian signals, component amplitude values are normally distributed. Signals may be specified as functions of time or frequency. If they are periodic, a Fourier series will do this, otherwise a Fourier integral is required. A Fourier series consists of a sum of sine/cosine terms of various magnitudes and frequencies. These frequencies are harmonics of the periodic signal. In the case of a sound signal, the harmonics are called tones. Each harmonic can be represented by a line on a spectrum diagram. A non-periodic signal has a continuous spectrum.

If information is put into words, a signal will convey meaning in the form of the spoken or written word. The spoken word is easily converted into a voice-shaped current or wave, and this has led to the widespread use of electrical signals that are analogues of speech. The speech signal, like the sound it represents, is continuous in time and magnitude. Symbols are not spatially continuous and so convert naturally into signals that are not continuous in time, but consist of pulses of varying magnitude. A number n can be represented by a pulse of n units of amplitude or by binary pulses forming codewords that represent n units. Binary pulses form the type of signal used in digital computers. Television pictures appear to be continuous, but the broadcast signals that provide this illusion are, at present, a mixture of continuous and discontinuous components. Signals that are continuous in both time and magnitude are classified as analogue signals. Signals that are discrete in both time and magnitude are classified as digital. Signals that are time-discrete and amplitude-continuous are often described as digital.

The continuity of sound does not prevent electrical audio signals being transmitted as time-discrete samples of the original, which may be used to recreate the original sound at a receiver. If a tone is sampled at more than twice its frequency (the Nyquist rate), the original tone can be recovered from the signal at the receiver by low-pass filtering.

The magnitude of a signal at any instant can be represented by a binary (digital) codeword. This value could be communicated as a sequence of timed electrical pulses along one channel or as a simultaneous set of pulses, one pulse to each channel. Converting the continuum of values allowed in an analogue signal into a finite set of values implies quantization. The size of the quantum decreases as the number of different values increases. This leads to reduced quantization error and a more accurate digital representation of a signal.

Analogue-to-digital conversion enables sound and data to be telecommunicated digitally. Economic digital TV broadcasting will be possible in the near future. At present, digitization of all the basic forms of signal (audio, video and data) occurs only in point to point transmission.

A receiver must convert an electrical signal back into sound or vision and present the information as an acceptably accurate replica of the original. The criteria of acceptability depend on subjective judgement tempered by economic factors. Telecommunication engineering is concerned with seeking ingenious solutions to the conflicting pressures of high fidelity and low cost. The factors that affect fidelity, so far as telecommunications is concerned, are the distortion and noise in the communications channel.

If the output of a channel is a non-linear function of the input, non-linear distortion will occur. This has the effect of changing the relative amplitudes and phases of the Fourier components of the transmitted signal, giving harmonic and phase distortion and producing signals that are sums, differences and products of these components (intermodulation distortion). Thus the received signal is degraded. Various applications can tolerate different amounts of degradation; for example, voice telephony is much more tolerant than stereophonic music broadcasting.

Noise is any unwanted signal that interferes with the wanted signal. If the wanted signal power is high compared with the noise power, or if the signal and noise are very different in form, then the noise may be tolerable. When listening to music, for example, a quiet, steady hiss is much more tolerable than another tune; telephone line crackle is more tolerable than cross-talk. The noise in these instances is not overpowering, and is so different from the wanted signal that the hearing system is able to filter out most of it.

If electrical signals are sufficiently different from the channel noise, an electrical receiver can be designed to filter out much of the noise. Unfortunately, all electrical processors add more noise to the signal. High transmitted signal power and cable screening can increase the received signal-to-noise ratio (SNR) if the noise enters a line link from outside. However, thermal noise, generated in the line conduction process and the electronic equipment, produces channel noise that cannot be reduced easily. Such noise takes the form of a random signal that can be specified in terms of its spectral power density. In practice much noise is 'white', a description used to indicate its uniform spectral density over its frequency range, its bandwidth normally being limited by the channel.

Some order or non-randomness is required if a signal is to carry any worthwhile information. A receiver detects this order or pattern. The random nature of thermal noise suggests that the only information that noise can convey is its power level and how this is spectrally distributed.

Channel bandwidth (B) and transmitted signal power (S) offer alternative means of improving channel capacity (C) according to the Shannon–Hartley theorem, which states that $C = B \log_2(S/N + 1)$ b/s. Improving the information-carrying capacity of a noisy channel requires devising signals that a practical receiver can distinguish from noise. This requires the addition of more information, over and above that required for the actual message. Adding extra information inevitably means increasing the bandwidth of the signal—more bits per second implies higher frequency.

Channel capacity is the maximum rate at which information may pass through a channel without error. The maximum channel capacity is limited to $1.44S/\rho$ by the increase in white noise power (spectral density ρ) that occurs when the channel bandwidth increases.

Codes used for encoding information before transmission (source codes) are designed to be as efficient as possible, so that the channel capacity is used economically. If spare capacity exists, redundant information can be added to enable transmission errors to be detected and, perhaps, located. This redundant information takes the form of parity-check bits, which are combined with the data according to the code rules. The total bit rate, obtained by adding channel coding to source coding, should not exceed the channel capacity if the channel code is to be effective in dealing with errors at the receiver.

If there is sufficient link bandwidth, a number of different signals may be multiplexed, either on a frequency-sharing or time-sharing basis. In frequency-division multiplex (FDM), each channel is allocated a slot of the available frequency band for use during the whole transmission period. The frequency band of the signal, the baseband, is shifted into frequency slots by modulation.

An alternative method of multiplexing, which can be used only with pulse signals, employs time sharing. In time-division multiplexing (TDM), each channel is allocated the whole of the link bandwidth for short periods or time slots. The shorter these periods, the more channels there will be. However, short time slots imply narrow pulses, which require wide link bandwidth. In both FDM and TDM, more channels therefore require more bandwidth. Higher bandwidths become available towards the infrared band, so glass-fibre links are ideal for point-to-point multi-channel transmission. The high immunity to interference, absence of thermal noise and low distortion that they provide are important bonuses. The allocation of frequency bands for cable telephony, and carrier frequencies for radio communications, are both applications of FDM.

At ELF and VLF, radio waves travel round the Earth as though in a waveguide bounded by the ground and the ionosphere, LF and MF waves can propagate similarly or as sky waves reflected by the ionosphere. Reflected waves are subject to fading due to ionospheric variations. HF radio also uses sky waves. VHF and UHF radio works with direct and/or ground-reflected waves. SHF and EHF signals use space waves and line-of-sight operation. Highly directional aerials must be many times larger than their operating wavelength, so dish aerials can operate only in the microwave bands.

Frequency shifting or translation is much used in telecommunications, not only for FDM, but also for radio transmission. Suitable aerials for baseband transmission are difficult to construct, so the baseband is shifted to higher frequency bands where the radio wavelengths are shorter. Wire aerials propagate most efficiently when their effective lengths are great relative to the wavelength of the carrier wave. Frequency shifting is achieved by a process in which a high-frequency continuous carrier wave is modulated, usually by varying its amplitude or frequency with the baseband signal.

Amplitude modulation (AM) is normally achieved by multiplication of a continuous carrier frequency (ω_c) with the baseband signal, a representative frequency of which might be simply ω_t. This produces signal components consisting of a carrier of frequency ω_c and two sidebands of frequencies $\omega_c \pm \omega_t$. Transmission of all of these components results in doube-sideband AM. Since the carrier contains no information, power savings can be made if the carrier is suppressed. Transmitting two sidebands is, in effect, sending the same information twice, so further power saving can be made by transmitting only one sideband. Most importantly, however, bandwidth saving is achieved by sideband suppression because the channel bandwidth requirement for single-sideband (SSB) AM is only half that of double-sideband (DSB) AM. The choice of AM system depends on the particular application, DSB being preferred for domestic radio broadcasting and SSB for telephony. Vestigial-sideband transmission is a compromise between the two, which is used in TV broadcasting where the baseband is too great for DSB to be used.

Frequency modulation (FM) is achieved by changing the frequency (ω_c) of a variable-frequency oscillator with the baseband signal. A single tone of frequency ω_t produces signal components at frequencies $\omega_c \pm n\omega_t$. When n is large, we have wide-band FM, which uses more bandwidth than AM in order to achieve greater base bandwidth and higher SNR. Since the available bandwidth increases with frequency, FM operates at about 100 MHz whereas AM works at about 1 MHz. At the modest bandwidths used at 1 MHz and below, receiver thermal noise is small compared with atmospheric noise, so inefficient aerials can be tolerated since these will not reduce the SNR. At VHF and above, this situation reverses, so efficient aerials must be used. Fortunately, the shorter wavelengths ease the design of highly efficient antennae.

Digital signals may be carried by sine waves of higher frequency than the pulse rate. This type of continuous-wave modulation is called shift keying and it has its amplitude and frequency forms. However, various kinds of phase shift keying (PSK), such as differential and quadrature PSK, have proved to be of greater practical advantage.

Signal sampling can be regarded as pulse amplitude modulation (PAM). When a periodic pulse carrier of frequency f_s is modulated by a tone of frequency f_t, the Fourier components of the carrier give rise to tones at f_t

and $nf_s \pm f_t$. PAM is not used for radio transmission because frequency shifting is more efficiently accomplished with AM. PAM could be used in line transmission, but in practice pulse code modulation (PCM) is preferred. In PCM, each pulse amplitude (sample) is converted into a binary-coded pulse train by an analogue-to-digital converter (ADC). An ADC outputs quantized sample values of an analogue signal.

The main task of a receiver is to recover the baseband signal from the modulated carrier signal—a process of demodulation. This is a simple task when PCM is used because the receiver simply has to detect the presence of a pulse and then use a digital-to-analogue converter (DAC). Cheap DAC integrated circuits are available. Sensing the mere presence of a pulse is much easier than sensing its actual amplitude, and has the further advantage of eliminating all noise below a threshold slightly less than the binary pulse amplitude. Pulses can be easily re-formed, thus overcoming channel distortion. This means that low-noise, high-fidelity communication is possible over long distances with PCM, if line repeaters are used. The occasional errors caused by noise in a PCM link can be detected by simple parity checking.

The accurate recovery of AM and FM signals, in the presence of noise, is more difficult than the recovery of PCM. Cheap integrated phase-locked loop (PLL) circuits are available that perform both amplitude and frequency demodulation. Wideband FM gives reception with a much higher SNR than AM primarily because of its greater bandwidth. A secondary improvement arises if amplitude limiting is used at the receiver to remove the noise, which, in broadcasting, mainly takes the form of spurious amplitude variations. The SNR obtained with PCM is superior to that of FM.

Line telecommunication between many terminals requires a switched network. The use of PCM for voice transmission on the public switched telephone network (PSTN) has not only made possible high-quality TDM operation, but has also facilitated time-slot interchange (TSI) switching. TSI, achieved with digital computing techniques, has superseded mechanical switching. The digital exchange is an example of the convergence of telecommunications and computing technologies. The digital exchange is really a computer embedded in a telephone network, with much of the operation of the exchange carried out under software control. This enables connections to be made very quickly, increases flexibility, eases administration and makes available services that were previously only available on private branch networks.

In addition to the PSTN, there are two other public networks: the telex system, which is used almost exclusively for business, is the most extensive but is a slow data network, and the packet-switching sytem (PSS), which offers much higher data transmission rates. There are also many private networks for both voice and data signals.

Large organizations require internal telecommunication networks that are linked to public or leased lines. These private branch networks (PBNs) are fully or partly automated and usually digital in operation. Such networks are designed for human communication. Local area networks (LANs) provide the equivalent service for computers and their peripherals.

Human conversation takes place over relatively long periods. The procedures used to establish a connection follow roughly defined rules, some governed by the hardware (dialling and call tones) and others by the subscribers' customs. Computer 'conversation' tends to take place in short bursts and requires carefully defined rules or protocols for every simple step for establishing and closing down the link. Complete computer messages are normally broken down into packets before transmission so that efficient use can be made of a network. Voice packet switching is also being developed. International agreement on standards and protocols is being strived for in order to achieve equipment compatibility and overall economy of operation.

Mobile radio telephone development has taken place over many years. Very-large-scale integrated (VLSI) circuit technology has facilitated the commercial application of cellphone systems linked to the PSTN. The restriction imposed by the hard-wired handset is removed by the cordless phone, which uses short-distance radio linking between a transceiver handset and a base transceiver wired to the PSTN. The cordless phone, together with the radio pager and telepoints, provides a cheaper personal mobile service than the cellphone. Both systems use radio for local links and rely on the PSTN for trunk calls. Personal communications networks (PCNs) are widely regarded as the point to point telecommunications development of the near future. The PSTN grew as a system for interlinking premises; the PCN will provide a national network linking people rather than premises by means of portable radiophones. These will operate at 0.25–1.0 W transmitter power with millimetre wavebands in contiguous, radio-linked cells of various size.

High-quality sound programmes are now provided at VHF using FM. The future of TV broadcasting would seem to lie with microwave satellite telecommunications. The problem of providing a good-quality TV service in the UK cannot be completely solved by using terrestrial links: there are line-of-sight and interference problems at UHF that limit the number of high-definition TV channels. The use of microwave carriers and satellites opens the way for more high-definition TV channels and, at the same time, solves many of the problems of audio and data global communications. The technology is in place, but propagation difficulties are severe at such frequencies.

Very-large-scale integration (VLSI) of electronic circuitry has made available cheap processing power, which is enabling the increasing exploitation of frequency re-use and companding techniques. Replacing old systems, which are still producing revenue for the service provider and an acceptable service for the consumer, is an inherently slow procedure. Handsets and computer terminals are frequently updated, but the millions of miles of wire already hung in the air and laid in the ground or on the sea bed are not likely to be completely replaced with glass fibres for many years.

Many future developments in telecommunications will be related to data transfer. The equipment now on the market fulfils most of the demands of voice communication, and digitization of the PSTN improves the quality of the telephone service. The picture-phone is already in the course of advanced development, but its commercial appeal is in doubt. Value-added data services (VADS) will provide the main area of future expansion in the move towards a paperless and cashless society. Electronic data interchange (EDI) uses a data communications network to eliminate much of the paperwork between organizations and companies. Routine tasks, like ordering and invoicing, will become more automated. Services operate, both nationally and internationally, over the suppliers' private networks and via the PSTN.

Electronic funds transfer at point of sale (EFTPOS) is a facility that is replacing cheques and cash with a plastic card that is passed through a reader at a check-out. A smart card will replace the many different standard plastic cards used at present. The standard card carries its information on a strip of magnetic material. The smart card has an integrated circuit built into it, comprising a microprocessor and a RAM. It can take in and process data, store it and then deliver it in response to appropriate signals. The smart card will replace the office door key, act as a clocking-in device, provide the authorized log-in key for computer access, operate vending machines and payphones as well as act as an EFTPOS credit card and cash card. Shopping by phone will be a growing trend. The service already exists on viewdata systems such as Prestel and others. Telebanking is also offered on a viewdata basis.

Telemetry (measurement at a distance) is well established in industrial organizations and likely to grow. Environment control and security alarms come under this heading, as do chemical processes that are monitored remotely over private or public networks. Picking up a record of the consumption on a customer's premises of utilities like electricity, gas and water, without using human meter readers, will be possible when the integrated services digital network (ISDN) standardizes the connection of the metered premises to the public network. With duplex connection, device operation as well as reading can be directed over a line, giving a telecommand facility. So, for instance, the supplier could cut off a service to any individual premises in the event of an emergency or a failure to settle bills. Calling in by phone to home to activate a cooker or the central heating system will be possible, just as one can now operate a telephone answering machine remotely via the PSTN.

The three major networks in current existence (PSTN, PSS and telex) have the drawback of being separate networks. Many of the overheads incurred in operating the three could be reduced by blending them into one integrated service. The user who requires all the services has the inconvenience and expense of separate physical connections. The ISDN offers all these services on one network. The growth in demand for data services is expected to accelerate, but the demand will come mainly from the business community and special interest groups, with the general public some way behind. The public demand can be met, at present, by the existing services allied to the PSTN. However, the will to implement the ISDN as a universal service is there, so it will have an expanding part to play in the future of telecommunications.

Appendix A Power and frequency ratios

Power ratios (PRs) occur frequently in system calculations and measurements, usually in the form of the ratio of output power to input power (P_o/P_i). In practice, power ratios can cover a very wide range from perhaps a few millionths to several millions. Such large ranges can be most conveniently represented on a decade or logarithmic scale. Figure A.1 illustrates this idea.

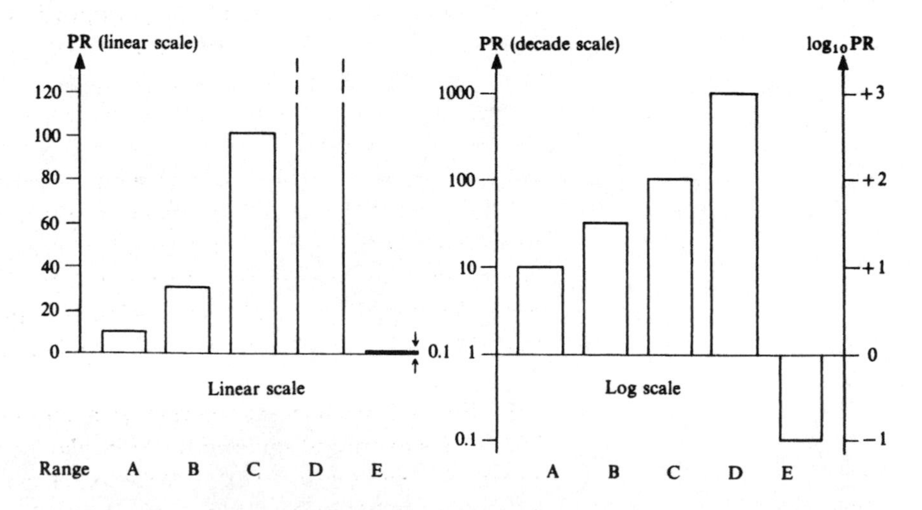

Figure A.1. Linear and log scales for power ratios.

In Fig. A.1, range D is too large to be easily accommodated on the linear scale but not on the decade scale. The latter makes it easier to cope with small ranges like E, which is 0 to 0.1 on the linear scale and hardly shows.

It is also convenient to calibrate the decade scale in terms of logarithms to the base 10 to give a log scale as shown on the right of Fig. A.1. PRs expressed as $\log_{10}(P_o/P_i)$ are given a unit called bel (B). One-tenth of this unit, the decibel (dB), is about the minimum discernible change in power ratio that can be detected by the human senses so unit subdivision has been generally adopted, the bel being too coarse a unit. Hence PR in dB is $10\log_{10}(P_o/P_i)$. A PR of 2 is 3 dB, for example, while 1 dB represents a PR of 1.26.

In electrical terms $P_o = V_o^2/R_L$ where R_L is the load resistance and $P_i = V_i^2/R_i$ where R_i is the input resistance. Hence

$$\begin{aligned}\text{PR(in dB)} &= 10[\log_{10}(V_o/V_i)^2 - \log_{10}(R_L/R_i)] \\ &= 20\log_{10}(V_o/V_i) - 10\log_{10}(R_L/R_i)\end{aligned}$$

$20\log_{10}(V_o/V_i)$ is known as the voltage ratio is dB. This ratio is equal to the power ratio only when $R_L = R_i$. Similar considerations apply to current ratios.

PRs greater than unity (positive dB values) indicate an increase in signal power from input to output, that is *gain* or *amplification*. PRs less than unity (negative dB values) indicate a decrease in signal power, that is *loss* or *attenuation*.

A power level, rather than a power ratio, may be expressed on a log scale if some convenient reference level for P_i is used, such as 1 W or 1 mW. Then the power level would have the units dBW of dBm respectively.

The dB is particularly suitable for sound pressure level (SPL) comparisons because the human ear responds to a wide range of pressure levels, the sound sensation being roughly proportional to the log of the stimulus at a given frequency. The weakest SPL that a normal human ear can detect is about 20 μPa (micropascals). A pascal is equal to one newton per square metre. The loudest tolerable sound has an SPL of about 20 Pa, giving an SPL span of six decades or 120 dB (think of pressure being analogous to voltage). An increase of 6 dB represents a doubling of SPL, but an increase of about 10 dB is required for a sound to seem twice as loud. The smallest sound change that can be perceived is about 3 dB.

Perceived loudness does not simply depend on SPL because the ear is not equally sensitive at all frequencies. It is most sensitive in the range 3 to 4 kHz and this sensitivity is more variable at low SPLs as shown in Fig. A.2.

In Fig. A.2 each curve represents a contour of equal loudness. This loudness is expressed quantitatively in *phons* by the relative SPL in dB at 1 kHz. For example, a 30 dB, 100 Hz tone has the same perceived loudness as a 10 dB, 1 kHz tone, that is 10 phons. By way of further example, a 62 dB, 4 kHz tone would have a loudness of 70 phons.

The subjective response of the human ear is catered for in the loudness control provided with many Hi-Fi audio amplifiers. When the loudness switch is pressed, the bass and treble frequencies are accentuated. This facility is recommended for listening at low sound levels.

Notice that frequency is also plotted on a log scale in Fig. A.2. Changes in frequency are perceived as changes of pitch. A decade is perceived as a very great change, so smaller multiples than 10 are more convenient for pitch comparison. A doubling of frequency is perceived as an eight note

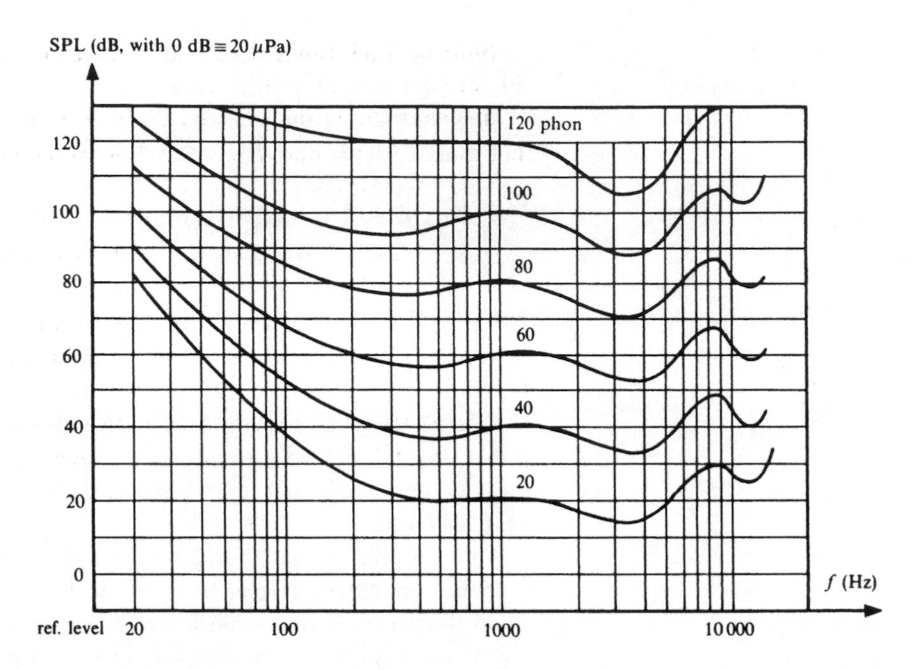

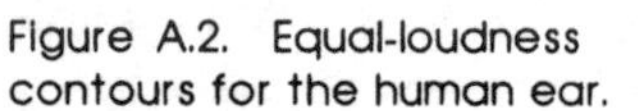
Figure A.2. Equal-loudness contours for the human ear.

(one octave) interval in a *major* or *minor* note sequence or *scale*. These scales consist of a mixture of six tones and two semitones whose arrangement determines whether the scale is major or minor. The frequency ratio corresponding to a tone interval is about 9/8. Half this interval is a semitone, whose frequency ratio is the geometric mean of 9/8 which is

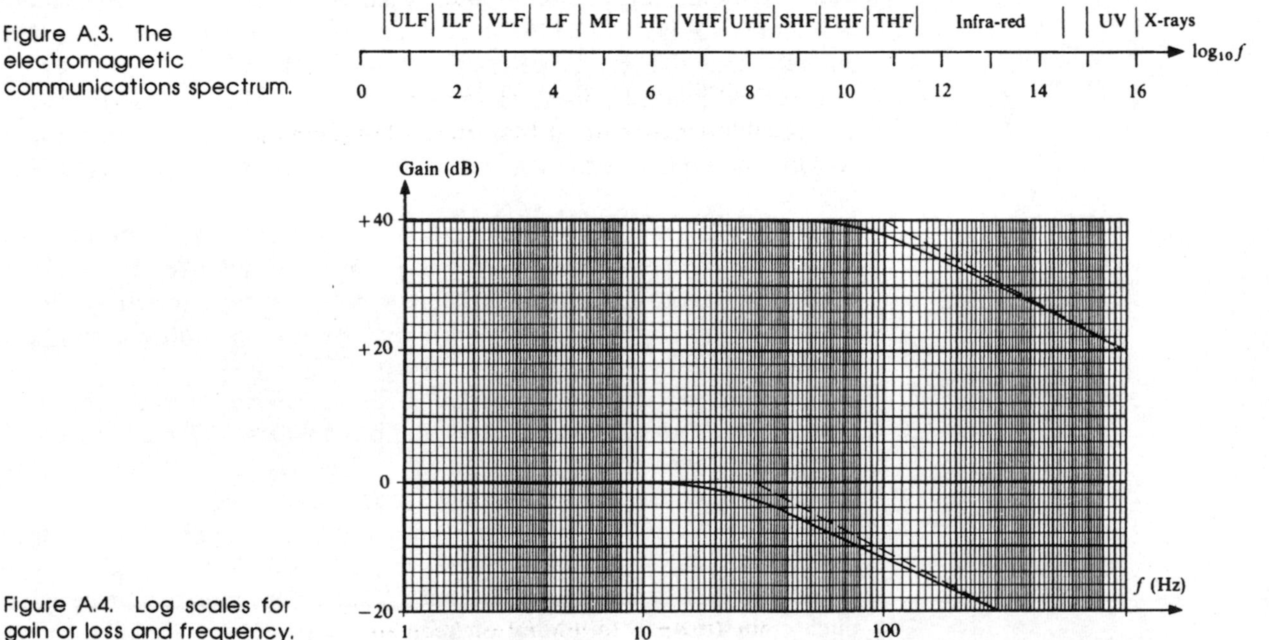

Figure A.3. The electromagnetic communications spectrum.

Figure A.4. Log scales for gain or loss and frequency.

$\sqrt{(9/8)}$ or 1.06. Since 1.06^{12} is 2, this means that there are 12 semitone intervals between notes which are one octave apart. Many keyboard instruments have the keys laid out in repeating patterns of 12 keys, with adjacent keys producing notes at semitone intervals. Playing adjacent keys in sequence produces a 12-note *chromatic* scale.

A log scale would be very appropriate for illustrating the electromagnetic communications spectrum (refer to Table 2.1), as in Fig. A.3.

The operating frequency of electronic equipment covers a very wide range from a few hertz (Hz) to several gigahertz (GHz). Individual instruments or systems may have a frequency range extending over several decades so it is convenient to represent their frequency response on a logarithmic scale. Some typical gain versus frequency responses for electronic equipment are shown in Fig. A.4 by way of illustration.

Appendix B Trigonometric identities and series

The basic trigonometric identities are

$$\sin(A \pm B) = \sin A \cos B \pm \cos A \sin B$$
$$\cos(A \pm B) = \cos A \cos B \mp \sin A \sin B$$

From these we can deduce

$$\sin(A - B) + \sin(A + B) = 2 \sin A \cos B$$
$$\cos(A - B) + \cos(A + B) = 2 \cos A \cos B$$
$$\cos(A - B) - \cos(A + B) = 2 \sin A \sin B$$
$$\tan(A \pm B) = (\tan A \pm \tan B)/(1 \mp \tan A \tan B)$$

Putting $A = B$ we get

$$\sin(2A) = 2 \sin A \cos A$$
$$1 + \cos(2A) = 2 \cos^2 A$$
$$1 - \cos(2A) = 2 \sin^2 A$$
$$\tan(2A) = (2 \tan A)/(1 - \tan^2 A)$$

which leads to

$$1 = \cos^2 A + \sin^2 A$$
$$\cos(2A) = \cos^2 A - \sin^2 A$$

The trigonometric series are

$$\cos x = 1 - x^2/2! + x^4/4! - x^6/6! + \cdots$$
$$\sin x = x - x^3/3! + x^5/5! - x^7/7! + \cdots$$

The exponential series is

$$\exp x = e^x = 1 + x + x^2/2! + x^3/3! + x^4/4! + \cdots$$

from which we can deduce the trigonometric and exponential relationships

$$\cos A \pm \mathrm{j} \sin A = \exp(\pm \mathrm{j}A)$$
$$2 \cos A = \exp(\mathrm{j}A) + \exp(-\mathrm{j}A)$$
$$2 \sin A = \exp(\mathrm{j}A) - \exp(-\mathrm{j}A)$$

The logarithmic and binomial series are

$$\ln(1 + x) = x - x^2/2 + x^3/3 - x^4/4 + \cdots$$
$$(1 + x)^n = 1 + nx + n(n-1)x^2/2! + n(n-1)(n-2)x^3/3! + \cdots$$

Appendix C CCITT recommendations for data communications

The Comité Consultatif International Téléphonie et de Télégraphie (CCITT) recommends telecommunication standards for global adoption. Those of particular relevance to data communications are the 'V' series, which is concerned with data communication over voice networks, and the 'X' series, which deals with communication over data networks. The scope of these are listed for reference.

The 'V' series

V1	Equivalence between binary symbols
V2	Power levels
V3	International alphabet no. 5
V4	Signals for alphabet no. 5
V5	Modulation and signalling rates on public switched networks
V6	Modulation and signalling rates on leased lines
V10	Electrical characteristics of unbalanced interchange integrated circuits
V11	Electrical characteristics of balanced interchange circuits
V13	Answer-back unit simulator
V15	Acoustic couplers
V16	Modems for medical analogue data transmission
V19	Modems for parallel data transmission using signalling frequencies
V20	Modems for parallel data transmission on the PSTN

V21	300 baud modem standard for use on the PSTN
V22	1200 baud modem standard for use on the PSTN
V22 bis	2400 baud modem standard for use on the PSTN
V23	600/1200 baud modem standard for use on the PSTN
V24	Modem standards list
V25	Automatic calling and answering equipment
V26	2400 baud modem standard for use on point-to-point equipment
V26 bis	2400/1200 baud modem standard for use on the PSTN
V27	4800 baud modem standard for use on leased lines
V27 bis	4800/2400 baud modem standard for use on leased lines
V27 ter	4800/2400 baud modem standard for use on PSTN
V28	Electrical characteristics of unbalanced interchange circuits
V29	9600 baud modem standard for use on leased lines
V31	Characteristics for interchange circuits controlled by contact closure
V32	Modems for high-speed transmission over PSTN or leased lines
V35	Data transmission at 48 kb/s using 60 to 108 kHz band circuits
V36	Modems for tranmission at 48 kb/s using 60 to 108 kHz band circuits
V40	Error indication with electromagnetic equipment
V41	Code-independent error control systems
V50	Standard limits for quality of data transmission
V51	Maintenance of telephone-type circuits used for data transmission
V52	Characteristics of distortion and error rate measuring equipment
V53	Limits for maintenance of telephone-type circuits used for data transmission
V54	Loop test for modems
V55	Impulsive noise measuring instruments for telephone-type circuits
V56	Comparative tests for modems used over telephone-type circuits
V57	Comprehensive data test set for high signalling rates

The 'X' series

X1	International user classes of service in public data networks
X2	International user facilities in public data networks
X3	PAD facilities in public data networks
X4	Structure of signals for alphabet no. 5 code used in public data networks

X20		Interface between asynchronous DTE and DCE used in public data networks
X20	bis	Compatible interface between asynchronous DTE and DCE
X21		Interface between synchronous DTE and DCE used in public data networks
X21	bis	DTE for interfacing synchronous 'V' series modems
X24		Definitions of interchange circuits between DTE and DCE
X25		Interface between DTE and DCE for packet terminals in public networks
X26		Same as V10
X27		Same as V11
X28		DTE/DCE interface for accessing PAD on public network
X29		Exchange of control information and data between DTE and PAD
X32		DTE/DCE interfaces for packet terminals accessed via the PSTN
X50		Multiplexing scheme for interface between 48 kb/s synchronous networks
X50	bis	Parameters for signalling in above scheme
X51		Multiplexing scheme for interface between 10-bit synchronous networks
X51	bis	Parameters for signalling in above scheme
X52		Asynchronous-to-synchronous conversion
X53		Number of channels on multiplex links at 64 kb/s
X54		Allocation of channels on multiplex links at 64 kb/s
X60		Common channel signalling for switched data applications
X61		Signalling system no. 7
X70		Signalling systems between asynchronous networks
X71		Signalling systems between synchronous networks
X75		Signalling systems for packet switched networks
X80		Signalling systems for circuit switched networks
X87		User facilities in public data networks
X92		Connections for public synchronous data networks
X95		Network parameters in public data networks
X96		Call progress signals in public data networks
X200		Reference model of OSI for CCITT applications
X210		OSI layer service definitions
X213		Network service definition for OSI
X214		Transport service definition for OSI
X215		Session service definition for OSI
X224		Transport protocol for OSI
X225		Session protocol for OSI
X244		Protocol identification procedures to estabish virtual call
X250		Techniques for communications protocols and services
X400		Message handling service for text communications

Appendix D American Standard Code for Information Interchange

The seven-bit ASCII code (alias international alphabet no. 5) has $2^7 = 128$ characters that are used in computer keyboard links. The code covers graphic (printable) characters such as letters, numbers, punctuation marks and arithmetic signs, and control characters such as 'carriage return' (CR), 'start of text' (STX), 'data link escape' (DLE), 'device control' (DC) and others, which appear mainly in columns 000 and 001 of Table D.1 (these are non-printable characters). A parity check bit is normally added to the seven-bit code to form a complete byte.

Table D.1. The ASCII code

	$b_7b_6b_5$							
$b_4b_3b_2b_1$	000	001	010	011	100	101	110	111
0000	NUL	DLE	SP	0	@	P		p
0001	SOH	DCI	!	1	A	Q	a	q
0010	STX	DC2	"	2	B	R	b	r
0011	ETX	DC3	#	3	C	S	c	s
0100	EOT	DC4	$	4	D	T	d	t
0101	ENQ	NAK	%	5	E	U	e	u
0110	ACK	SYN	&	6	F	V	f	v
0111	BEL	ETB	'	7	G	W	g	w
1000	BS	CAN	(	8	H	X	h	x
1001	HT	EM	)	9	I	Y	i	y
1010	LF	SUB	*	:	J	Z	j	z
1011	VT	ESC	+	;	K	[	k	{
1100	FF	FS	,	<	L	\	l	|
1101	CR	GS	–	=	M	]	m	}
1110	SO	RS	.	>	N	^	n	~
1111	SI	US	/	?	O	-	o	DEL

Appendix E RS and Centronix connectors

The Electronic Industries Association (EIA) issued, in 1969, a recommended standard (RS-232C) that specifies voltage levels, control signals and connector pins for *serial* data linking between such devices as computer terminals, printers and modems, that is cables and signals used between data terminal equipment (DTE) and data communication equipment (DCE). A later improved standard, RS-449 (released in 1977), retains many of the functions of its predecessor, but improves on distance and speed specifications.

The *Centronix parallel interface* is preferred for computer/printer links because data is transferred in bytes rather than bits. The *general-purpose interface bus* (GPIB), which is also known as either the *Hewlett Packard interface bus* (HPIB) or the *IEEE-488 bus*, is used in many other data communication tasks. An RS-232C, 25-pin, D-type connector is shown in Fig. E.1. A similar connector is used for Centronix interfaces. Parallel connection is practicable when the link is short; for distances in excess of about 3 m, serial links are recommended.

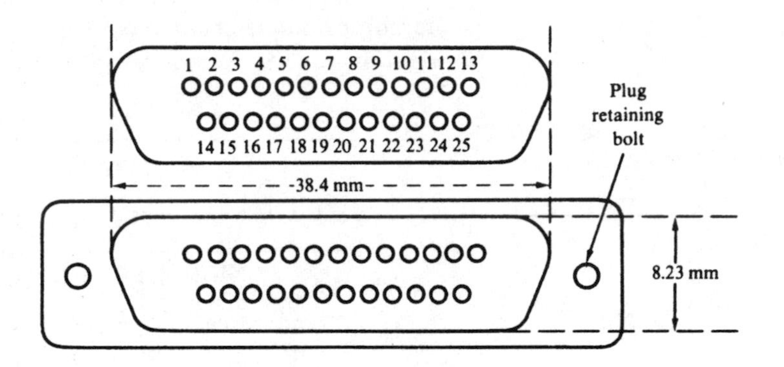

Figure E.1. RS-232C 25-pin D-type connector.

Table E.1. Pin functions of RS-232C and Centronics connectors

Pin no.	Circuit function (RS-232C)	Signal (Centronix)
01	Frame ground	STROBE
02	Transmit data	DATA 0 (LSB)
03	Receive data	DATA 1
04	Request to send	DATA 2
05	Clear to send	DATA 3
06	Data set ready	DATA 4
07	Signal ground	DATA 5
08	Data carrier detect	DATA 6
09	D.c. test voltage	DATA 7 (MSB)
10	Voltage	ACK OUT
11	Equilizer mode	BUSY
12	Secondary data carrier detect	PAPER OUT
13	Secondary clear to send	SELECT
14	Secondary transmitted data	AUTOFEED
15	Transmitter clock	NC
16	Secondary received data	0 V
17	Receiver clock	CHASSIS GND
18	Divided receiver clock	+5 V
19	Secondary request to send	GND
20	Data terminal ready	GND
21	Signal quality	GND
22	Ring indicator	GND
23	Data rate selector	GND
24	External transmit clock	GND
25	Busy indication	GND

Appendix F PC Communications – Using Modems and FAX Cards

F1 Modems

Modems are used to transfer data over the PSTN. This data may come from Viewdata services such as Prestel, Email or bulletin boards, or from computers. Modems would be used to transfer files between remote PCs.

The procedure used to establish a link via modems is covered by the Hayes Standards. For example, the procedure used by a V22 bis modem is as follows. The called modem, set in the 'receive-call' mode, detects a ring and answers, after two seconds, by transmitting a 2100 Hz tone for 3.3 s. After a further 75 ms it transmits binary ones at 1200 b/s, known as the 'unscrambled binary signal' (UBSI). The sending modem then responds within 150 ms after receiving USBI. After another 450 ms, the sending modem transmits alternate double zeros and ones at 1200 b/s (the SI signal) for 100 ms to identify itself as a 2400 b/s modem. After sending SI, it switches to a burst of scrambled binary at 1200 b/s (the SBI signal). The answering modem detects SI and replies with 100 ms of SI to identify itself as a 2400 b/s modem. It then sends SBI for 500 ms before switching to scrambled binary at 2400 b/s, indicating its readiness to receive data. Meanwhile, the sending modem, after receiving SBI, sends scrambled binary at 2400 b/s for 200 ms after which it is ready to send. The modems at each end of the data link must follow the same standard procedure.

The early 300 and 1200 baud modems use the V21 and V22 standards respectively. In the mid-1980s, manufacturers began producing modem ICs to the V22 bis (2400 baud) standard. Prestel transmits at 1200 baud and receives at 75 baud and some early modems are permanently set at this V23 split-rate standard. The V26 to V29 standards cover the 2400 to 9600 baud modem standard rates. Many commercial modems are now switchable between these transmission rates.

Error control and data compression are two other features of a modem specification. These features can be either built into the modem (the hardware) or included in communications software packages. It can be counter-productive to use hard and software features simultaneously: one only should be chosen.

Error Control

Noisy links require error control, especially at the higher baud rates, and the Microcom Network Protocols (MNP) 1 to 10 cover the standards for hardware error control. The CCITT introduced V42 in 1988, which is equivalent to MNP 2 to 4. Both sets of standards use cyclic redundancy or sum checks on data packets.

Most communications software packages provide a selection of error control protocols for file transfers that can be used instead of the hardware protocols. The main software protocols on offer are: Crosstalk, Smartcom, XModem, YModem, ZModem and Kermit. These protocols, each of which offers various advantages, check files only, whereas MNP and V42 check everything, such as commands and screen data.

Data Compression

Software data compression is used to save storage space. Since compressed files take less time to transmit, this technique also reduces transmission costs. Alternatively, file compression can be built into the modem. In this hardware alternative, the data to be transmitted is examined for redundancy and, by reducing this redundancy, the information rate is increased. MNP 5 to 10 include data compression and error control, and many of their features have been included in the V42 and V42 bis standards. Error reduction is achieved by adding redundancy in a controlled manner that will reduce the data rate. However, an overall improvement is achieved by using these protocol standards.

MNP 5 uses two different compression algorithms: run-length encoding and Huffman encoding. The former identifies repetitive characters. After a character has been repeated more than three times successively, a count of the next repeats of that character is sent instead of the characters themselves. Huffman encoding uses the fact that some characters, such as the letter e, are sent more often than others. Instead of sending all the ASCII characters as 8-bit characters, the most frequently used letters are sent in four bits and the less frequently used letters in up to eleven bits. The ASCII code for carriage-return (CR) is 0001101, so six successive CRs, using an even-parity MSB transmit as:

100011011000110110001101100011011000110110001101

With compression, this becomes:

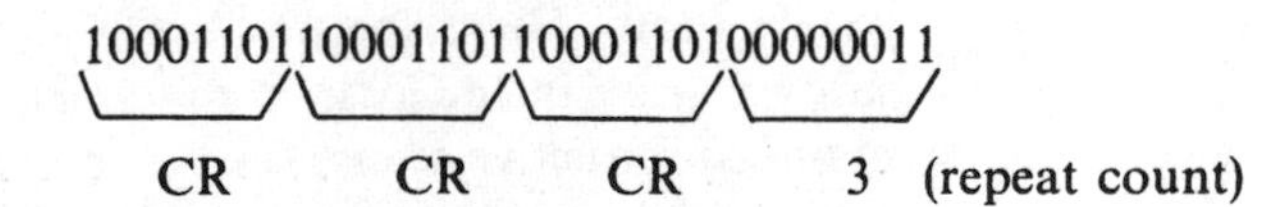

With Huffman coding the sequence:

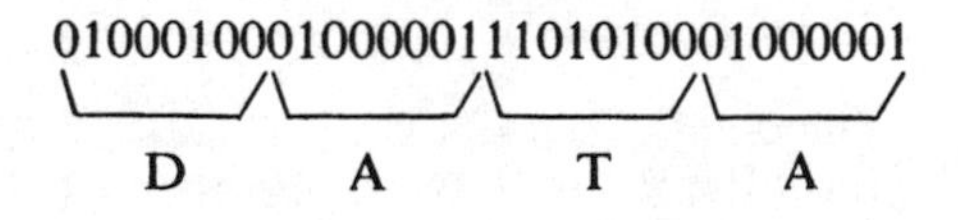

compresses to:

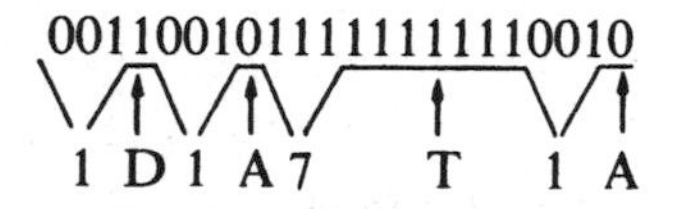

The arrow indicates the data position and the count digits show that D and A are one digit long, whereas the less frequently occurring T is seven digits long.

If it is required to transfer data from one PC to another within a range of a few metres, a null-modem cable connecting the PC serial ports is all the hardware required. Such a cable links the output pin (pin 2) of one PC to the input pin (pin 3) of the other PC and provides a common earth link: only three wires in all. The communications software takes care of the file transfers as it would if a modem had been used. There would be no point in using error correcting or compression protocols in this application.

F2 Facsimile

FAX is normally used to send electronically-scanned documents over the PSTN. Since the output from a scanner is a graphics file, a PC with a graphics card could receive FAX data using FAX circuitry instead of a modem. Such circuitry is commercially available on a printed circuit board known as a FAX card. FAX cards and modems are not compatible because different modulation methods are used in the two systems. Commercial modems incorporating FAX cards are available. The modem/FAX cards can be connected to the internal expansion slots of the PC if available, otherwise stand-alone machines, connected to the serial port (RS-232C) of the PC, must be used.

A PC can generate graphics files and send these as FAXes via a FAX card to other computers or FAX machines. Most FAX cards at the present time use the Group 3 FAX standard and are switchable to 2400, 4800 or 9600 baud transmission rates. A PC with a scanner, FAX card and printer can double as a FAX machine. The total cost may be greater, but a PC offers processing facilities which are unavailable in a FAX machine.

F3 Teletext

Teletext cards can be added to a PC in the same way as modem or FAX cards. They enable data from the Ceefax and Oracle services to be retrieved and stored. These cards require a TV aerial input instead of the telephone input used in modems and FAXes.

Bibliography

It is practically impossible to provide a complete list of the many worthwhile books that have been written on telecommunications. This list is not intended to be complete—it is simply a list of books that the author has found useful at this introductory level. Each has its strengths and weaknesses, and their various styles will have diverse appeal. There is, therefore, no order of merit and there seems little point in trying to categorize into topics because most of the books are quite general. The publication dates given are not necessarily those of the latest edition. Further references to other books and specialist papers may be obtained from the books in the following alphabetical list.

Barden R. and Hacker M., 'Communication Technology', Delmar, New York, 1990.

Beauchamp K. G., 'Computer Communications', Chapman and Hall, London, 1987.

Black H. S., 'Modulation Theory', Van Nostrand Reinhold, London, 1953.

Brewster R. L., 'Telecommunications Technology', Wiley, Chichester, 1987.

Chilton P. A., 'Introducing X.400', NCC, Manchester, 1989.

Coates R. F. W., 'Modern Communication Systems', Macmillan, London, 1982.

Conner F. R., 'Antennas', Edward Arnold, London, 1982.

Conner F. R., 'Noise', Edward Arnold, London, 1982.

Conner F. R., 'Signals', Edward Arnold, London, 1982.

Cullimore I., 'Communicating with Microcomputers', Sigma, Wilmslow, 1987.

Das J., 'Review of Digital Communication', Wiley, Chichester, 1988.

Davidson C. W., 'Transmission Lines for Communications', Macmillan, London, 1978.

Dunlop J. and Smith D. G., 'Telecommunications Engineering', Van Nostrand Reinhold, London, 1984.

Evans B. G., 'Satellite Communication Systems', Peter Peregrinus, London, 1987.

Freeman R. L., 'Telecommunication Transmission Handbook', Wiley, Chichester, 1981.

Green D. C., 'Radio Systems for Technicians', Longman, Harlow, 1985.

Halsall F., 'Data Communications, Computer Networks and OSI', Addison Wesley, Reading, MA, 1988.

Haykin S., 'Communication Systems', Wiley, Chichester, 1983.

Haykin S., 'An Introduction to Analogue and Digital Communications', Wiley, Chichester, 1988.

Haykin S., 'Digital Communications', Wiley, Chichester, 1988.

Housley T., 'Data Communications and Teleprocessing Systems', Prentice-Hall, Englewood Cliffs, NJ, 1987.

Keiser N. F., 'Local Area Networks', McGraw-Hill, New York, 1989.

Kennedy G., 'Electronic Communication Systems', McGraw-Hill, New York, 1984.

Killen H. B., 'Digital Communications', Prentice-Hall, Englewood Cliffs, NJ, 1988

Langley G., 'Telecommunications Primer', Pitman, London, 1990.

Lathi B. P., 'Modern Digital and Analogue Communication Systems', Holt, Rinehart and Winston, New York, 1983.

Lewis G. E., 'Communications Services via Satellites', BSP Professional Books, London, 1988.

Macario R. C. V., 'Mobile Radio Telephones in the UK', Glentop, London, 1988.

Marshall G. J., 'Principles of Digital Communications', McGraw-Hill, New York, 1980.

Martin J. D., 'Signals and Processes', Pitman, London, 1991.

Maslin N. M., 'HF Communications', Pitman, London, 1987.

O'Reilly J. J., 'Telecommunication Principles', Van Nostrand Reinhold, Wokingham, 1984.

Pettai R., 'Noise in Receiving Systems', Wiley, Chichester, 1984.

Purser M., 'Computers and Telecommunication Networks', Blackwell Scientific, London, 1987.

Roddy D. and Coolen J., 'Electronic Communications', Prentice-Hall, Englewood Cliffs, NJ, 1984.

Roden M. S., 'Digital Communication Systems Design', Prentice-Hall, Englewood Cliffs, NJ, 1988.

Schoenbeck R.J., 'Electronic Communications', Merrill, New York, 1988.

Shanmugam K. S., 'Digital and Analogue Communication Systems', Wiley, Chichester, 1985.

Slater J. and Trinogga L. A., 'Satellite Broadcasting Systems', Wiley, Chichester, 1985.

Smale P. H., 'Introduction to Telecommunications Systems', Pitman, London, 1986.

Stanley W. D., 'Electronic Communications Systems', Reston, Virginia, 1982.

Stremler F. G., 'Introduction to Communication Systems', Addison-Wesley, Reading, MA, 1990.

Taub H. and Shilling D. L., 'Principles of Communication Systems', McGraw-Hill, New York, 1986.

Wilson R. G. and Squibb N., 'Broadband Data Communications and LANs', Collins, London, 1986.

Young P., 'Electronic Communication Techniques', Merrill, New York, 1990.

Index